Cal VanderWerf:
Anchor of Hope

by
Douglas C. Neckers

President VanderWerf in his Van Raalte Hall office, 1966

Manuscript Edited and Formatted by: Joe Faykosh

Cover art by: Hailey Connell
Photo: Hope College Archives, Geoffrey Reynolds

Library of Congress Control Number: 2016933539

ISBN: 978-0-9973096-1-4
First Printing: March 2016

BioSolar Publishing
27800 Lemoyne Rd, Suite I
Millbury, OH 43447
BioSolarPublishing.com

Visit DougNeckersExplores.com

Printed in the USA by
Morris Publishing®
3212 E. Hwy 30----Kearney NE 68847
800-650-7688----www.morrispublishing.com

Table of Contents

Spero in Deo: Anchor of Hope

Foreword

As any good former commencement speaker can tell you, an erudite bard of wisdom will proffer enough Latin for the first minute, enough Whitman or Shakespeare for the second minute, a few jokes for another minute, a school song or two which will make a point or two, and a five sentence take-home message in a speech that can last no more than ten minutes.

That is if one's audience are the graduates of a state university.

If one's audience is a major private university collection, *veritas* must show up somewhere, or *lux,* too, though *perpetua* is left for a funeral.

If one is speaking to graduates of a Protestant church-related liberal arts college, the Latin has to be little or less, because the Reformation threw Latin out of the reformed churches, though the word *Deo* must always be front and center.

I doubt Albertus Van Raalte had 21st century commencement speakers in mind when he ordained Hope College his "anchor of hope." But, were he alive today, he might be happy with all those who use a small smattering of Latin and then preach to his listeners and take an hour to do it. Words were the currency of his day, and the Hope of the 19th century grew from their delivery, and a healthy dose of *Deo*, according to the gospel of the rebels from 19th century Netherlands. My impression of Van Raalte ranges from admiration for one who managed, early on, to benefit from an American dream, to being querulous about how he managed to begat such hard heads as the settlers in Western Michigan turned out to be. Van Raalte was well educated and evolved into a businessman, leader, and an educator. His comprehensions were local and immediate, as well as wide-ranging and futuristic. He managed to settle his group in the woods because then they could build houses, in addition to founding a college. He bought the land so he could sell to newcomers, and he bought it cheap from tax sales and foreclosures. Public schools came soon and higher schools later. Hope College graduated its first class in July 1866, as a place where the young people were to be educated in a Christian institution. Van Raalte felt education should be taken to the people, not the other way around. But it is now essential to realize that he came from a background in the Netherlands, where

1

church, state, and school were all connected. It was the American dream, the freedom to make one's own way, that drove him.

The Hope College that Cal VanderWerf inherited had lost much of Van Raalte's vision and had become, partly by necessity, non-competitive. In order for it to survive, Hope College needed a revolution. Cal never called it a return to Van Raalte's principles—but he could have and probably should have. Van Raalte's principles drove Cal VanderWerf during his tenure at Hope College, as well as in life.

Anchor of Hope started as a biography of a relatively minor player in American mid-twentieth century higher education. Cal VanderWerf was the eighth president of a small liberal arts college located in Holland, Michigan, serving from July 1963 through June 1970. Hope College was also Cal VanderWerf's alma mater.

Cal VanderWerf served as president during the turbulent 1960s. In those seven years, he took a college that could not meet the academic expectations of the day and revolutionized it. He overwhelmed an intact faculty (less than 33% of whom had terminal degrees in the fields) with new appointments of Ph.D.s with ambitions to make major imprints on their chosen fields while serving in a non-graduate school setting. Many of them did just that. By the time he was done, at least four future college presidents had been hired, several national and international award-winning scientists had begun practice, other scholars had been attracted, and students whose financial capital would impact the college well into the 21st century were placed on pathways of success. President VanderWerf suffered significantly for his effort. His physical health was never the same, and his psychological energy was spent. But the college survived, was reformed, and built anew. Therein came the seeds for this analysis.

Anchor of Hope also started from a worldview: I knew the Hope College of his day. The political Hope that survived him never really appreciated VanderWerf and his energies, his contributions, or his personality. He was a university professor and department head. He had published multiple research papers, consulted for major industries and philanthropies, had directed dissertations research, and had a sterling record as a teacher and scholar at the University of Kansas. The Kansas campus and town revered him: He and his wife were strongly involved in events that led to attention towards racial equality in Lawrence, Kansas. This was merely one moment in a career dedicated towards finding the best in scholars, regardless of skin color.

2

What I have found are documents from the Hope archives, the national news, and hundreds of personal interviews, each confirming Cal VanderWerf's impact as the greatest president in Hope's history. He overcame resistance: his academic presence and personal force was strong, and he had a dream to make Hope College a place where an educated man or woman could find hope.

At this stage, I have to get personal. My mother, Doris van Lente Neckers, was the only person to finish a college degree in the allotted time in a family with a mother and father whose education stopped at grade school. She managed to get through Hope with an AB degree and a major in English by working three jobs during the Great Depression and living at home. Hope College was within walking distance of her house. When she finally managed to get three boys in school, she determinedly used that education and that degree to become a teacher. Though she lived in the hills of western New York, she managed to impact a number of good students whose backgrounds were as limited as hers. She was a religious woman—she lived her dream and her beliefs.

Hope was also Cal VanderWerf's hope. His father died when he was nine and he was raised in a single-parent household where three older sisters had much influence. Yet, through Hope College, he became a university professor and later college president. Neither his example, nor my mother's, is unique.

What this biography has become as it has evolved is an anatomy of the hope that comes with quality education in America. That value drives students toward achievement. Education, though it was not always that way, now points young persons in the direction of success in the world's most competitive arenas, armed with the basics, and says "Go to it."

Cal VanderWerf remade Hope College so that it could manage a new Hope for generations of students who had options, but would value the local, intense small-college presence that Hope had always been. He did so by understanding the value structures of the emerging university and collegiate environment of the late-20th century and demanded those in an institution shaped by the Depression and World Wars. With his leadership and direction, Cal VanderWerf led Hope College to realize its potential.

Acknowledgements and Sources

Many people at University of Kansas were gracious with their time and professional help. There are too many to name them all individually. I acknowledge, with gratitude, the archives,

3

archivists, and the collections at the University. The alumni office helped me with identifying, and finding, the persons whose efforts, with the VanderWerfs, started the Lawrence League for the Practice of Democracy. Their files are particularly extensive in dealing with the difficulties of racial segregation and the integration of the community of Lawrence, and the University.

If a person is quoted in the book, I either interviewed that individual in person, or talked to him/her on the phone. The K.U. Alumni staff pointed me to former students and colleagues. These included both Robert D. Coghill, who was one of those responsible for the commercialization of penicillin, and alumnus in 1919. Most, but not all, of Cal's former chemistry faculty colleagues are dead but some of their wives were/are still living.

It is a particular joy to acknowledge the courteous generosity of Rich and the late Sue Givens who allowed me to use a bedroom at their home – a home originally built in Lawrence by Cal and Rachel VanderWerf – while I traveled around the community interviewing Cal's former colleagues. Rich sent me the picture of the house Cal and Rachel had built. I remember it as being a cool place on a typically hot August day in Lawrence because even then there were shade trees. I met Cal and Rachel there in August 1960.

Marnie and Bill Argersinger and Helen and Paul Gilles told me many stories, along with others, and pointed me in the right directions. Marnie, since deceased, said "No matter what you hear, Wilt Chamberlain did not integrate the theaters in Lawrence, Kansas! We, the Lawrence League, did that long before him!" Harriet Wilson, attorney Paul Wilson's wife, was generous with her time. Mrs. Wilson told me about her husband's trip to Washington alone, in April 1952, for the oral arguments in defense of the State of Kansas in the case that today is known as *Brown v. Topeka Board of Education*. She also told me of going back to Washington with her husband, on the train, with a six-month old baby, to hear the second argument in December 1953. He had told her that it was probably going to be important so she should go with him. In his prior trip, in 1952, he had gone alone spending the 28 hours on the train from Topeka to Washington preparing to defend Kansas law at the Supreme Court. He was the youngest in the attorney general's office, and had the least to lose politically in that defense – so he was chosen for the case.

The Lawrence women, and others, pointed to the stresses rising to the surface that led to the organization of the Lawrence League for the Practice of Democracy. Cal's particular friendships

4

with Hobart Woodie, Sr. and Jr., chemistry department storeroom keepers, led to his involvement with the possibility that black children might be bussed to Lincoln Elementary School across the river to accommodate workers being moved to Lawrence during the War to be employed at Sun Flower Powder works. Rachel VanderWerf and Mrs. Alta Wilburn, their friend, colleague, and Holland associate, told me many stories about the ugliness of segregation in Kansas, supposedly a free state.

Physical chemists Bill Argersinger and Paul Gilles, along with Jacob Kleinberg, and particularly Cal's dear friend, Arthur Davidson, made racial equality a subject of discussion on the "hill" at K.U. Though they might argue vigorously about the values of their own research areas and expertise, they were of one mind when it came to racial equality and justice. The founding energies of the Lawrence League came from Rachel VanderWerf, and Cal, as well as the head the Y.M.C.A. in Lawrence at the time. Rachel was a professional herself, and on the staff of the Y.W.C.A. in Columbus before she and Cal married. By the time of *Brown v. Board*, more than 500 persons, black and white, assembled to hear Thurgood Marshall at K.U. in April 1954 for the annual Lawrence League event. Marnie Argersinger told me that they gave Cal no chance to get Marshall in Lawrence at the time. But he did.

I also thank former N.F.L. star Curtis McClinton for a long conversation about his taking freshman chemistry with Dr. VanderWerf as his tutor (As we approach Super Bowl 50, it should be recalled that Curtis scored the first AFL touchdown in Super Bowl I). It was unknown today (and ignored by Reformed Church historians) but Cal invited Deane Waldo Malott, Kansas Chancellor in the 1940s to join the Hope Board of Trustees in 1968. Malott, with whom Cal had argued about the appointment of I. Wesley Elliott, a young black chemistry major who Cal was encouraging to go on for graduate study, was a highly distinguished academician. However, Malott was, according to Rachel and Margery Argersinger, political, so the appointment of a black was an issue for Kansas, in Malott's opinion. But he approved the Elliott appointment eventually. Elliott could only work in the chemistry store; he did not get to serve as a teaching assistant.

I also thank University of Kansas history professor Bill Tuttle, Jr. for his detailed insights on racial equality in Lawrence and at the university. Bill's father was a Hope alumnus ('26) who became a physician. His father served in World War II, and for a significant time.

Unfortunately, the situation in Holland when the VanderWerfs moved there in 1963 was not a lot better—in fact, in some respects, it was worse for blacks because the prejudices there were buried beneath Dutch Calvinist piety. Many years ago, I had the good sense to ask Rachel VanderWerf to record her thoughts about Cal, their lives together, and to share those thoughts with me. I have two rather long tapes of Rachel speaking, after Cal's death, about their many journeys and experiences. Rachel had encouraged me to write a nomination for the Scientific Apparatus Makers Award for Cal. From that nomination process, I received the many names of VanderWerf students from the 1940s and 1950s. Rachel also shared information about Cal's friends and colleagues at Kansas and in the scientific community. I knew some, but there were many others that I did not know.

I also benefitted from long conversations with Gene and Vic Heasley – two brothers from Burnips, Michigan, who went to Hope College and then to Kansas, where both studied in Cal's laboratory. Vic was my mentor at Hope and more important to my learning organic chemistry than were any of my professors. I've told Vic dozens of times over how I appreciate this, but this time it is written for all to see.

Special thanks goes to my Ph. D. mentor at Kansas Earl Huyser. He and the late Albert Burgstrahler were the young faculty in organic chemistry when I went to K.U. in the fall 1960. I benefitted significantly from Earl's information about his work after his undergraduate studies at Hope College ('50), at the University of Chicago and at Columbia University. There's much in the book about him and his career.

By the time I met Cal he was a marked citizen in the organic chemistry community. Many knew him or of him. Melvin Spencer Newman, his professor at Ohio State and to whom I talked often, shared with me about his life growing up as a Jew in a racially less than tolerant America. The Ohio State University was badly intolerant until the 1960s. Mel was Cal's Ph. D. mentor at Ohio State and told me, though not by name, of the six chemistry faculty voting against his appointment, and his tenure, because he was a Jew.

Milton Orchin, a graduate school classmate of Cal's and Mel's first student was another person who helped me understand the laboratory VanderWerf. One of the more delightful and thrilling conversations I had about the VanderWerf family was with Samuel Massey, a black chemist who originally came from Fisk University to study chemistry with Henry Gilman at Iowa State University.

Massey came to Bowling Green to speak to our minority graduate students one fall, and I had a chance to hear his story.

Cal's colleagues at Kansas – Bill McEwen, Jake Kleinberg, and others collaborated with him scientifically and those collaborations were successful based on the science as reported in the journals. The work with positive iodine compounds with Kleinberg spilled into McEwen's labs. He and student Jerry Knapsyk published the first work on the photodegradation of aromatic iodonium salts. That, in the late Jim Crivelo's hands at GE Research, with Hope alumna Julia Lam Lee's able technical assistance, led to the development of photocationic polymerization. There would be no stereolithography as we currently know it without that discovery.

Before I say a word about Hope College, let the message be heard I love the place and hold it in deep personal respect. But it had some serious shortcomings when I went there as a young faculty member. So I read as much as I could about Holland, the Reformed Church and Hope's history. Wynant Wichers' book on the first 100 years of Hope College, *A Century of Hope,* was carefully researched and written; it is packed with information. Aleida J. Pieters' book, *A Dutch Settlement in Michigan,* was quite analytical and of a much earlier vintage. What surprised me from Wichers' reports was how much Phillips Phelps Jr.'s tenure at Hope and Cal's tenure at Hope were similar—even to the point of property on Lake Macatawa.

I rummaged through old college catalogs from archivists at several schools founded at the same time or before Hope College. The curricula of all of these institutions were, in general, similar. In reflection, what does one study in college when the only knowledge of consequence resided with the ancient languages? One's own literature, or at least the study of one's own English/American literature, was just starting. Mathematics study was a part of most college curricula. There was a touch of astronomy and physics; and an extensive discussion on human hygiene, because this was before the introduction of community sewer systems. British scientists like Lyon Playfair were encouraging the installation of sewers in Glascow, and London. Hope's curriculum included chemistry taught from the books of the day (*Draper's Chemistry*), and this generated interest among some students. There was the ever present study of the Bible. But that too was not just at Hope. Religious study was the norm in the 19th century, and was present in every curriculum I found.

This biography would not have been possible without the valuable correspondence, minutes, notes, letters, articles and other dated matter assembled at the Hope/Holland Archives. That these items were mostly recovered from the Van Raalte fire (1980) is further testament to the good sense of Hope's administration at that time. Though those who originated that project are long since departed, the collection is now in the able hands of Geoffrey Reynolds. Geoff is gracious, helpful, and an accommodating professional. He allowed us to scan copies of the *Hope Anchor,* helping refresh my memory on the campus scene. I am indebted to his help.

I was disgusted by the deprecation of current Holland writers for VanderWerf, and the work he did for the College. Much they wrote was hearsay, the result of sloppy research, if they did any before writing, and just plain wrong. Two historians said it was "even alleged that President VanderWerf had something to do with Wilt Chamberlain at the University of Kansas." A trip to the library and a read of *Life* magazine could have shown these persons that Cal's roll in that recruitment coordinated with his concern for the rights of minorities in Lawrence, and the University of Kansas. The athletic department leadership at Hope was a problem for the VanderWerf administration. It was satisfied with where it was, and did not like a mere chemist interfering with its territories. The fact that young men often selected a school more because they could play one or more sports at the collegiate level went by them. Cal wanted to use this as a recruiting tool for young men. Cal's works were social good works that did not count for much among the religious of the Reformed, and Christian Reformed traditions.

My analysis of the Hope chemistry department's productivity included a listing of all of Gerrit Van Zyl's publications and a report on Harvey Kleinheksel's Ph. D. work. My discovery of Cal's early work on phosphorus chemistry, more where the support for this came from, was stimulated by something he said about nerve gasses in his inauguration speech at Hope, November 8, 1963. I had no idea he knew about this because the structures of sarin, tabun and soman were embargoed until quite recently. Students had no reason to know this information, and I did not know it until I read the structural information as given by Gerhard Schrader, to the British interrogator, Colonel Robert Tilley, in Elberfeld in 1945 from the Archives at Duxford years later I looked everywhere I could to find if Cal was first to recognize the stereochemistry of the phosphorous atom in the nerve gas sarin. I suspect he was but a Freedom of Information Act request for a critical Office of Naval Research report was not fulfilled.

Libby Hillegonds shared numerous insights particularly about Cal's several illnesses during the Hope years. She also told me how hard Cal worked to get Bill to become Hope's chaplain. Bill did this for about twelve years, finally deciding it was time to marry and bury again, rather than counsel. So he left Hope to become the very successful senior pastor of the 1st Presbyterian Chuch in Ann Arbor, MI.

I am particularly indebted to my College classmate Lee H. Wenke. Lee and I had many, many long discussions about the Hope of our mutual experiences. While Vic Heasley liked to point out that I transferred to Hope, so I really did not get its full charisma, character and ethos, Lee knew full well that I understood that more than most. He shared his collection from the Holland Sentinel and his own recollections. It was Lee who dubbed Cal VanderWerf, Hope College's greatest president.

This project has gone on since the early 1970s. So I have read a lot, studied more, and formed opinions, too. The opinions are just mine but the references are, I hope, copious and detailed.

NEWSPAPERS
Holland Evening Sentinel
Lawrence, KS Journal World
Hope College Anchor
Toledo Blade
Jamestown, NY Post-Journal
NY Times
Washington Post
San Francisco Examiner
Chautauquan Daily
Harvard Crimson
Detroit Free Press
MAGAZINES:
TIME
Newsweek
Life
Colliers
Saturday Evening Post
Bowling Green State University Alumni News
Hope College Alumni News
University of Rochester Alumni News
Bucknell University Alumni News

9

University of Kansas Alumni News; Chemistry Department Newsletter
INTERVIEWS
Interviews referred to are those occasions when I had a specific conversation with the individual mentioned about Cal or Rachel VanderWerf, or about both. This book took almost 40 years to come to fruition, so many of the persons with whom I had detailed discussions are dead.
DISSERTATIONS
Neckers, J. W.
Kleinheksel, Harvey
VanderWerf, C. A,
Ter Meer, Fritz
Ter Meer, Edmund
Schrader, Gerhard
Immerwahr, Clara

Acknowledgements and the Hope community
      Most of what I submit about Hope comes from material in the archives, personal recollections, and a few personal interviews. One was with Professor Ken Weller. He was a co-chair of a faculty committee to help them address retirement options when TIAA CREF was introduced. As I am writing this, I received from Hope a report of Investment Options for my TIAA-CREF accounts, then worth more than $133,000,000. TIAA-CREF was started for Hope faculty by President VanderWerf in 1967. Among the many things this did was to provide Hope's faculty and administrators with a mobile retirement plan. The "Hope plan" that preceded it was a self-insurance and only vested after a faculty member had spent 25 years at the College. Ken Weller told me that he and Bert Ponstein set up a retirement counseling service for Hope faculty at the time TIAA-CREF was begun. I did not talk with Ponstein, but remember fondly the evening at a faculty meeting when he got up and said "Mr. President, there's this thing called tenure." Ponstein with just a Master's degree, likely felt threatened by all "those Ph. D.'s from prestigious places" showing up at the College. Cal VanderWerf did not just revamp the faculty—he about doubled it in size in just four years. After all, if one is going to build a "Harvard of the Midwest," one is not going to do that with a few graduates of Holland Christian High School with Ed. D. degrees from Western Michigan.
      One of the matters that shows up in the *Anchors* of the day (1958-72) is an alleged lack of communication between VanderWerf,

the president and the faculty/students about matters at the Board of Trustees level. I was teaching at the place through all this, but cannot for the life of me divine what it is that was being talked about. None the less immediately after VanderWerf left, the Board appointed a non-voting faculty member to meet with it.

There was a contention about compulsory chapel that was student driven during Cal's years. But the non-life threatening question of whether persons have to go to church regularly in the morning was just that – non-life threatening. I went to Chapel regularly as a Hope student because it was required and do not guess it hurt me much: the things I remember are those occasions in which school officials looked foolish in Chapel. As a senior, I was in the Chapel choir so I got to sit in front and sing a hymn, but rarely an anthem. The main sound was of the theater-ticket machines, as every student took a ticket on which to register his/her attendance. The students were right in wanting to go Chapel less, but the religion department needed something to do since there was no Sunday services at the Chapel for them to supervise. Cal was between a rock and hard place—an inconsequential one as young persons were dying in Vietnam, which was probably the real reason for the intensity of the confrontation. The religion faculty did complain about the scientists being away from campus "making money." Yet, every Sunday, each of them was off praying for dollars, their own, in some Reformed Church, in the area.

Another issue that was mostly nonsense was that the chemistry department was favored and built by the President. President VanderWerf did nothing for the chemistry department that he did not do for the rest of the campus. What bothered the humanities faculty of some tenure was, apparently, that those Ph. D's from prestigious universities spent more of their time working in their fields – and successfully – than in the coffee kletz sharing gossip.

Another more interesting take about these years was produced by Kennedy and Simon, *Can Hope Endure? A Historical Study in Christian Higher Education.* As in many cases of personal relationships, this book gives pause to the knowing reader because of the company it keeps. Allan B. Cook, a campus chaplain, is quoted; yet, the students of his day did not give him a lot of credence. Dave Marker was quoted in detail about the Reformed Church; yet, the long departed Marker was a practicing Episcopalian and a member of Grace Church in Holland. I guess one finds one support for one's argument where one can find it. Gordon Van Wylen, the president that followed Cal, had the luxury of faculty leadership that could

11

carry that mantel and a working committee structure. His idea that deans continue teaching was an excellent one, but not original with him. Places in the West – UCLA, Utah and many others – had practiced this for years.

I also read every other book I could find written about Hope College, by anyone—Hope College faculty, administrator, detractors, or friend—in order to understand the place that I had known since I could remember. George Marsden's writings on university secularization stimulated at least two telephone conversations. He thought that Hope's direction was an unusual anomaly. His longest and most well-known work deals with the Secularization of the American University.

Unfortunately, those that wrote about Hope's recent history were either residuals – those who were left – or those that came after. So, a school that was mostly academic in the eyes of the majority of its alumni and faculty, was portrayed from the perspective of a small, by this time, minor Protestant denomination by those that wrote about the College. Just one example makes the point – on January 6, 1968 President VanderWerf wrote chair of the Board, Hugh DePree an 18-page letter describing the things that had to happen for Hope to survive financially (see appendix). The College was in difficult, though not yet impossible, financial straits, and he was working with old financial managers, no development staff, and paying bills President Lubbers' administration had generated. There had to be changes, he said. At this time, the Board consisted of the sixty preachers he had inherited. Among other things, Cal said that if it did not change, the Board had better get someone else to run the place: eventually, that's what they did.

Other Reformed Church books include: Lynn Japinga's *Loyalty and Loss, The Reformed Church in America 1945-1994*; Elton Bruins and Karen G. Schakel's edited volume, *Envisioning Hope College, Letters written from Albertus C. Van Raalte to Philip Phelps, Jr. 1857 to 1875*; and Robert Swierenga's *Family Arguments in the Dutch Reformed Churches in the 19th Century*.

Anyone who seriously identified with the liberal arts college ideal knew that few in the community cared to be a Harvard, or any other research university, for that matter. Hope, like its ideals – Oberlin, Carleton, Harvey Mudd, Amherst, Dartmouth – wanted to be the best it could be, for the students who studied in its corner of the world.

Swierenga blames Lubbers as well as VanderWerf for Hope's secularization. But the point is obvious from the perspective

12

of the holy and sanctified looking backwards, Hope had gone to hell and it would take a re-identification with *the faith* to right all these wrongs. Fortunately, a Christian Reformed engineer showed up in the person of Gordon Van Wylen. Van Wylen had written a classical text for engineering undergraduates on thermodynamics – basically "the going and coming of hot and cold treated scientifically."

So VanderWerf, the son of a Dutch Reformed minister, and Van Wylen the Christian Reformed opposition, ran the institutions from different perspectives. According to Swierenga, Van Wylen arrived in order to "emphasize teaching in the manner of the historic Christian faith." Bruins, more or less a sole source for this printed word, said Van Wylen "saw something special in a combination of seriousness about Christianity and ..."

Thank goodness for that!

The Van Wylen Hope College went out its way to distance itself from the VanderWerfs. He sprinkled the Board with alumni with direct connections to the faculty. He brought additional members of the Christian Reformed Church to the administration, and sold the Marigold Lodge to the DePrees. Fortunately, a few of those that wished the College to be the "Harvard of the Midwest" hung around and the place prospered academically. When Mike Doyle and Sheldon Wettack both left, I lost touch with Hope in every way. It was only resurrected when Paul Schaap gave large funds to the college.

Hope today says its "graduates are educated to think about life's most important issues with clarity, wisdom and a deep understanding of the foundational commitments of the historic Christian faith."

I trust them; it is, after all, my alma mater, too. All documents and photographs are either from Cal's private papers or the Hope College Archives, with the assistance and permission of Geoffrey Reynolds.

Prologue

Cal VanderWerf was the most significant president in modern Hope College history. During his presidency, from 1963-1970, Hope transitioned from a college-centric institution into the more common department-centric school of postwar America. Though his tenure was cut short by the tensions of Vietnam, Civil Rights, and a generation of the querulous young who found their elders' political leadership wanting, under his leadership, the school entered the modern age. He added faculty in every field Hope knew to build a very energetic "Cal's Hope." Young faculty, most of whom had earned the highest advanced degree in their field, took their specializations more seriously than they did the college identity. So Hope emerged from the small institution that had survived the Depression and a World War as a college with enormous academic strength among four-year colleges.

Cal's presidency built Hope College into a place where it would not just survive, but lead in intellectual experiences for American undergraduates in the final decades of the twentieth century. Undergraduates were the currency of Cal's faculty, and that faculty began to use such resources wisely and conservatively.

Those old enough to remember recall that, following the total surrender of the Germans and the Japanese in 1945, the world emerged from a century of death and mayhem, international crimes, and genocides, to one where peace was at least attempted. The United Nations was chartered in San Francisco, and international problems were addressed in that forum. Because of discoveries during the war, the average life expectancy rose worldwide from around forty years in 1900 to over seventy-five in 2000. This was aided, in part, because there were lessor wars, but mostly because, among those deadly side-effects of wars, armies and battles, were infections and diseases. Antibiotics discovered during the war mostly eradicated deaths from bacteria. Vaccinations prevented diseases like smallpox, diphtheria, polio, mumps, whooping cough, and measles and deaths in childhood were also reduced. Public sanitation in cities became commonplace in civilized nations. Better sanitation made life safer for everyone. In diagnostic medicine non-invasive tools like M.R.I. took some of the guesswork out for those that became ill. Worldwide travel came via the jet plane so people who used to be told about others now could meet them face-to-face. Care in the air came from the invention, and predictions of future weather on the ground came from the universal

14

use of radar. By today, perhaps sadly, digital communication has almost replaced face-to-face meetings.

An enhanced national interest in the sciences in the 1960s had been one result of the war. Much of the world's growth was based on discoveries during the awful wars of the twentieth century. The advances that extended so many lives can almost one by one be attributed to scientific research carried out in support of the armies at battle. Much of this came from mobilization of the university professoriate in America in the sciences during that conflict. That transition transformed universities staffed by those prepared mostly in teaching classes in chemistry and physics departments, among others, to universities leading by their research. This change, the development of the nation's research universities, was irreversible.

After the war, the developments were transitioned to the civilian sector such that even the politicians of the day could claim that science was on the leading edge of their national futures. By 1946, even Winston Churchill was ruing Britain's inattention to technology in its institutions of higher education, and by the mid-1950s, President Eisenhower was declaring that the United States had a shortage of scientists and engineers. The vestiges of World War II in the US and Russia were not only positive. Militarists begat a chemical weapons and nuclear arms races that we still live with. The weapon development of the 1940s and 1950s, if unchecked, could destroy the world and everything in it. Atomic scientists started a doomsday clock counting down the seconds toward nuclear annihilation while armies armed with nuclear weapons looked at one another across short distances in Europe, in Korea, and in many other potential hotspots worldwide.

World War II changed higher education in America forever. Thousands more students enrolled because their tuition was paid by the G.I. Bill. University professors engaged by the U.S. government for their creative, scientific, intellectual skills during war kept that momentum going in peacetime. The nation wanted that, too, so that never again would America be so far behind other nations in science that our national safety was threatened. The National Science Foundation was set up in the national interest in 1950. Laboratories in chemistry classes in colleges left test tubes behind for ever-seeing scientific instruments; physics labs forgot about inclined planes and dealt with spectroscopy, computation, and eventually computers. Biologists became chemists while engineers became biologists. On the whole, those who taught science in universities saw their duties much more consumed by research interests, their skills, and creativity at same. Almost as important was their ability to convince others of that they had worthwhile ideas. Research funds replaced tuition as the most important financial resource for scientific study in universities and the entrepreneur scientist became commonplace, too.

Liberal arts colleges, with no graduate programs in the sciences, were not part of this growth. Almost without exception, faculty who taught chemistry, physics, and biology in liberal arts colleges, most of which had remained small, were not research scientists. They were the school teachers of somewhat more advanced students. Their job was the pre-graduate school preparation for those students who aimed for scientific careers. If they had a job outside the classroom, it was a second job to support their families. Gerrit Van Zyl, Hope's chemistry icon, worked with his wife in a downtown boutique. Harvey Kleinheksel, the second of Hope's chemistry duo, worked with his wife's family in Dykstra Funeral Home. Small college faculty served pre-health professional communities as well and liberal arts college faculty were loved by their students and their colleges for the jobs that they did. The graduate colleges wanted none of that to change. But, in the era following World War II, the skills of the prior generation of small college teachers in the sciences were less and less relevant. They were already obsolete. The young people emerging on the scene, if they were employed by a small school, wanted their piece of the research pie, too.

When the Russians launched Sputnik, President Kennedy declared the United States would put a man on the moon before the end of the 1960s. While World War II continued to simmer at some level in southeast Asia, the national attention was no longer only on the war. Kennedy redirected America's national psyche to putting a man on the moon. The war in Vietnam was a point of total transition for the United States as it tried to accept an uneasy peace with a still-undeclared war against an almost unknown enemy. The political organizations shook and stumbled and some of its leadership caved under its weight. As irresponsible as it seems from the distance of years, one potential president even talked about unleashing the power of nuclear weapons in Vietnam, while the actual later president dropped thousands of tons of bombs on a poor helpless country who simply shrugged its shoulders, and continued to maim young Americans.

Cal's tenure at Hope lasted just seven pressure-packed years, at which point he lost support of his Board of Trustees and was fired. But in many ways he was Hope's twentieth-century founder. President VanderWerf, who everyone called Cal, put Hope in a

16

position to not just survive in higher education in the twentieth century, but to lead all small colleges. He brought contemporary science instruction to Hope through convincing young scientists, many of whom were to be at the top of their fields later, to start careers there. Ten young scientists were added in chemistry and physics during his years, and though all but two left by 1988, their impact on the school was institution-altering, and significant for higher education as well. Four of these individuals became presidents of small colleges or foundations. Two others were trend setting leaders in emerging areas of technology as department and institution leaders.

The Hope College that was a liberal arts college started in 1866 by the Dutch Reformed Church in America in the western frontier disappeared forever in the VanderWerf administration. Though a subsequent president would try and reclaim it, finances were such that could not happen. The Hope that had struggled financially from the beginning continued to do so for a while. President Philip Phelps, Jr. had, in the 19th century, thought beyond a four-year school. He envisioned a Hope that would evolve to a university, and set financial planning on a path to make that happen. Had Phelps' plans succeeded, the 1960s would have been very different for the College. Hope College would have been Hope Haven University, and scientists perhaps almost all from the Dutch Reformed Church tradition would have brought to it levels in research that would have singled it out in the eastern Midwest for scientific excellence. John Sheehan, from Battle Creek, might have gone to Hope Haven instead of the University of Michigan on transferring from Battle Creek College. Hope Haven might have had an alumnus who synthesized, later, penicillin. (On the negative side, Sheehan also developed the commercial synthesis of the world's most powerful non-nuclear explosive, R.D.X. So, many sunken battle ships would have been on Hope's conscience, too.) But that was not to be. The eastern Dutch Reformed Church failed to support its future. Though Hope avoided extinction, its plans to become a university were killed, forever, by decisions in the early 1880s.

The problem Cal VanderWerf faced in 1963 was identified by the North Central Accrediting Association in a report that had been put in motion before he arrived. "Hope College has done very well with a school whose students are much better able than its faculty. In the current climate, that will have to change. Hope needs a revolution, and an evolution and President VanderWerf is the person to orchestrate that."

17

Hope College had rested too long, the review committee said, on its Reformed Church heritage. It had used being poor on several levels for good religious Dutch immigrant farm kids as a recruiting tool that had now gone well beyond its limits. The emerging generation of college students was going to expect much more from its higher education than just a college that was there. Often, potential Hope students had parents who were college educated. Now the Reformed Church kids had choices of where to go to school. Given what was also going on in higher education then, Hope had to become competitive or lose out for the best students. The critical question in 1963 was not if Hope College could survive in an environment dominated by ever-larger state universities, for which the US government was providing ever growing research resources. The question was whether any small college without graduate programs could remain relevant in the emerging Sputnik generation.

Fortunately, two far-seeing university administrators at the Alfred P. Sloan Foundation, Everett Case and Larkin Farinholt saw this as handwriting on the wall. Through a one-time series of grants to twenty small colleges, they set a change in motion that saved most, if not all, of the twenty. Hope had the most to gain: they had a scientist president, and, based on the very hard work of two science alumni, built a Hope College proposal into the largest grant in College history. Sloan, and VanderWerf, saved Hope's scientific future.

Cal VanderWerf inherited, as any president at the time would have, an institution where either a revolution was required in what it could sell in its future, or it would no longer retain relevance in, at least, its strongest academic units – the physical sciences. The question of which fork in the road was to be taken was open in 1963.

Had Cal not had the academic recognition in a basic science that he had, and had he not been able to hire young scientists because of that reputation, Hope would not have survived in anything approaching its present from. Many other schools attracted more able students and had more resources.

Cal needed strong Board support at the beginning of his tenure as president because he had to hire new faculty, and he received it. Hope College was far behind and he had to change its direction. Even though Bill Vander Lugt, Dean of the Faculty, said in the North Central report, "the only academic unit at the College that was known outside the Reformed Church" in America was the chemistry department, the recognition was in comparison to the rest of Hope College. Hope's faculty recognition in chemistry was that of a feeder school able to send good students to graduate schools. As the North Central report attests, this was more the result of its students than the faculty. Scientists in universities would build their academic reputations on the graduates of Hope and other small colleges. If Hope's faculty had a research reputation of any consequence, it was because of Gene Van Tamelen, later professor of chemistry at Stanford, an undergraduate Gerrit Van Zyl funded and promoted.

The National Science Foundation was feeding the appetites of scientists at big state universities with research grants, and those grants needed people to be carried on successfully. Hope's faculty was recognized as servants for the scientists of the state universities, and they played this role very well. But this was not the role of the future nor could it be the role of any liberal arts college that expected to maintain currency in the heady early days of federal support for scientific research. With success for the academician in the university came money and recognition. State governments were advocating a college education for students available within fifteen miles of their homes, too, so undergraduate enrollments in the basic sciences were booming at big state universities. These enrollments needed teachers; and Hope graduates as teaching assistants served that purpose. But, and Hope was caught in this bind, postwar science was based more on the ability of scientists to control the instruments that did the work than it was on the wet lab work of the scientists themselves.

An important question that had to be addressed, and not by Cal VanderWerf alone, was could a faculty be built in any four-year school that 1) could compete for the limited resources that were available from federal sources for programs in small colleges, and 2) could those faculty, with his mentorship, develop leadership positions in their fields nationally and internationally so they, in turn, could help increase the government's attention to smaller schools. This in turn required their ability to find funds for the instrumentation of the then contemporary research laboratories.

I am happy to report that Hope had done this by the mid-1980s. Hope, because of the faculty Cal VanderWerf hired, and the traditions set in place in the 1960s became the leading proponent of research with undergraduates in America. By 1970, it could be part of, and latter even lead, a greater academic world because of its research strengths. For some years, its physics department was widely recognized as being the strongest in a non-Ph.D. program setting in the United States. Biology also became very strong; chemistry maintained its national position; and the College built programs in engineering, computer science, geology, nursing and other areas of direct interest to students with scientific aptitudes.

A legitimate questioner could ask "Why all the focus on the sciences?" and many on the then-campus did ask that question. I would have answered the question with "It's always been, by its own admission, Hope's academic strength. It needs to maintain that." Cal answered the question then by saying that "Science appeals disproportionately to male American undergraduate students so, in order to keep the enrollment of males up, the sciences needed to be maintained." He sought parity in the male female ratio – and it was less than that in 1963, and declining.

Today I would answer "Scientists make things that can be sold to others. Science is a producing profession, not a service profession. Strengthening the sciences strengthens the overall American economy."

With Cal's expertise, and because of Hope's history, the VanderWerf administration built undergraduate mentored self-study into a College mantra that stands much more broadly a half century later. I came on this scene in the fall of 1964 as an aggressive, arriving from Harvard, newly minted Ph. D. in organic chemistry from Clymer, New York. Seven years, to the day, after I enrolled as a nervous, insecure freshman at Bucknell University on September 15, 1956, I walked into Paul Bartlett's office at Harvard a freshly minted Ph. D. in organic chemistry looking for my research assignment. The incongruity of this is evident, but so is the American dream. A year later, I was back at Hope as an assistant professor of chemistry. My affection for Hope was born-in but my affection for organic chemistry was learned, and well. Hope was the only college about which I heard as a child, and though I chose to go to Bucknell University out of high school because I needed to break that mold, I came back home as a college junior.

When I went to Hope as a faculty member, my Uncle Jim Neckers said, famously, "They'll make you mad." My Hope alum father, Carlyle, said "You're going to a place to work with such an awful retirement program?" Sue, my wife, said "Do we have to go back there? I prefer the University of Wyoming." But my instincts had it right: I had been inculcated with the excitement of research in organic chemistry, and there was no stopping me from that. I felt that working at Hope could be useful for me at the beginning of my career, and it was. I also, along the way, did some pretty good things for the College and for the students with whom I interacted directly and indirectly.

But these are parts of the rest of the story. In what follows, I first want to outline what Hope was and from where it came. Then, I want to set President VanderWerf's background before Hope reached out to him. Next, I want to deal with what led up to the convulsions in the academy in the 1960s. Then, I want to measure those convulsions at Hope, giving benefit to the judgment of the years. Finally, I want to wrap this all up with a bit of fiction as in what might have been that was not.

I present a heavy dose of the history of chemistry because the Hope College story is not complete without that. I offer a bit of analysis of the really important contributions of its most sensational chemistry alumni, and the most important president in Hope College's history, Cal VanderWerf.

Chapter 1

Growth of an Educated America

Hope College was founded to educate future clergy for the Dutch Reformed Church (in America). Missionaries from that Church were to travel the world and preach the gospel. Hope's founders were dreamers. Some, like Albertus Van Raalte, were practical men; others, like Philip Phelps, Jr. were too ambitious. But, now, after decades of ups, downs, and struggles, Hope is doing very well more because of the mistakes its clergy founders made than because of their original objectives.

In the 19th century the clergy of mostly Protestant denominations carried "higher education" to the American people. Higher education was a product of something broadly termed as the "Enlightenment," as in the Reformation started by Calvin and Luther. These movements took understanding and salvation from eternal death and damnation out of the hands of a few in the Roman Catholic Church and placed them firmly in the hands of a believer. "By grace, thou art saved," Paul wrote, of the priesthood of all believers. For the people to be able to see what was necessary for their own salvation, the Bible needed to be read to them in words that they could understand and those translators of the holy writ were the clergy. In the Dutch Reformed Church and many other reformed denominations, the sermon took center stage. The Dutch Reformed Church became clergy-centered, in spite of having been reformed to grant parishioners freedom from hierarchy, it became almost as structured as the Roman Catholic Church though without bishops and a pope. Complications necessarily arose, but the basic step taken by both Luther and Calvin was to elevate salvation from the church hierarchy and place responsibility for it on the individual. In the United States, higher education was tied to the protestant churches from its beginning. Though the universities secularized reasonably early in their histories, Harvard College, Yale College, King's College (now Columbia), New York University, Rutgers, the University of Chicago, and many other schools trace their origins to one or more Protestant denominations.[1]

Essential morality was always a target of religions. Though this was made complex later, the Ten Commandments rooted the

---

[1] For a detailed analysis of the secularization of the American university see George Marsden, *The Soul of the American University: From Protestant Establishment to Non-Belief* (Oxford University Press, 1994).

Judeo-Christian traditions: "Have no other gods before me; I am a jealous God punishing those who hate me, and loving thousands of generations of those that love me and keep my commandments; take a weekly rest; honor your father and mother; do not murder, commit adultery, steal or commit false testimony against your neighbor. Do not covet your neighbor's wife, ox, ass, servants or anything that belongs to your neighbor."

In the Christian tradition, Jesus taught his principles of service to others in the Sermon on the Mount (Matthew 5-7): "Blessed are the poor in spirit, those who mourn, the meek, those who seek righteousness, the merciful, the pure in heart, the peacemakers, and those persecuted for righteousness sake."

Christ also taught principles of Christian modesty (Matthew 6:5): "When you pray, you must not be like the hypocrites. For they love to stand and pray in the synagogues and at the street corners, that they may be seen by others. Truly, I say to you, they have received their reward. But when you pray, go into your room and shut the door and pray to your Father who is in secret. And your Father who sees in secret will reward you."

He instructed on how to pray to the one God (Matthew 6:7): "And when you pray, do not heap up empty phrases as the Gentiles do, for they think that they will be heard for their many words. Do not be like them, for your Father knows what you need before you ask him. Pray then like this," directing them to the Lord's Prayer.

The added acclimation of Jesus' direction to the eleven apostles surviving after Judas (Mark 16: 15-16): "Go ye into all the world and preach the Gospel" in search of believers who shall be "baptized and saved." This translated, in the 19th century to the marching orders for the zealous, the young, and the converted; numerous Hope College alumnae and alumna only were able to study there if they agreed in advance to become missionaries.[2]

In the 19th century death from disease was always present. There were no antibiotics and sewer systems were non-existent. Scourges and epidemics often took the very young. There was a strong sense of community of those that survived it all. We see this in the history of America's earliest settlers, especially among the pilgrims. The ideas carried to America by a very small, but dedicated group of Calvinists and their leader William Bradford, who

---

[2] Victoria Henry, "The Dual Calling of Missionary Wives: A Look at the Reformed Church of America in China, 1917-1951," (The Joint Archives Quarterly; Hope College, Vol. 25, (1) 2015), 1-5.

eventually led the Plymouth Colony were of a strong, clergy centered religious faith. They had seen the deaths of most everybody in their families in 17th century Yorkshire. Thus, the young Bradford, being even more enthusiastic than most, joined religious separatists before the Pilgrims moved to Leiden.

At least one biographer compared the motivations of Hope College founder Albertus Van Raalte, who had brought a small band of settlers to western Michigan, to the Plymouth leaders, this time as the second "leader to a Plymouth colony, but in the 19th century in western Michigan." Van Raalte also had seen many people die because of epidemics and had experienced a near-battlefield conversion that led his being enthusiastic for salvation of others on terms he was to define. He developed contrary ideas to the Reformed Church in the Netherlands, and was never officially ordained, but became instead a "separatist" who identified with the poor in the Dutch provinces of Gelderland and Overijsel.[3] In September 1847, Van Raalte embarked with a group of about sixty Dutch from the country-side for New York. He taught himself enough English to survive when they reached New York. After a winter in Detroit, most, but not all, of the band went westward and settled in Ottawa County, Michigan. Van Raalte's gang (though he himself was university educated) was comprised mostly of poor working families. Van Raalte had incredible imagination though, and began an "academy" in 1848, just one year after the beginnings of the Holland settlement. He engaged the Dutch Reformed Church in the East for support, and hired a principal, John Van Vleck, in 1851. With that principal and some teachers, the academy started to have success teaching the settler boys English, reading, writing and arithmetic.

Van Raalte purchased thousands of acres of western Michigan land for under $2 an acre, and donated five acres of land for a building spot for the school, with the caveat that he would appoint the board himself. Van Raalte took care to specifically separate the Reformed Church clergy from the board, appointing members who would provide "a decisive hand to take charge of this important responsibility to the young." Van Vleck Hall, built on Hope's campus to house students and for instruction, was erected on the site where it still stands with $12,000, most of it donated by churches in the east. The Van Raalte Academy was placed under the Board of Education of the Dutch Reformed Church in 1853. After the

[3] Albert Hyma, *Albertus C. Van Raalte and the Dutch Settlements in the United States* (Grand Rapids, MI: Eerdmans, 1947), 15, 23.

donation of more land by Van Raalte,[4] Hope College separated from the Academy in 1866. These decisive steps in the history of the College impacting its future for generations took place almost immediately after the school was established. It does not necessarily follow that the Church received (because it was authorized by the State of New York to hold property in Michigan) a much larger role than it might have had Van Raalte waited a time before beginning the school. But had he been able to establish his presence, financially, more at the beginning, the organization of Hope College could have been quite different.

Van Raalte's insights though recognized by many, were sometimes not appreciated by his own people. Holland remained conservatively attached to its Dutch heritage(s) for generations. This disabled some its own future and even frustrated Van Raalte, who began threatening to return to the Netherlands as early as 1856. He feared the people did not support his projects, specifically at the Academy. Van Raalte, and later Philip Phelps, Jr., wanted greatness, and struggled to keep this dream alive. The Dutch Reformed Church, headquartered in New York City, and in upper Hudson River was reticent to provide that much support to those "westerners" And the Holland locals were not that supportive.

Pitches for money have to be correctly focused, or they do not always convince even the generous. Consider a recent example: When former president of The Ohio State University Gordon Gee sought money for his university a few years ago, he was quoted in the New York Times: "Sometimes, when I see the pitches other nonprofit institutions are using, about how they're facing hard times and they need money, I wonder why they think anybody would want to support something that's going under. People do not want to spend their hard-earned money on keeping something afloat. They want to support something aspirational, something that's going to thrive. You have to

---

[4] The name Van Raalte, means "from Raalte"; Raalte was a small town in Overijsel. Many of the towns that figured in Albertus Van Raalte's history in the Netherlands, Ommen, Deventer, and Zwolle, are within a few kilometers. As far as his record is concerned, Albertus Van Raalte did not live in Raalte. But other Van Raalte's did. A Jewish family immigrated in the late 1890's founding Van Raalte Lingerie, and later Van Raalte Mills in New York and New Jersey.

give them an image, a grand vision of what you are trying to do."[5]

Van Raalte's vision was educational excellence in the strongest Christian tradition. His motto, albeit in Latin, a forbidden language for many Dutch Reformed, was *Spero in Deo*. He was poor, but educated and he had a dream and that was the hope that through God he might achieve academic significance for his College.

The Setting: 1856

Education before the beginning of the 20th century favored the wealthy. It was not that the majority of Americans were purposely excluded from a college education. Most simply did not live long. Those who did live into adulthood often found their educational backgrounds insufficient. Many young persons needed more education to qualify for college admission than routinely available in rural communities, but there were no public schools of consequence that taught anything but reading, writing and arithmetic. So private schooling became almost a necessity, and this was prohibitively expensive. There also were not always teachers available who were, themselves, well-educated. Though public education started to be free in many states after the Civil War, it did not become compulsory everywhere until the mid-20th century, and only then after contentious arguments about public support for parochial schools, particularly Roman Catholic schools. There was also the problem of discrimination later dealt with in *Brown v. Board of Education* (1954). But an increasing number of young Americans did not have access to quality schools. The public schools of the late 19th century also had to fight with the practice of "work out": young persons that had to work for another (mostly farmers) to pay for their own food. There was not much time for schools for most Americans. Later, James Bryant Conant's post-Harvard study of the *American High School Today* (1959) took the public education movement apart. Conant thought there should be comprehensive public schools that would provide for the general education for all future citizens, offer good elective programs for those who wished to use their skills in the job market after graduation, and that would prepare those students that were able, a stepping stone to post-secondary education. In 1959,

---

[5] *New York Times* interview with Gordon Gee, president of The Ohio State University, 2009 ("The Man With Nerves Like Sewer Pipes, *The New York Times,* February 9, 2014.)

when the Conant report was issued, about 30% of the US population was proceeding towards a Bachelor's degree.[6] But not in Van Raalte's day – and that day saw the founding of Hope College.

Public School Education as it Evolved

The concept that sets public education apart in the United States is that of equal opportunity. When compared to European school systems, political equality like that which Jefferson championed, widened by the late nineteenth century so that increasingly students that had the ability to compete would be given a chance to do so. American's attitudes no longer just required making do on the frontier. "Equality became, above all, equality of opportunity—an equal start in a competitive struggle," Conant wrote.[7]

"By 1900, the power of the twin ideals of equality of opportunity and equality of status" led the American people to "believe that more education provided the means by which these ideals were to be realized." As a result, along with institutional expansion among colleges and a declining need for child labor, enrollments in secondary schools expanded greatly.[8] Conant cites the figure of 17 year-olds enrolled in school, from 35% in 1910 to over 70% at the time of his book (1959). During the same period, the percentage of youth attending college jumped from 4% to 35%.[9] Later, the need for a college would become almost a given. Though, in the cases of the real trend setters, Robert H. Jackson, Bill Gates, Steve Jobs, and others—education was just in the way of their talents and objectives, those Americans that could achieve it were bettered by a college education.

In Europe, in contrast, common schools stopped at elementary levels. Adolescents were sorted, often on the basis national examinations, and sent off to entirely different trainings depending on the outcome of their examinations. Some training led immediately to apprenticeships or other forms of employment, while the most prestigious, and smallest segment, prepared students for the university entrance examinations. Secondary schools that tried to serve all of a community's children simply did not exist. Conant notes

---

[6] James Bryan Conant, *The American High School Today* (New York: McGraw-Hill, 1959).
[7] Conant, *American High School Today*, 5.
[8] Conant, *The American High School Today*, 6-7.
[9] Conant, *American High School Today*, 6; and Floyd M. Hammack, *Future of the Comprehensive High School* (New York: Teacher's College Press, 2004).

that at the time of his book, "three-quarters or more of the youth go to work at fourteen or fifteen years of age."[10]

The United States, responding to more democratic forces, evolved a secondary school system where vocational and college preparatory education took place under the same roof. *The Cardinal Principles of Secondary Education* report, issued in 1918 by the U.S. Office of Education and under the sponsorship of the National Education Association, strongly influenced thinking about our high schools.

Conant summed up the tension between a nation moving towards progress and one who had to make significant strides in its education, writing in 1945: "One of the highly significant ideals of the American nation has been equality of opportunity. This ideal implies on the one hand a relatively fluid social structure changing from generation to generation, and on the other mutual respect between different vocational and economic groups, in short, a minimum of emphasis on class distinctions. It is of the utmost significance for our future that belief in this ideal be strengthened."[11]

Education was paramount, particularly at the high school level, for restoring a fluidity to our social and economic life. This type of education, Conant believed, was important for preparing the future generation for an economic system based on private ownership but also to train them in the ideals of social justice. Students would also, it was hoped, learn the virtues that would define their nation.

In Floyd Hammack's analysis: "Conant clearly had a sense that 'academic talent' was a finite quantity among any cohort of youth. Though he cited, with approval, the expansion of secondary and collegiate enrollments, he thought that truly democratic schools would offer a variety of curricula, leading to different occupational careers. Students would be sorted among these courses and programs according to their performance, inclinations, and ambitions."[12]

Conant, as president of Harvard, systematically began targeting students based on academic talent and ability, and less on the means to attend. With the development of placement and intelligence tests, Conant sought to cultivate the 15-20% of youth that he thought were sufficiently prepared for college, either in the academic or vocational setting. He did the same with the Harvard faculty in the early years of his presidency. It had become

[10] Conant, *The American High School Today,* 2, 7.
[11] Hammack, *Future of the Comprehensive High School,* 11.
[12] Floyd Hammack (1998), "The Future of Comprehensive High School Education." Retrieved from https://files.nyu.edu/fmh1/public/memorandum.html.

exclusionary and elitist he said, and he opened faculty positions to the competitive forces of the American academy we know today.

Hammack wrote that Conant believed that "Not all students were to be expected to want to pursue a college preparatory curriculum. However, the (public) school had an explicit responsibility of keeping open lines of mobility and of reducing the barriers to mobility."

Toward the end of his long and storied career, Conant defended his principles, writing, "The comprehensive high school has been defended on social and political grounds as an instrument of democracy, a way of mitigating the social stratification of society. Such has always been my argument."[13]

Van Raalte's vision of Hope College from 1847-1872 did not have the benefit of Conant's years of analysis and criticism, but in his *Anchor of Hope* he laid the groundwork for a bright future for his charges "based on education, and eventually higher education." In the hands of Philip Phelps, Jr. this vision took root in a form for undergraduates, as we know them now, of a liberal arts college.

In a similar way, another American leader was also trying to democratize education in a most different setting where the objectives were entirely different. By 1878, John Heyl Vincent, the founder of Chautauqua Institution, was also addressing a more generally available university education. Vincent's dream begat what later became a Great Books reading program. His attention, as much as anything, was on women – homemakers and mothers. "Education, once the peculiar privilege of the few, must in our best earthly estate become the valued possession of many."[14]

Vincent founded the Chautauqua Literary and Scientific Circle, writing that its mission was "to promote habits of reading and study in nature, art, sciences, and in secular and sacred literature," and "To encourage individual study, to open the college world to persons unable to attend a higher institution of learning."

This idea spread after, and because of, the involvement of William Rainey Harper then a professor of Hebrew at Yale, and by the turn of the century, over ten thousand such reading circles had been formed. The Chautauqua Literary and Scientific Circle (C.L.S.C.) was more than a book club, or a university extension program, it became an authentic beginning of adult education.

---

[13] Hammack, "The Future of Comprehensive High School Education."

[14] Leon Vincent, *John Heyl Vincent – A Biographical Sketch* (New York: Macmillan, 1925).

Carnegie's instigation of libraries in small communities helped as well. Eventually all of this stimulated a significant growth in an interest in higher education, and in an interest in institutions of higher learning particularly among those at home. It still would take until after World War II for states to get into higher education in ways beyond those schools that received land grants after the war.[15] Eventually, excellent state college and university starts transfixed the nation and today, many of their reputations surpass that of their private competitors.

The Chautauqua Circle reading, or book, list consisted of what the educated people of that generation thought important for learned individuals to retain; Greek history, Greek literature; English history; English literature; Bible history; and a bit of astronomy. American and Canadian history were, naturally, slipped in, too. J. R. Green's "A Short History of the English People" was the first book selected for reading by the C.L.S.C. leadership. This book shares Green's fascination for the people around Oxford, where he was born and lived his whole life, but it is really a story of English people and not the language or the history. Over a course of years, geology also showed in the reading list, as well as writings by Longfellow, Hawthorne, and Washington Irving. *A Beginner's Handbook of Chemistry*, by John Howard Appleton was on the 1888-1889 reading list and Chautauqua Course in physics on the 1889-1890 list, demonstrating the breadth of information available to readers across genres and professions. Even though the lawyer who argued for *Plessey v Ferguson* (1892), Albion Tourgee, lived in Mayville, New York (the county seat of Chautauqua County), the first mention of the U.S. Constitution in C.L.S.C. books shows up in "The Story of the U.S. Constitution" by Francis Newton Thorpe in the 1891-1892 edition of books. Perhaps unsurprisingly, at Chautauqua and in Chautauqua County, there were no blacks. It was a white enclave, isolated from everyone, even its Chautauqua County neighbors. One of the more curious books on the C.L.S.C. historical list is *Science and Prayer* by W.W. Kinsley and William Cleaver Wilkenson, and edited by W. D. and P. L. McClintock, an 1893-1894 entry. Content did not matter to Vincent, and later, to Harper, as much as intent. The intent was to provide an educationally broadening experience for Americans, mostly women.

The founding of Chautauqua Institution, and its intellectual

---

[15] For a list of the books used on the Chautauqua Literary and Scientific Circle reading list, visit: http://www.ciweb.org/phocadownload/literaryArts/CLSCBookList.pdf.

parenthood by Vincent and for a few short years, William Rainey Harper, also coincided with the beginning of what American colleges now call the liberal arts and sciences. Harper, though a serious scholar, was the consummate salesman and even convinced large numbers of literates spread over the country to sign up for summer courses in Hebrew! Harper was said to be, like Van Raalte, "A Christian and a scholar," after having been converted at a Baptist camp meeting in Granville, Ohio, where he was an instructor at Denison University. He connected with Vincent because in the summer of 1882, he had set up a Baptist summer school, on Chautauqua Lake in western New York at place called Point Chautauqua during his vacation. John Heyl Vincent's Chautauqua, for Methodists, met across the Lake at the former Fairpoint, and was by this time, known as the Chautauqua Assembly. Vincent decided he could not beat Harper, and had to join him. So, as Harper's biography, written in 1941 says, "Vincent knew, like John D. Rockefeller to the petroleum people, that if he didn't get Harper under his tent, he – Vincent – would lose."[16] So, Vincent, afraid of a competition that never materialized from the Baptists, hired Harper to be the "Principal of his school" at Chautauqua.

Harper's tenure at Chautauqua was short, and eventually contentious. His tenure may be idealized in other's review of his work, who said he was, "A Young Man in a Hurry," and Chautauqua's summer schools bloomed in his hands. Someone said "He could read a railway timetable aloud and convey the impression this was an important document." Several hundred students took his classes.

John D. Rockefeller was one of America's paradigms of the "rich industrialist." He was by then coming under pressure from all sides to provide funds for education. The president of what became Colgate Divinity School in Rochester, NY, Augustus Strong, wanted to "take possession of New York City for the Baptists" and was asking Rockefeller for $20 million to do that. Because, that president said, "there were already more than enough one-horse colleges (in small country towns, like Holland, Michigan) to stock the world." But Rockefeller could not be convinced – by Strong anyway, and Morningside Heights went to the Columbians.

Chicago's treasurer, on the other hand, was much more

---

[16] There are many biographies of Harper, such as Milton Mayer's *William Rainey Harper, Young Man in a Hurry*, but the online archive of Harper College is excellent and accessible.

modest. He asked Rockefeller for a mere $100,000 which would do nothing but build a great high school, according to Colgate Divinity's Strong. Strong wanted Rockefeller's New York money to build a great university in Morningside Heights – where Columbia now stands – that was to be a militant Christian college closed to "infidels." Yale also made an offer in 1886. At that point, Chicago's trustees immediately elected him president. Rockefeller had heard Harper lecture, and, though he did not commit right away to support his efforts in Chicago, he made the point of his affection in a telegram sent Harper in 1892.

Harper's ideas took off. He used his insight to promote the idea that America needed a great research university. There were fine institutions, according to Harper (Harvard, under Charles Eliot, being one). But, though Eliot's report for the Academic year 1900-1901 at Harvard talks about multiple academic opportunities, more than 50% of the graduates received Bachelor of Arts degrees. Chicago's literature said that "until Harper, there was no institution that was 'setting the pace' to provide a daylight of higher understanding." By the turn of the century, the German research institutions were well on their way to seeding what became, around 1903, the beginnings of I. G. Farben. Harper felt American universities had to catch up and he hired American scientists educated in German universities. These scientists built his great research university for him.

Scientists, in Harper's idealized version of "university" (the word was only coined around 1835!) were to lay the ground work for discoveries of general use, but not focus on anything practical. Football, fun and fraternities were out as near as possible, and there was no concentration at all on undergraduate enrollment. What Harper did was to build the first great research university: "Harper's university would water the tree of knowledge at the roots." [17]

At the beginning, there were no funds for Harper's salary at Chicago. So he took the position at Yale where he taught Hebrew to the masses, and gave Bible lectures in downtown New Haven. Until, as it turned out, with Chautauqua growing, he collapsed from overwork after the summer of 1889. But the pursuit of Rockefeller by Chicago went on relentlessly. Then the "University of Chicago," a euphemism for a Baptist outpost for higher education with an

---

[17] Thomas Wakefield Goodspeed, *William Rainey Harper: First President of the University of Chicago* (Chicago: University of Chicago Press, 1928).

affiliated seminary, in Morgan Park, Illinois, crashed and burned. By this time, Harper was an instructor there, too, so they actually made him its president.

The first gift to the University of Chicago by Rockefeller was only $600,000; "only" because, in the end, he would contribute close to $80,000,000. The $600,000 was accompanied by a matching demand; that $400,000 be provided by Chicago's other supporters. Harper held his ground about the university that it was to be a *university of the highest character* targeting research. Undergraduate programs could come later. He said openly that he was not sure he would go to Chicago (unless Rockefeller gave him a blank check). That blank check began to happen when Harper was elected President of the University of Chicago on September 18, 1890; he opened the University in October, 1892 with $2,000,000 from all quarters, ten acres of land donated on the South Side of Chicago by Marshall Field, an ardent scattering of Baptists, and William Rainey Harper. Harper's *modus vivendi* in addition to pushing research, included individualized learning, publication and extramural teaching (as in extension schools) to the lowest chronological level. Undergraduates were to concentrate their studies in one or two individual subjects or majors. Fifty years before, Robert Maynard Hutchins began to clamor for a revival of the human tradition, Harper wanted to select the world's great books and publish them. So, he founded the University of Chicago Press.

The First Great Research University

Harper was fond of saying that one could have a great university without classrooms, but one could not have a university without research. He created the first great research university where every faculty member was a scholar; if one was not a scholar, there was no job for that person at Chicago. Harper worked himself into a frenzy, killing himself with work by 1905. His effort would remain to be finished by others, but his legacy had been established.

It took Harvard until the era of Conant (1932-54) to firmly plant the idea of academic excellence that we know of it today for a great university among its departments. Until Conant's presidency, at least according to his biographer,[18] there was significant inbreeding in the faculty and only Harvard alumni were hired for Harvard's teaching faculty. It has been said that the Nobel laureate in organic

---

[18] Herschberg, J. C.; *James Bryant Conant: Harvard to Hiroshima and the Making of the Nuclear Age* Knopf, 1993.

chemistry, Robert Burns Woodward, was first denied tenure by his Harvard colleagues. Conant saw Woodward's potential though, and instituted an outside review team to study the case. On the second vote, tenure was accorded. Chicago, in contrast, managed to hold its academic principles together through thick and thin though, seeding what became the nuclear generation through its physics department in 1942. The wrench of the humanities, and Eliot's Harvard College offering only a B.A., continued to play out at various levels at other universities, however, and to large extent still does today. The ideal for the undergraduate's degree is a degree in the liberal arts and sciences, a Bachelor of Arts.

So, what were the liberal arts?

Woodrow Wilson, while president of Princeton in 1907, and speaking to the teachers of New York City described the liberal arts in his speech "The Meaning of a Liberal Education":

"I suppose that what perplexes every man to-day in every walk of life is the extraordinary complexity of modern life as compared with the life in the midst of which our grandfathers found themselves, as compared with the life in the midst of which the generation immediately preceding ours found itself. The life of the present day is incalculably complex, and so many of its complexities are of recent rise and origin that we haven't yet had time to understand just what they are or to assess the values of the new things that have come into our life. Not only is life infinitely complex in our day as compared with the previous age, but learning is correspondingly complex. In the old days of the fixed curriculum of the college and the school one could say with a degree of confidence that the elements of these curricula did contain the main bodies of knowledge, by specimen at least. But who can say that any curriculum that can be packed into the years of school life and the years of college life combined   contains all the elements of modern learning? Modern learning has been so drawn into a score of consequences, has been so extended into a system of uses, that it is a sort of mirror held up to life itself, and the man of affairs now seeks in the laboratory, in the quiet places of counsel, from the scholar, those main elements which shall guide him in accomplishing the particular material tasks which lie immediately under his hand. So that life and learning are equally complex, and they are interlaced with each other, they are related as never before. There is not the scholar on the one side with his door closed and his window open, and on the other side the manufacturer and the man of commerce beating the seas with his ships and searching the distant markets of the world for new stuffs. That is not the contrast which exists to-day. The man of learning has on his table a telephone that connects him with all the activities of the world, and his windows look out on smoky chimneys; he feels that he is one of the many servants to carry on the great tasks of to-day, whether they be material or intellectual. So that these complexities interlock and are the same complexities, the complexity of knowledge and the complexity of life.

It goes without saying that there is an equal complexity of economic effort, of employment, and therefore an infinitely greater difficulty than there used to be in calculating the future orbit of any young person. When you say a young person must be prepared for his life-work, are you prepared, is he prepared, are his parents prepared, to say what that life-work is going to be? Do you know a boy is going to be a mechanic by the color of his hair? Do you know that he is going to be a lawyer by the fact that his father was a lawyer? Does any average and representative modern parent dare to say what his children are going to be? My chief quarrel with the modern parent is that he does not know, and that he hands that question over to the youngster whom he is supposed to be advising and training.

With this complexity, what has the modern school attempted to do? It has attempted to do everything at once. It has said: Here are a lot of boys and girls whose future occupations we do not know and they do not know. They must be prepared for life. Therefore, we must prepare everybody for everything that is in that life. We have not found it amusing. We have not found it possible. We have attempted it and we know we have failed at it. You cannot train everybody for everything. Moreover, you are not competent to teach everything. There is not any body of teachers suited in gifts or training to do this impossible thing. Neither the schools nor those who guide them have attempted to make any discrimination with regard to purpose or to settle upon methods which will promise some degree of substantial success. That is the situation we are in.

I do not wonder at it. I think it is hardly just to blame those who have brought this situation about, because this change in modern life has come upon us suddenly. It has confused us. We are in an age so changeful, so transitional, I do not wonder that this confusion has come into our education, and I do not blame anybody. I do not see how it could have been avoided, how we could have avoided trying our hands at a score

36

of things hitherto not attempted to determine at least if they were possible or not. Therefore, this is not a subject for cynical comment, this is not a subject for criticism. It is a subject for self-recognition. The present need is that we should examine ourselves and see whether this be true or not; and, if it is true, ask ourselves whether the air has cleared enough, and whether our experiment has gone far enough, to make a definite program, to make a radical change, in the things we have attempted. This is the moment for counsel. The thing that is imperative upon our con-science is that we should ask ourselves whether it be possible to do it differently and better.

If we are going to do it differently or better it is imperative that we should distinguish between the two things. It is imperative that we distinguish between education and technical or industrial training. And before we distinguish between these two it is necessary that we distinguish between the individuals who are going to take the one and the individuals who are going to take the other. There is no method in American life by which the state or any public authority can pick out the persons to be educated in the one way or the other. The vitality of American life, and the vitality of all democratic life, lies in self-selection; it lies in the challenge put upon all to make up their minds as to what they want and what they intend to do with themselves. It is absolutely essential that we should start with that or we can never have any system of education."[19]

Wilson saw his day as a time to differentiate between the MITs of higher education and the Princetons – one school taught practicality and the other school educated men, and for Wilson there had to be some difference. What he may not have seen was the day in which his destiny and all the destiny of mankind would be in the hands of a few like himself – a few presidents of universities, and the president of the United States.

A fellow proponent of the possibilities of university education, Conant was a Harvard undergraduate when Wilson was president of Princeton, majoring in chemistry. Following his doctorate, he moved his science to Washington where he worked with others in chemical warfare for the Army. Finally, he was assigned to Willoughby, Ohio, where he directed the manufacture of tons of Lewisite, monochlorovinyl arsine. This toxic gas was invented

---

[19] Woodrow Wilson, "The Meaning of a Liberal Education," (*High School Teachers Association of New York*, Volume 3, 1908-1909), 19-31.

originally by Ralph Lewis, professor of chemistry at Northwestern University, following developments outlined by Father Julius Nieuwlands, in his Ph. D. dissertation at Catholic University in 1905. Nieuwlands showed that arsenic trichloride, when treated with acetylene, produced three compounds all of which were exceedingly toxic arsenic derivatives. Conant and his crew separated these three compounds into the most toxic (named Lewisite after Ralph Lewis) and developed a commercial method for its separation. This was practiced on an old auto maker's site in Willoughby. When the war ended, eighty tons of Lewisite was on the way to France.

Conant was one of four scientists that interfaced with the United States government immediately prior to World War II where he served on the first National Defense Research Committee (N.D.R.C.). The N.D.R.C. began operations in June 1940, and it was under N.D.R.C.'s direction that the establishment of the great American academic/government research complex came into being.

Almost because Conant had worked at the Army Research Labs in Washington during World War I, the vestiges of that are felt to this current day.[20] Twenty-seven years later, at the end of World War II, Conant was directly involved in the decision to use the atomic bomb on people in Japan. This cost him dearly, but he defended the decision to the end. And his relationship with one his sons in a story told recently by his granddaughter, Jennet Conant, was irretrievably destroyed.

"It was a tragedy, in the end, for my grandfather and the other Los Alamos scientists, who saw the tremendously powerful new force they helped create used to incinerate hundreds of thousands of innocent civilians. It was never their intention to cause the radiation epidemic that followed, yet they did.

It was certainly never their purpose to let nuclear weapons loose on the world, initiate an arms race and allow petty dictators and terrorists the means to cause mayhem at the push of a button, yet that is precisely what happened. As a result, my grandfather took no pride in his wartime service, once confessing to Newsweek, "I no longer have any connection with the atom bomb. I have no sense of accomplishment."

He devoted the latter part of his life (until his death in 1978) to trying to secure international control of nuclear weapons and did not want to recognize himself in the role of destroyer. But the verdict

---

[20] For the stunning story, see the New York Times' "Zeroing In on Mystery of an Old Site Called Hades," March 17, 2012.

of history, which tormented him in his final days, will deal most heavily with the politicians who waged war, not the scientists who provided them with the latest weapons.

Science has no choice but to move relentlessly forward; it is society that always seems to be slipping backward."[21]

Robert H. Jackson – another 19th century icon

The college story of Robert Jackson, Supreme Court Associate Justice, has been made a legend by John Q. Barrett. Jackson learned to read from a list of great English literature provided by his high school English teacher as almost his sole higher educational experience. Jackson took a year of post-graduate high school education in Jamestown High School where he came upon a Great Books list given to all students by English teacher, Ms. Mary Willard dated 1893.

Willard's list of thirty great books, though probably a little more detailed, was nothing like the Chautauqua Literary and Scientific Circle list of John Heyl Vincent or Harper. Her list was a small sample of books from the English-speaking world comprised mostly, though not entirely, of books by English and a few American men. Neither Vincent's list, nor Willard's, were as comprehensive as Robert Maynard Hutchins' and Mortimer Adler's *Great Books of the Western World* published as collective volumes first in 1952. This effort was – according to Hutchins – an attempt to provide the complete reader with a "liberal education" because "the great bulk of mankind has never had a chance to get a liberal education.[22] So the list included some science, though it is doubtful the normal reader could have comprehended Einstein's *Relativity: the Special and General Theory,* or Niels Bohr *Discussion with Einstein on Epistmemology,* or Schrödinger's *What is Life?* any more than the average reader can read *Scientific American.* Nonetheless Great Books included these works, and Lavousier's *Treatise on Chemistry,* too. Hutchins' University of Chicago owed much to the first Chautauqua in Chautauqua County, creating a lineage of learning that made both vital. It was there, at Chautauqua that Harper led a summer school, starting in 1893, in languages as esoteric as Sanskrit and Chaldee.

Vincent said it this way: "Chautauqua will exalt the

---

[21] Jennet Conant, "My Grandfather and the Bomb," *Los Angeles Times*, May 2, 2005.
[22] Ashmore, Harry Scott. *Unseasonable Truths: The Life of Robert Maynard Hutchins.* Boston: Little, Brown & Co., 1989.

profession of teacher until the highest genius, the richest scholarship, and the broadest manhood and womanhood of the nation are *consecrated* to this service."[23]

His explanation of the movement: "We expect the work of Chautauqua will be to arouse so much interest in the subject of general liberal education that by and by, in all quarters, young men and women will be seeking means to obtain such education in established resident institutions... Our diploma, though radiant with thirty-one seals—shields, stars, octagons—would not stand for much at Heidelberg, Oxford, or Harvard...an American curiosity.... It would be respected not as conferring honor upon its holder, but as indicating a popular movement in favor of higher education."[24]

In western New York, there was little that Jackson could afford – so he educated himself with the help of a superb teacher and thirty great books. Endowed with strong intellectual, organizational skills, and a clear writing style, he rose to one of the most influential justices in the history of the Court.

We have presented bits about great persons, and their thoughts, of the pre-20th century American college education; Vincent, Harper, Wilson, Conant, Jackson. For much of our history, who went to College was determined by what was available, and though high achieving, ambitious, and able men like Jackson were successful, the efforts to expand education focused on making college available where people were. In Holland, Michigan, that would become Albertus Van Raalte's *Anchor of Hope*.

---

[23] Oyen, Henry. *The Founder of the Chautauquas – The Varied and Helpful Career of John Heyl Vincent* in *World's Work: A History of Our Time, Vol. XXIV.* May 1912.

[24] The Papers of John Heyl Vincent are located at the Southern Methodist University Perkins School of Thelogy and are available online: http://www.smu.edu/Bridwell/Collections/SpecialCollectionsandArchives/~/media/Site/Bridwell/Archives/BA30125.pdf.

Chapter 2

Anchor of Hope

Van Raalte and Hope's beginnings

Van Raalte managed to get Holland, Michigan, and Hope College, started pretty well. In many ways, he was the sole driver of education in western Michigan and the initiative was not misplaced, though the follow-through was really quite poor, especially in terms of advanced education. Money was always an issue. The western Michigan Dutch churches were simple minded with obstinate clergy and hard-headed congregants. The Van Raalte gang were outliers as far as the eastern Reformed Church was concerned and Hope College was almost their afterthought. Though leaders in eastern Reformed churches had been in the United States long enough to be established, and they had resources, and were also philanthropic at least at some level, documents in different places suggest the General Synod of the Reformed Church (known as the Protestant Reformed Dutch Church then) in the mid-19th century did not give Van Raalte's "anchor of hope" much credence or a lot of help. Van Raalte looked to them, and maybe gave away too much of the vision, but their help in return was less than enough.

Van Raalte was, however, very clever. He insisted that the school be separated from the Church at least at some level through the composition of its Board of Trustees. At least one-third of the board, which could be up to thirty-five members but not more, was to consist of laymen. The General Synod of the Dutch Reformed Church was the body that had "corporate authority, according to the laws of the State of New York, to hold property for educational purposes in the State of Michigan, and had ultimate control."[25] It had the power to accept or reject nominations for members of the Council. When Hope College was chartered in 1866, the General Synod was the chartering agent and it guaranteed upwards of $30,000 for Hope though it remains unclear if the sum was ever paid.

Schuyler Colfax

Van Raalte's Board was not intended to be political, but attracted the attention of an important political figure. On Van

---

[25] Articles of Incorporation, Hope College, 26th day of April, 1866; filed office Secretary of State, May 14, 1866; copied – as certified – by Secretary of State, State of Michigan, 24th day of April, 1896.

Raalte's first Board of Trustees was former Vice President (serving under Ulysses Grant) Schuyler Colfax had been replaced by Henry Wilson for Grant's second term because he lost the support of the Republican Party after the Credit Mobilier scandal came to light. Note his signature on an envelope from the Vice President's chamber while in office, and on the first Articles of Incorporation of Hope College, May 14, 1866.

There is no evidence that Colfax had much input at Hope or that he wanted input. But Van Raalte's idea was correct: Colfax was a mover and shaker in South Bend, and his presence on the Board was a business decision for the pioneer Van Raalte. A college as an institution of higher education needed well-known people from the community on their side and Colfax certainly fit that description. Van Raalte also knew well the control mania of the Dutch Reformed Church. He knew that money counted so some of his intent with his first Boards had to be to include people who had at least access to funds, and at least the Eastern Church had some of those. No matter the intent, the Articles of Association of the College could not be amended without a vote of the General Synod of the Reformed Church. Both the president of the college and the executive secretary of the General Synod also were *ex officio* members of the board without term.[26] In the end, this Church connection, particularly in the context of the "Theology Department," almost sunk the new college.[27] Regardless of Van Raalte, Hope College became a college of the Dutch Reformed Church.

Board composition remained a matter of concern, surfacing over and over again even into the VanderWerf days. By that time, board membership had grown well beyond the intended thirty-two members and was a burden, if not a problem. In fact, Bob De Young, a Vice President who served under many presidents, maintained that President VanderWerf's most long-lasting contribution to the College was in reducing the number of Board members from over fifty-five to eighteen during the third year of his presidency. By the time he did this, Board meetings cost Hope College more in travel

---

[26] The definitive work on Hope's first one hundred years is Wynand Wichers' *A Century of Hope* (Eerdmans, Grand Rapids, 1966).
[27] Wichers, *Century of Hope,* Chapters 6 and 7.

and accommodations for members than Board members contributed as individuals to the Alumni fund.[28]

The history of the College for its first 100 years has been written by others, with Wichers' *A Century of Hope* the most detailed. The beginning of the College was rocky. A school or academy was started first, and the Academy (high school) had to be extracted from the district "public" school. Then, Hope College had to distinguish itself from the Academy. There were untold arguments with the eastern Reformed churches about money and jealousy about money being spent in Holland that was, it was said, letting the "westerners" off easy. After the initiative to start a college in Holland began, there were arguments with John Van Vleck, a minister who knew no Dutch, who was named President in 1859, about Van Raalte's attempts to extract support for the College in the east. Van Vleck, from Geneva, New York, lasted long enough to build a hall, Van Vleck Hall.

Phillip Phelps, Jr.

Arguably, the most important first academic steps taken by Hope College was the recruiting and appointment of Philip Phelps, Jr., as President. The Reformed Church in the East viewed the West as a missionary charge, and Phelps came to Holland to take that charge and push the west with its "hope and aspirations" to its next higher educational level. Phelps was a Phi Beta Kappa graduate of Union College (now University) in Schenectady, New York, who had tremendous dreams for his new charge. He bought a farm, called Suydam's farm, on the Lake Macatawa where the current Waukazoo is, and intended this farm would grow products that he could sell to support the College. At a time when travel was difficult, Rutgers, the Dutch Reformed Church school, was too far removed in the East from the opportunities in the West. How could it educate future clergy for service in the West? New Brunswick seminary, located on the Rutgers campus was too dominant and distant to train the clergy for the Western churches, too. So Phelps saw an opportunity and had dreams of a university in Holland grown under his command. His notion was the name "Hope College" was a passing name that would soon transition to Hope Haven University.

President Phelps recruited the first freshman class and Hope College was born when the Reformed Church sent the Vice President

---

[28] Interview, Robert De Young, Vice President of Development for Hope College – 1968-2005, conducted in 2013.

and Secretary of its Board of Education to a commencement on June 23, 1863. Hope College was originally dealt with by the Church as being under the Particular Synod[29] of Chicago, which proposed to the General Synod that an endowment of $85,000 be raised in order that "Holland Academy obtain a charter under the laws of Michigan." These numbers were met.

Hope's Incorporation

On May 14, 1866, Hope College was incorporated under the terms of the Michigan College General Law. The original documents of Hope College stated:

> "the <u>character</u> and <u>object</u> of the college and the corporation are to provide
>
> the *usual* literary and scientific course of study, in connection with sound, evangelical, religious instruction, according to the standards of the Reformed Protestant Dutch Church, as based on the Holy Scriptures, although the College is denomination in character, yet students shall be admitted to all its advantages without reference to their ecclesiastical connections subject only to the general rules and regulations of the Institution."[30]

The charter expressly retained the option of creating a Theological Department, for the "training of missionaries and ministers of the Gospel," and a "Normal School" for the training of teachers. The Grammar School, or Academy, was to remain a part of Hope College, also under its Council. The charter expressly linked the General Synod with the functioning of Hope College, financially and in terms of other support.

A Dutch lawyer from Detroit, Theodore Romeyn, drew up the first governing documents – the Articles of Association – in 1871. These were likely written as much by Philip Phelps as anyone and are impressive for their depth. The academic sector of the document understandably focused heavily on the classics, considering it as a college founded in the mid-19th century and it really did not differ

---

[29] In the Reformed Church hierarchy, the Particular Synod was the body above the Classis (which oversaw each congregation's Consistory) and below the General (or national) Synod.

[30] The College archives have a typed copy dated April 24, 1896; not the original dated May 1866. According to Michigan law then, a corporation could officially last only for 30 years at which point it had to reapply for certification. So the Hope/Holland Archives copy of the Articles of Incorporation is for the 30 year recertification of the College in 1896.

much from many other colleges begun at the time. For example, the first catalogue of the University of Kansas, also begun in 1866, listed a curriculum much like that proposed by Phelps.

Phelps' curriculum

Phelps' curriculum, as listed in the 1871 document, was also an almost verbatim copy of the Union College curriculum of the time. The mathematics in the curriculum is today completed in the first two years of high school but it was clearly a part of the expectation that every student have mathematics.

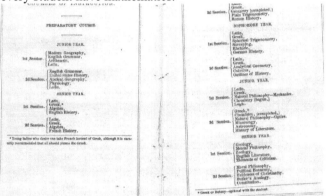

General course curriculum includes chemistry

Particularly striking is the presence of "chemistry" in the first list of courses, and Hope College would later excel because of its chemistry program. *Draper's Chemistry*, published in 1859 was targeted by Phelps in the original list of courses, and two semesters of its study were included in the junior year in the first Academic Department's general curriculum for all students. John Draper was an Englishman who came to New York to accept an appointment at New York University. He is widely credited with being America's first professional chemist, credited with having taken the first photograph and inventing what became photochemistry. Most significantly, he was the first president and founder of the American Chemical Society (1876). *Draper's Chemistry* is a nearly 500-page tome, with some realistic science in it, and a good number that is not. But for the day, it was better than other texts the students could have studied. Almost as an aside, as Hope grew, there was the science department. It was out of sight and out of mind. The clergy did not understand the sciences and science faculty were not active in campus or church politics. Philip Phelps, John Heyl Vincent, and particularly

46

Chautauqua's William Rainey Harper admitted that the sciences, and chemistry in particular, were an essential part of the education of a liberally educated man. But I seriously doubt any one of them ever read a chemistry text – they knew chemistry was good, but were not sure just what it was.

The Hope chemistry department, when it was delineated as such, did well for the school and at several levels. First, a succession of leading faculty starting with Almond T. Godfrey succeeded by Gerrit Van Zyl, made placing their boys in graduate schools that fit their abilities a priority. And so long as Godfrey first, and then Van Zyl, stayed on the side of campus away from management and met their obligation, they were left alone.

Theology

The Articles of Association of Hope College also prescribed a Theological School and a Preparatory School, at a detail that is significant for its foresight. The Theological School became a serious issue in the 1870s mostly because it was too expensive. President Phelps' dream for his farm also did not pan out because the land on Lake Michigan's shore was too sandy and poor for growing crops. In an omen of what was to come, the churches also met less than half of the money pledged. By 1877, when complaints in the West that the Reformed Church in the east had not met financial obligations pledged as early as 1850, the Theological School was suspended. The Reformed Church had seen this coming as early as 1868, when it began complaining that the College should be, by this time, self-supporting. Nonetheless, when the financial crisis came to a head in 1877-1878, out of nearly $53,000 in promissory notes from the Church to Hope College, only $28,000 was in invested funds. Van Raalte died in 1876, and the Reformed Church, after numerous attempts to get Phelps to right the ship, fired him. They sent a team comprised of the Education Secretary of the Reformed Church, E. W. Bentley, and two elders, Peter Danforth and G. Van Nostrand, to see if Hope College could be saved. The Theological School was closed, and Hope College chastised for investing so much in real estate, because property values had fallen so drastically. The advisors also focused on the Council of the College, saying it was too dominated by those from the West. The Board of Trustees was subsequently reduced to sixteen with each Classis having two representatives not four, and the General Synod with five members (terms were for five years). The programs of the school were reduced to two tracks – the Preparatory School and the College, though the option was left open for a School of Theology in the future.[31] Phelps refused to give up the presidency though, as much because the school owed him money as for any other reason, and continued to live in Van Vleck Hall rent-free. He later took positions at two churches in New York eventually but not until 1886.

Near Failure

This history of Hope has been entirely written by Wichers, and not important to repeat here. Recognizing the genius of Van Raalte first, and then President Phelps' contributions is important, however. Whatever followed that in terms of the consolidation of the College, Western Seminary, and the closing of the Preparatory School could not have happened without the creative beginnings orchestrated by these two men.

In a recent brief analysis of the VanderWerf years, Robert Swierenga suggests that "by paying more attention to academic credential in hiring, than religious faith, this president and his predecessor, Irwin Lubbers, undermined the relationship of the College to the Church."[32] The relationship of the College to the church, from Phelps to VanderWerf continued to be contentious, if not acrimonious, well into the mid-20th century.

There were always three issues:

1. Money – the College wanted/needed more support from the churches; the churches did not have it.
2. Control – the College wanted as much freedom as possible; the Church wanted to control as much as possible.
3. Philosophy – Call it right wing/left wing; piety vs. sociology; whatever. The Midwestern church particularly was much more conservative than most of the faculty of the College, and the students.

Early Boards of Trustees

The first Board, 33% of which were to be non-clergy, had nineteen members of whom thirteen were clergy. Albertus C. Van Raalte and Philip Phelps, Jr. were members of the Board for terms not to exceed thirty years, which was the statute of limitations for the charter of the institution according to the terms of the Michigan constitution of that day. So was the Corresponding Secretary of

---

[31] This actually happened, but not until 1885.
[32] Robert Swierenga, *Holland, Michigan* (Grand Rapids, MI: Eardmann's Publishing/Van Raalte Press, 2014), 520-523.

Education of the Reformed Church, a member of the Particular Synod of Chicago (appointed for thirty years), and another number not to exceed thirty-five members. All had to be approved or confirmed by the General Synod of the Reformed Church. Board presidents though, with the exception of Lodewecus D. Viele, in 1877-8, were always from the clergy. After Phelps was gone and Van Raalte dead, no layman served as President of the Board until Ekdal Buys who served as Board president from 1961-1966. Buys was among those who hired President VanderWerf; he was followed by Hugh DePree (1966). It was DePree who fired President VanderWerf.

Hope's First Formal Constitution

Hope's first formal constitution, adopted June, 1871 vested the Board of Trustees with supervisory responsibility for the College. College presidents were to be appointed by the General Synod that determined the salary and kept the money for its payment. Others have dealt with questions regarding theology and Hope College.[33] The bibliography in the Wichers' book is particularly useful.

Phelps' Curriculum

President Phelps had big ideas. He was also said to have been responsible for "mismanagement and financial reverses," leading to his termination. But what Phelps clearly set forth, in plain English in the Constitution of 1871, was a course of study that was much more academic, than were counterparts in many other nascent schools of the time. The tension between academic preparation, and a job, did not occur much at the time. Later it would, but it did not, in theory at least occur, at the beginning.

Here was the course curriculum from the 1870s:

FRESHMAN YEAR
First term: Mathematics – Algebra (Davies); Latin, Cicero de Amiecitia –Arnold's Latin Prose Composition; Greek – Xenophon's Memorabilia – Arnold's Greek Prose Composition; Rhetoric Original Essays – Delivery of selected pieces – Vocal exercises in the gymnasium; Hygeine Health and mental education; Modern Languages Dutch.

---

[33] See Wichers, Wynant, *A Century of Hope*, Eerdmans, 1968; also Bruins, Elton J.; Swierenga, Robert P. *Family Quarrels A History of the Reformed Church in the 19th Century* Eerdmans, 1997.

Second term: Mathematics – Plane Geometry; Latin, Livy – Prose composition; Greek – Homer's Iliad - Prose Composition – Ancient geography; Rhetoric continued History Ancient History; Modern Languages Dutch.

Third term: Mathematics – Solid Geometry; Latin, Livy, continued – Prose composition, concluded; Greek – Homer's Odyssey - Prose Composition, continued – Ancient geography, continued; Rhetoric continued History Ancient History, continued; Modern Languages Dutch.

SOPHOMORE YEAR

First term: Mathematics – Solid Geometry; Latin, The Odes of Horace; Greek – Thucydides - Prose Composition, continued –; Rhetoric Trench on the Study of Words – Delivery of Selected Speeches – Vocal Exercises History Schieffelin's Foundation of History; Sacred Literature Harmony of the Gospels; Modern Languages French and German; Didactics Holbrook's Normal Methods of Teaching.

Second term: Mathematics – Plane and Spherical Trigonometry; Latin, The Odes of Horace; Greek – Thucydides - Prose Composition, continued –; Rhetoric Trench on the Study of Words – Delivery of Selected Speeches – Vocal Exercises; History - Schieffelin's Foundation of History; Sacred Literature Harmony of the Gospels; Modern Languages French and German; Didactics Holbrook's Normal Methods.

Third term: Mathematics – Surveying; Latin, Terence – Roman History; Greek – Sophocles, Greek Antiquities - Prose Composition, continued –; Rhetoric as before History Modern History; Sacred Literature Harmony of the Gospels; Modern Languages French and German; Didactics Page's Theory and Practice of Teaching.

JUNIOR YEAR

First term: Mathematics – Conic Sections and Analytic Geometry; Latin, The Satires of

Horace; Greek – Aeschylus – Greek literature Rhetoric – Blair's Rhetoric Analyses – Delivery of Original and Selected Pieces History; Modern History– Sacred Literature Historical

Introduction to Sacred Scriptures; Modern Languages French and German; Chemistry –

Draper's Chemistry; Intellectual Philosophy Upham's Mental Philosophy – Hamilton's MetaPhysics.

Second term: Greek – Demosthenes on the Crown – Greek lit; Rhetoric Blair's Sacred Literature As before Modern Languages French and German; Chemistry *Draper's Chemistry*; Intellectual Philosophy Upham's Mental Philosophy – Natural Philosophy Statics, Dynamics.

Third term: Latin, Juvenal – Roman Literature; Rhetoric Essays and Discussions; History –

Modern History; Sacred Literature; Modern Languages French and German; Intellectual Philosophy– Upham's Mental Philosophy; Natural Philosophy Hydrostatics, Hydrodynamics, Pneumatics, Acoustics, Optics; Natural History Botany.

SENIOR YEAR

First term: Greek – Plato – Lectures on Greek philosophy; Rhetoric Original Essays and Speeches; History– Guizot's History of Civilization; Sacred Literature– Evidences of Christianity. Modern Languages French; Natural Philosophy Electricity, Magnetism, Galvanism. Natural History Geology; Astronomy Olmstead's Astronomy; Moral Philosophy Wayland's Moral Science; Political Philosophy Political Economy. Aesthetics Kames' Elements of Criticism

Second term: Latin Cicero's Tusculan Disputations; Greek testament – Lectures on Greek Philosophy; Rhetoric as before; Sacred Literature Evidences of Christianity concluded; Modern Languages– French concluded; Natural History Geology concluded; Astronomy– Olmstead's Astronomy, concluded: Moral Philosophy– Wayland's continued; Political Philosophy Political Economy; Aesthetics –Kames' Elements continued; Logic Whately's –Logic.

Third term: Rhetoric Original Essays – Vocal Exercises; Hygeine – Oral Instruction in Sanitary Science; History– History of Philosophy; Philosophy –Wayland's concluded; Political Philosophy Constitution of the United States.

Phelps was not equipped personally to teach much, at depth, in his curriculum so he must have relied on good text books; and there were almost no others on the faculty to help him because the founders had shackled him with a Theology Department that seemed more expensive than the rest of the institution in total. Nor is it clear if he got any help from Shiefflelen in history, or Suydam in natural history. But it is almost a moot point, because some of his dreams never transitioned to being a reality. The school ran out of money and its own devices – donations, student fees, endowments, legacies,

Reformed Church Board of Education – never had a chance to gear up and support it at the level of Phelps' dream. The Reformed Church closed the Theology Department in 1878. Phelps was fired, and the College struggled until the 1890s. It would never be the same – a pattern had been established that basically said "We are small, simple, and poor. We will have to make due with a heavy emphasis on _____." What filled the blank would be up for grabs. That worked for the obviously able students that were looking to Hope, because of their heritage and history in the Church, but for those that did not want to be ministers, they were forced to find something else to study.

Sewers

The third term senior year course on oral hygiene is a characteristic of the day. Cities were, in the 1860's, just beginning to think about sewer systems. For example, in the case of Chicago, sewers were built to drain human waste into Lake Michigan, but built incorrectly in the 1850s. A heavy storm in 1885 flushed sewage out beyond the drinking water intakes. The resulting typhoid and cholera outbreaks killed more than 10% of the population. From that point, the flow of Chicago's sewer canal was reversed, and it has since drained down the Mississippi River.[8] The Scot, Lord Lyon Playfair,[34] addressing his peers in the British House of Commons in 1855, talked about cleanliness, and the lack thereof. Playfair was one of the first British chemists, and studied – wrote a dissertation – with Justus Liebig in Giessen in 1839. Subsequently, Liebig was invited to England by Prince Albert and, in 1846 gave a series of lectures on the chemistry of animal fats to the Scottish Wool Growers, at Playfair's invitation. The result was a laboratory in London, to which the young chemist August Wilhelm Von Hofmann was sent. Hofmann began training a group of youngsters in the techniques of chemistry[35] in his laboratories, and in 1852, a fourteen-year old William Henry Perkin discovered "royal purple," or *mauvene* while searching for a synthetic route to quinine.[36] The sale of role purple

---

[34] Reid, W. *Memoirs and Correspondence of Lyon Playfair: First Lord Playfair of St. Andrews,* Castle and Co., London, 1899; pp 433-4.

[35] Chemistry, the word, probably traces its 17th century origins to the word 'alchemy' with which it was interchangeable. Practical chemistry, as in doing planned laboratory experiments, was a 19th century phenomenon. The origins are German. R.B. Woodward and William Doering achieved the authentic synthesis in 1943; see Doug Neckers' "The Art of Synthetic Chemistry" at dougneckersexplores.com.

[36] Woodward and Doering achieved the authentic synthesis in 1943; see Doug Neckers' "The Art of Synthetic Chemistry" at dougneckersexplores.com.

by Perkin's company began the synthetic dye industry in England and by the late 20th century had given rise to the largest English corporation – ICI – Imperial Chemical Industries.

The stories of filth in Britain are particularly poignant – Handel's offering of Messiah in London, starting around 1750, was a charity event to provide funding for the Foundling Hospital. A foundling was a fetus born to a mother unable to care for it, and left in London sear the river front.

The Newtonian physics of Phelps' curriculum remains at some level in college physics curricula of this day; though obviously, the detail and presentation would be different Newtonian physics was physics before quantum theory. Joseph Henry, who discovered the relationship of electricity and magnetism, was appointed professor of *natural philosophy* at Princeton in 1839. This followed the invention of the telegraph by Samuel Morse in 1831. Michael Faraday, an Englishman, discovered many of the facets of electrochemistry that continued to fascinate chemists including those who taught Gerrit Van Zyl at the University of Michigan in the 1920s. Faraday, with Henry, discovered the principle behind the electric power generator in the 1840s. Edison translated into a light bulb in the 1880s, and formed GE to develop the light bulb as a commercial lighting entity.

The American Chemical Society's Division of the History of Chemistry referred to *Draper's Chemistry* a 400-page effort published by Harper Brothers in New York in 1846. Draper delivered a well-received lecture to medical students at New York University in 1844 that was published as a result of the student's request, further drawing interest to the burgeoning field.

The Hope curriculum was a classical one for many years. The textbooks changed as new ideas and studies emerged. Draper died in 1876, and Ira Remsen came forth with a series of texts. Remsen was of Dutch and Hugenaught parentage, and had taken his Ph. D. in Gottingen (Germany) in 1870. Later the President of Johns Hopkins in Baltimore, he wrote books and published papers with substantial regularity. Remsen's thesis from Gottingen was on the chemical compound, saccharin. Soon, the Remsen's chemistry texts were being incorporated into Hope's curriculum:

COURSE OF STUDY.

Modern—Edgren's French Grammar; some French Author.

RHETORIC.—Essays, Debates, Orations.

HISTORY.—Mediæval and Modern History.

NATURAL SCIENCE.—Remsen's Chemistry.

SACRED LITERATURE.—Greek New Testament, and Harmony of the Gospels.

By 1922-1923, the course of study had Godfrey's imprint and was outlined in the College Catalogue in the manner familiar for these documents going forward.

54

MATHEMATICS.—Hardy's Calculus.

MATHEMATICS APPLIED.—Olmsted's Natural Philosophy.

LANGUAGE.—

Latin.—Stickney's Cicero's De Officiis; Sloman's Terence; Seneca's Moral Essays.

Greek.—Odyssey or Lyric Poets; Humphreys' Aristophanes' Clouds; Allen's Prometheus of Aeschylus; Literature.

Modern.—Joyne's Meissner's German Grammar; some easy German Author.

RHETORIC.—Bascom's Philosophy of Rhetoric; American Literature; Essays, Discussions, and Orations.

HISTORY.—Studies in History; Lectures on the Constitution and History of the United States.

NATURAL SCIENCE.—Chemistry, one term; Wood's Botany, two terms; Sedgwick and Wilson's Biology.

METAPHYSICS—Porter's Elements of Intellectual Science.

SACRED LITERATURE.—Butler's Analogy.

Hope's beginnings were no more stumbling, nor awkward, than were those of many other church related colleges begun in the mid-19th century. It was shackled by the attitudes of those that served on consistories in the Church. There's little reason to believe, given the reasons why the Holland Dutch emigrated though it could have been any different.

Chapter 3

Between the Wars

When Dean William Vander Lugt was interviewed by a North Central Association review committee, Appendix 5, during a re-accreditation visit in 1964, he said explicity that the department of chemistry was the one department in the school that had academic standing outside the Reformed Church. In some ways, Hope's emergence was almost directly tied to the beginnings of doctoral education in the sciences in America.[37]

In that report, Vander Lugt used the chemistry department as example of what the rest of the College should be and said: "An all-campus contribution is necessary if faculty are to make their maximum effort for the College in order to make for a smooth functional relationship between faculty and administration; but other departments (than chemistry) fail to make their department as strong as it can be. Hope's chairmen are extremely sensitive to the development of the college as a college and do not push ahead with the development of their own departments... Except for chemistry. In all honesty we must admit that much of the reputation the College enjoys in the academic world is a great deal due to the reputation of the chemistry department."

In short, Vander Lugt said that the Hope faculty and their department chairs spent more time minding the College's business than their own academic reputations, or that of their departments'. This was to the detriment to the institution to which they professed intense loyalty, especially when an ambitious alumnus like Cal VanderWerf returned to lead it as an academic institution. Regardless of merit, and there were some, this lack of attention to one's discipline would prove Hope's near downfall in the 1960s.

Cal VanderWerf showed up, and changed the course because – for the good of the School, according to the North Central review, Appendix 6, - he had to do so. But the transition was difficult for older faculty. To them, Hope was their life and in it they were totally invested. Cal's point to the faculty was that to hold Hope College accountable as an academic entity, the faculty's attention must be paid more to their own academic reputations and to that of their departments. This did not play well at all levels, but he held to the point, and when he found resistance, he chose other routes. Like a

---

[37] See "North Central Report" in the Appendix.

good general, he went around the opposing army – not through its middle. He brought in young faculty whose concerns were more their professional reputations than running the college. As elsewhere, young faculty, some then to work at Hope, identified more with their fields of study, than they did with their institutional home. For young scholars and future scientists, focus on academic rigor was paramount.

In higher education in chemistry, German universities had established the patterns. In the late 19th century, the universities collaborated with German industry to build the world's first academic industrial complex. German intellectuals also defined "chemistry," and physics, as the disciplines of the time. The dye or color (farben, in German) industry was the first industry solely built on the basis of research. There was a practical reason for this. Coal, when coked, leaves a tarry residue that is packed with valuable carbon rich substances. In order to use all of the natural products begat by coal, the Germans developed, through research, industry that used the products found in its tar. Products were developed and sold that replaced other products found, but more expensively, in nature. Synthetic materials were of the same composition as the natural product, but could be made more cheaply and in larger amounts. So, because of the dyes made from coal tar, most everybody could have colored clothes. Industry, and the organic chemical businesses in Germany, stimulated the growth of other research and development. While industries managed their own research units working specifically on their projects, more fundamental and often new ideas came from the university community. In Germany it was hard to tell where industry left off, and universities began. They were that closely allied.

There were no university research labs of consequence, except perhaps those at the University of Chicago, in the United States before the end of World War II. Patents, based on work done in universities with industrial support were the norm in pre-World War I Germany, but this strong relationship between industry and university was feared in the US, to our detriment. The Germans developed their patent laws to build their companies, and their sciences. In the US, we were pleased that companies could patent their technologies because that could build out industries – we thought. But it needed not go too far. Our policy was dictated by end users, as opposed to developers, of new technology. Dye prices were controlled by the textile companies that needed imports to make continuing and new products, where the dyes were made was of less importance than what they cost. As long as there was a continuing supply, the dominant textile companies were happy.

Hope College, the institution, grew up with the German model of higher education in the sciences. And any history of Hope College that does not include details of the history of the development of chemical education in the world, as it was emerging in the United States, is incomplete. Hope faculty, from its earliest post-Phelps days, gave young people confidence in their futures if they used their intellectual skills in practical matters, and because of that, a degree with a major in chemistry had an impact on the College. Even preceding Dr. Almond T. Godfrey, a widely recognized physician/educator whose advice impacted many students in the early-20th century, there were Hope chemistry graduates taking the lead in science. William Morris Dehn (Hope 1891) was the first student to be granted a Ph. D. degree in chemistry from the University of Illinois (1902) and he was a Hope alumnus. Dehn surely could not have benefited substantially from any science education he got at Hope aside from reading books on the subject, but today he is a Hope alumnus recognized as such more by the University of Illinois' chemistry department even, than by Hope College.

The academic Hope College excelled at laying the groundwork for futures through the graduate schools in chemistry. When the author, as a potential chemistry major, was considering schools at which to obtain an undergraduate education, the ability to obtain graduate assistantships at universities following graduation was important. Chemistry was not the only career choice, but it was a place where Hope had proven its mettle. And earning a graduate assistantship meant that the successful student would find his advanced education paid for by others.

One must look at the lives of many brilliant Hope alumni of those early days and ask, "Would they have become chemists had they had a broader array of choices?"

Each individual is clearly different—and "the road not taken" always remains, of course, a brighter road than the one traveled. But, given the circumstances at the time, Godfrey and later Van Zyl, gave *boer jungens* (Dutch for "farmer's sons" – alternatively sons of parents who mostly worked at manual labor) confidence that they could make a better life by using their mechanical abilities, and their intellectual skills through a career in chemistry. Cal VanderWerf graduated from Hope in 1937 as the Depression still gripped much of the country. There were virtually no jobs even for those with exceptional intellectual talents. Through the emerging value of the chemical sciences to humankind, this suggested that an investment in educational training for chemistry would be productive. Cal VanderWerf chose graduate school in chemistry in 1937 and was not only admitted to The Ohio State University with an assistantship but he would also be admitted to Harvard's chemistry program as well. This demonstrated his skill and promise with Rachel later wondering how different his career would have been had he chosen Harvard not Ohio State. But because they made him a first offer he honored his word and chose to enroll at Ohio State. Still, at a time when the Roosevelt government was investing heavily in the C.C.C. (Civilian Conservation Corps) and the W.P.A. (Works Progress Administration), it was not investing in the National Research Council or in any other programs supporting students to study in the sciences and engineering. The opportunities for a more general educational experience, even in chemistry, were severely limited. Later that changed. But in 1937, the opportunities were limited even for the best graduates. So Cal became a chemist.

World War II changed America's appreciation for science and engineering. After "the War," the place of chemistry in American higher education, and all the sciences, changed dramatically and irreversibly. America had awakened, and finally knew that the basic sciences would – at some level – dictate its future. One of the main things that changed was the available funds for those that wished an advanced degree in a science—any science. College graduates in chemistry from the earliest postwar days, were afforded financial support for advanced studies. These grants were almost entirely awarded to students attending state universities. Such schools were seeing increasingly large class sizes in first chemistry courses. State university enrollment in chemistry courses was driven by other programs popular with undergraduates like pharmacy and agriculture. The students paid for by 'teaching' assistantships helped the faculty of the chemistry departments at universities nationwide deal with the influx of all those new undergraduates. Hope undergraduates were conscientious, and careful. So most did a good job as teaching assistants. State universities in the Midwest, particularly the University of Illinois in Champaign/Urbana and Ohio State in Columbus, were the biggest beneficiaries of Hope graduates. But Van Zyl was skilled at spreading his boys around—so Wisconsin, Berkeley, Iowa, Kansas and other state universities all managed to attract Hope students. Certain outliers like Gene Van Tamelen (Harvard) and Earl Huyser (Chicago) chose different routes but the result was the same – a sound background in the principles of chemistry obtained from Hope College faculty of that day by exceedingly bright men gave young Dutch Americans, with the strong work ethic, a place to start on the road to academic successes and financial comforts exceeding their dreams.

Emergence of scientific specializations; more science in medicine

The emergence of chemistry at Hope was also, though indirectly, tied to a more demanding, scientifically based, medicine—the profession. Doctors, from the time medical certification became a matter for the attention of states, were required to complete undergraduate chemistry courses through organic chemistry (three years of chemistry at that time) to qualify for admission to medical schools. Medicine had a cache for Hope undergraduates from early on, too; in fact, more than would be expected for the size of the school. Some graduates, like Timothy Harrison and Marilyn Scudder, became medical missionaries. Others, and it is always dangerous to name names, like Vernon Boersma, Henry Vander Kolk, Jerome Wassink, and Phil Van Lente formed/form the strength of the medical community in western Michigan.

Graduate education in chemistry

Chemical education at the graduate level was tied directly to the value of the product (the graduate with an earned Doctor of Philosophy with a specialization in chemistry). It was here that Hope College graduates really flourished. By far, most alumni who received Ph. D. degrees in chemistry became industrial scientists. Some, like Howard Hartough, stopped at a Master's degree, and became successful industrial scientists. A few graduates became professors – some, like Gene Van Tamelen and George Zuidema became the best in the world in their respective fields. Paul Schaap is another case in point. He chose to work in university, because that was what young scientists did when he graduated from Harvard; but, in fact, he used his business talents to develop Lumigen, and make it a successful business. By the time he was forty-five, the academic Paul Schaap had become the entrepreneur and later philanthropist Paul Schaap.

Hope had carved out its own, good reputation through its graduates; and there was a sense of loyalty by the alumni to the school, so the College benefitted from the success of its alumni. Both the Van Zoeren Library and the Schaap Science Center came about through the intellectual curiousity, business skills, and the generosity of the chemists, John Van Zoeren and Paul Schaap. The facilities

60

named for them are the only buildings on Hope's campus solely built directly because of the education those two individuals were afforded at Hope. The Dow Center was also the result of the strong contribution of Hope graduates in chemistry to the success of the Dow Chemical Company. One can also argue that VanderWerf Hall is in this category.

Every other building on Hope's campus is named either for a former administrator, or for local, western Michigan philanthropists. Almost all of the latter made their money from historic, western Michigan business interests like furniture, and farming—Miller, Bultman, DeWitt, DePree, Lubbers, Phelps, Durfee, Kollen, Van Wylen, Peale plus Anderson/Werkman. Though Devos made his money in the mid-20th century, he was not a Hope graduate, and the college education which he did not have, had little or nothing to do with his business success. There was nothing wrong with this, mind you. Bless the development people that made these gifts happen. But the fact remains that few of the successful business people in Hope's history—except for those that majored in chemistry—have made much money or been that charitable.

From whence did the specialization – "chemistry" - come?

America and Americans did not invent research in chemistry. The German universities, research labs, and institutes did. Yet the comfort of human beings living in western countries is significantly greater now than it was at the beginning of the 20th century. This is mostly because humankind, at some level, has marshaled the efforts of science in its best interest. Even awful political events, like world wars, lead to growth and safety. Sulfa drugs were a product of the devastation of World War I; penicillin and the antibiotics a product of World War II. Much of this is directly attributable to universities and industries in the United States.

If the United States did not invent research, and research universities, who did? William Rainey Harper became the president of the first American research university, Chicago, in 1890. He almost immediately turned to the German educated chemists, J. Ulrich Nef, and Julius Steiglitz to form Chicago's chemistry department. Even today, the library of last resort in chemistry, as in when searching for information from an obscure 19th century Ph. D. dissertation, is the research library affiliated with the University of Chicago.

Nineteenth century German scientists did not know, or had no way of knowing, what they had unleashed. They stumbled step by step. The curious young persons out of university made Bayer, and B.A.S.F. If Bill Gates or Steve Jobs were to have a 19th/early 20th century role model, it would have to be Karl Duisberg, or Karl Bosch. Gates and Jobs exited formal education and drove their futures outside the stream, but Duisberg and Bosch found human miseries like diseases and starvation, after they had an education in chemistry that they could use to make pharmaceutical and industrial products. Of course it did not stop there. B.A.S.F. and Bayer also made fertilizers and bombs. When John D. Rockefeller said to Duisberg, "Let's make a deal. You take all the chemistry in the world; and I'll take care of all the petroleum," the Germans jumped at the chance. A decade of more later, Bosch's B.A.S.F. made could make 100 octane aviation fuel out of German coal. Rockefeller's petroleum engineers, could not even get to its biggest new supplies because the U.S. had been locked out of the Pacific by Japanese (Axis) military advances. America suffered from the choking decisions of the early 20th century business moguls because the value hunting and gathering more than research and study. That's why, when World War II broke out, we had a lot of making up to do.

German attention to chemistry led to German companies that sold chemical products. Eventually there came a German government that poured money back into the scientific community through a series of institutes, some of which were connected to universities. The picture, below, is of the German Kaiser (left) and Emil Fischer, Nobel laureate in chemistry, 1904 (right).[38] It shows a parade of the distinguished at the dedication of the Institute for Coal Research in Mülheim Rühr in October, 1913. The name was chosen for political reasons since no scientist working there then, and none since, specialized in the chemistry of coal. The Kaiser and Fischer surmised the people would be happy with people who slaved as they did, mining coal for the population. Though by 1918, the Kaiser would have abdicated, and Fischer, who lost two sons in World War I, would be be nearly dead, the Institute of Coal Research (which is today the Max Planck Institute for Coal Research, and later added an Institute for Radiation and Photochemistry) remains. The building still calls the unit by the original name, the Kaiser Wilhelm Institute of Coal Research.

The Kaiser Wilhelm Institutes became centers for research excellence and scientific leadership. At Dahlem, in Berlin, in the

---

[38] Richard Rhodes, *The Making of the Atomic Bomb* (New York: Simon & Schuster, 1987).

1930s, a Fischer student, Otto Hahn (working with Fritz Strassman and Lisa Meitner), discovered the fission of uranium U^{235} to barium and krypton. Those experiments led almost directly to nuclear fission, and the atomic bomb. Though the Germans did not develop that, they had the background and intellectual power to do so. What Hitler did not give the scientists was the political will to make that happen. The Kaiser Wilhelm and other labs managed to survive World War II.

The doctoral education of German young men that evolved when chemistry, the discipline, was being formed in the 19th century, was that through work in a laboratory with a senior scientist, the student would be trained as a research scientist. A research effort could lead to publication and a younger man would be ready for a job in German industry when it did. In most respects, this educational pattern still exists in the physical sciences. Young people working with mentors learn the skills and patterns of research, and are thus prepared for research careers of their own. Research work in chemistry means experiments carried out with, more or less, unknown outcomes. Of course, most experiments are well planned so that even a negative outcome provides information for the experimenter. But it was this attack of an unknown subject that chemists encompass in the word "research." That outcome of this original work would be published by German chemists, mostly in two German chemistry publications of the time—*Berichte der Deutchen Gesellschaft* or *Liebigs' Annalen der Chemie*. There were also emerging strong university programs in chemistry in the earliest twentieth century, all over Germany, and these were mostly associated with research institutes, as in the aforementioned Kaiser Wilhelm Institute for Coal Research. [39] Most of these were formed in direct connection with industrial collaoborators and laboratories, and after Paul Mendelssohn's time (1860s) when he paid the University at Heidelberg (Robert Bunsen) 260 guilden for his degree, these Institutes became breeding grounds for the growth and commerical development. German industry was, itself, created in the first generation by students that graduated with research backgrounds from the labs of A. W. Von Hofmann, or Adolph Baeyer and, earlier from the lab of Robert Bunsen. Graduates who benefitted from the mentorship of just two German chemists – Justus Liebig and Frederick Wöhler – started Badische Anilin und Sodafabrik

---

[39] Though the Nobel laureates, Emil Fischer (Humbolt – Berlin); Adolph Baeyer (Munich); Jacobus van't Hof (Berlin) and Fritz Haber (Berlin) were in the major cities, other universities were also emerging.

(B.A.S.F.), built Bayer, started AGFA, Hoescht, and built Weiler ter Meer. Another, from Hofmann's stay in London (William Henry Perkin) began what eventually became Imperial Chemical Industries. Felix Mendelssohn's son Paul, with Carl Martius founded AGFA. American companies, DuPont, Rohm and Haas, Sterling Winthrop, and eventually even General Aniline and Film (ANSCO) were direct beneficiaries of this German technical buildup.

War and its Aftermath -Reparations for World War I

Few things have impacted chemistry as much as reparation payments following World War I. Much of what the German nation paid after that war came from German chemical industries forced to make compounds at cost and send them to the victors where their companies could be sold for a profit. The United States, through the Alien Property Act (signed into law in early 1917) confiscated nearly 4700 German patents all from chemical industry. They also confiscated an aspirin plant near Albany from Bayer selling that eventually to the highest bidder, a patent medicine house called Sterling Remedy Company that led to Sterling Drug. At least 33% of the deaths from World War I were from bacterial infections, and scientists from B.A.S.F. returned after the war to immediately start testing against bacteria all new compounds made by the company–*streptococcis, e coli,* and several others. By 1932, Gerhard Domagk, under Heinrich Hörlein's direction, had discovered the sulfa drugs. Sulfa drugs were the first antibiotics

On the battle field soon after invading Belgium the Prussian generals ran out of shells. In September, 1914 because the British had blocked the ports in Argentina, using their base in the Falklands, no Chilean salt peter (potassium nitrate – KNO_3) could get to Bremenhaven. So the Kaiser called on his chemists and in a month or so, the engineers at B.A.S.F. had taken the already commercial Haber-Bosch process for the production of ammonia from the reduction of nitrogen in the air, used modifications arrived at by collaborations with the Leipzig physical chemist Wilhelm Ostwald, and found catalysts that could convert the newly synthesized ammonia to nitrogen oxides and nitric acid. This process moved forward with fertilizers made out of the air, replacing those made from salt peter. One of these was ammonium nitrate, from which shells could be made, and which has often been in the news since.[40]

---

[40] It was used by Timothy McVeigh in the Oklahoma City bombing of 1995; and blew up in a major industrial accident in West, Texas in 2013.

A German chemist with whom I was intrigued because of his Dutch name (though he was German) was Fritz ter Meer. Ter Meer was sentenced to seven years in Landsburg prison following the sixth subsequent trial at Nuremberg. Ter Meer was convicted of crimes against humanity for his role in budiling the Monowitz plant for the production of buna rubber, with the labor of concentration camp inmates from Auschwitz. With nitric acid, the German chemists at Weiler Ter Meer could make tri-nitrotoluene (T.N.T.), nitroglycerin, trinitrophenol (Picric acid) and trinitrobenzene. With that, the firm of Weiler ter Meer would develop its own position among the World War I leaders in German chemistry.[41]

Executives of Farben were tried in the sixth subequent trial at Nuremberg (August 14, 1947- July 23, 1948): *The United States v. Krauch et al*, the first trial of chemists as participants in the murder of civilians that came to a successful conclusion. The chemists were guilty of engaging slave labor in building their new plants in the East. Ter Meer was found guilty for crimes against humanity, and spent seven years in Landsberg prison. The Court found that he was aware that the Nazis were using prisoner labor from Auschwitz to build the Farben plant for the synthesis of Buna rubber. German scientists had develop a brute force approach to reduce crack coal first, and then reduce some of the residue to butadiene, which could be polymerized and thus from whence on could make an ersatz or synthetic rubber from coal. Prisoners were among the other innocent victims of their war crimes because they were marshalled from those incarcerated at Auschwitz to work on, and eventually build a plant to manufacture buna near there. Fritz ter Meer was the chemist that held the highest management position in the cartel called I. G. Farben, and a chemist that worked for him, Otto Ambrose, surveyed so he could pick Auschwitz as the site for the Farben buna expansion in the east. That decision was made in January 1941.[42]

Fritz Ter Meer's life still inspires protests long after his death in 1967.[43] But he developed a successful presence in I.G. Farben and,

---

[41] Weiler ter Meer was begun by Fritz ter Meer's father, Edmund ter Meer and was located in Uerdingen in the Rühr. I have made no attempt to trace the ter Meers to the Netherlands but they are obviously as suspected to have a Dutch heritage.

[42] For more on Fritz ter Meer's controversial legacy, read the Pittsubrgh Jewish Chronicle's November 14, 2006 article on his burial site "Bayer Honors War Criminal Fritz ter Meer": http://www.cbgnetwork.org/1695.html.

[43] For more on this case, read Richard Sasuly's *IG Farben* (New York: Boni & Gaer, 1947). Sasuly was a reporter covering the trial, though he had a checkered literary history following Nuremberg. Josiah DuBois' *Generals in Grey Suits* (1958) and Joseph Borkin's *The Crime and Punishment of I. G. Farben* (1977). Subsequently, more specific

at the end of the war, was its chemist of highest position on the *vorstand*, the board of operations. By the time he died, he had been returned to his former position, C.E.O. of a major chemical company former Farben company.

Graduate education in chemistry – The patterns

The major position established, in Germany, by those that pushed nature to its limits came in organic chemistry, and the commercial goal of converting coal tar, the residue left when coal is heated in the absence of air under reduced pressure, to useful organic raw materials. This crude tar was found rich in organic chemical compounds—compounds containing only carbon, hydrogen and nitrogen.

The Germanic traditions that evolved in chemical education at the graduate level are pretty much the traditions we practice in American universities today. A student, after being accepted into a graduate program, takes courses in various sub-disciplines of the science; sometimes also in math, physics, biology or pharmacy. The procedures for determining the qualification of the student to enter advanced study in the field are different from school to school. In the 1960s, for example, students had to evidence good backgrounds in the four then areas of chemistry, analytical, inorganic, organic, and physical, before being allowed into candidacy for the Ph. D. At that time, they were also expected to evidence at least a reading knowledge of German and French. Language requirements have mostly changed now.

After admission to candidacy, the student chooses a mentor or major professor, then he and his professor work out a course of study that, in a few years with success in the laboratory would lead to a thesis and a degree. The mentor student relationship must be close; some mentors interject themselves into a student's work on a daily, or sometimes even oftener, schedule. Others tend to retain a much larger separation on a daily basis, but provide diretion and advice in a regular way through meetings, reports, and sessions as requested by either the student or the mentor. In chemistry, because experiments are involved, there was a significant amount of teaching required some of which is just teaching a student how to do a particular experiment or procedure. Part of the learning process

---

books by Northwestern historian Peter Hayes, Stephen Lindner and others, have dealt with particular companies. Degussa, for example, commissioned Hayes to study its archives in 2002± in search of records about Zyklon B. Lindner has written a detailed analysis of Hoechst's Farben past.

involves the recognition of success. Experiments provide observations; they do not provide answers. Just learning the difference may be difficult for some. Publications are joint publications meaning all of who have contributed are listed as authors. But the work is, ostensibly the candidate's. Examinations at the time of the presentation of the doctoral dissertation to the graduate faculty of the university were in the German tradition oral; mostly they were a perfunctory discussion between the student, soon to be graduate and the examination committee. Though missteps happened because of improper preparation or, in some cases, from serious intellectual carelessness, mostly it is routine.

Though, in the American university, this procedure and process has been modified from the much more formal German tradition in which candidates and mentors/ faculty examiners wore tailcoats during the exam, and it still is a very much formal event.

Hope graduates in chemistry during World War II

Though World War II was more known as the war of the physicists than the chemists, it was widely contributed to in the United States by the work of a large group of chemists organized by the National Defense Research Committee (N.D.R.C.).

Harvard's Conant and others organized chemists into a series of subgroups the collected output of which has been reported "Chemistry, a History of the Chemical Components of the N.D.R.C. 1940-1946," edited by W. Albert Noyes. Conant enlisted Roger Adams, a professor of organic chemistry from the University of Illinois, to get the chemists mobilized quickly. As Noyes, Jr. pointed out in the preface of his postwar report, though the lay public was led to believe that chemists played a minor role while radar, the atomic bomb, penicillin, and D.D.T. became much more important. Yet more chemists worked on the atomic bomb than physicists. The manufacture of penicillin was almost entirely a chemical problem with Robert D. Coghill, a Yale Ph.D. chemist leading that effort. D.D.T., which greatly reduced the threats carried by disease ridden mosquitos particularly for fights in the Pacific was a synthetic organic compound prepared in a simple reaction from chloral, chlorobenzene and sulfuric acid. An advance with a west Michigan tinge was the development of the commercial synthesis of the powerful explosive, R.D.X. which has huge explosive powers approximately eight times more powerful than T.N.T. John Sheehan, originally from Battle Creek went to Michigan in Ann Arbor, majored in chemistry, and succeeded with the commercial synthesis of R.D.X. in Professor

Werner Bachman's labs in 1940. Bachman was Frank Moser's Ph. D. supervisor. Moser (Hope, 1932) helped Hope get the funds for its first major instrument – a nuclear magnetic resonance spectrometer.

Paul Bartlett carried out two projects funded by National Defense Research Committee at Harvard. One project was on chemical warfare agents including the arsenicals and mustard gas; the other was on the composition of D.D.T. and other insecticides related to it. This work was done well before any of the instrumental analytical techniques, so solving the latter project was a chore. Bartlett's lab also developed analytical procedures to differentiate between *cis* and *trans* Lewisite – a clever contribution made quite difficult before the advent of nuclear magnetic resonance spectroscopy. Lewisite was a World War I toxic gas some have called the Tears of Death. Noyes, Jr.'s report was published and we copy the cover page, and one of its committee structures below:

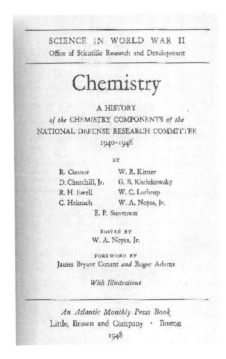

SCIENCE IN WORLD WAR II
Office of Scientific Research and Development

Chemistry

A HISTORY
of the CHEMISTRY COMPONENTS *of the*
NATIONAL DEFENSE RESEARCH COMMITTEE
1940–1946

BY

R. Connor   W. R. Kirner
D. Churchill, Jr.   G. B. Kistiakowsky
R. H. Ewell   W. C. Lothrop
C. Heimsch   W. A. Noyes, Jr.
E. P. Stevenson

EDITED BY
W. A. Noyes, Jr.

FOREWORD BY
James Bryant Conant *and* Roger Adams

With Illustrations

An Atlantic Monthly Press Book
Little, Brown and Company · Boston
1948

Three Hope College chemists need be spotlighted for their contributions during World War II – Howard Hartough, Cal VanderWerf, and Bill Arendshorst. The first two provided Van Zyl with numerous research ideas and both Gene Van Tamelen and George Zuidema worked on research problems that appear to have been sent Van Zyl by VanderWerf. Earl Huyser published on work that had to have originated with Howard Hartough. Arendshorst was later a Holland opthamologist who had majored in chemistry and this included his taking an advanced degree at Indiana University. Hartough was an expert in the chemical compounds found in crude oil. He paid particular attention to aromatic compounds that contained sulfur, and wrote at least two books on thiophene and its related compounds that were treated like the Bible in the Hope chemistry library.[44]

VanderWerf at Kansas had a program funded by the American Petroleum Institute (A.P.I.) that continued into the 1960s. A.P.I. represents, even today, all facets of the natural gas and petroleum industries. Many to-be Navy officers also studied chemistry at Kansas during the war, and Cal taught in these programs. He spent many years after the war working at China Lake, CA in projects for the U.S. Navy, and this work could have been directed toward the organic chemistry of optically active phosphorus compounds, of which the nerve gasses were examples. VanderWerf's A.P.I. support was for synthetic studies of nitrogen containing

---

[44] Howard D.Hartough, *The Chemistry of Heterocyclic Compounds*, (1952) Hartough and Meisel, *Compounds with Condense Thiophene Rings* (1954).

materials found in petroleum. With that project came a consultancy with Phillips Petroleum in Bartlesville, OK.

Doc Van Zyl did some work at Hope during 1941 under the auspices of a contract funded by N.D.R.C. that was directed by E. P. Mack at Ohio State University. The contract was said to involve solid state studies of R.D.X., though Van Zyl was not mentioned in the report filed by Mack, and reported by Noyes, Jr.

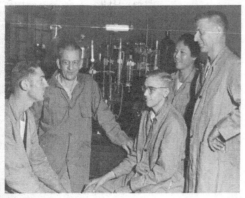

From left: Bob Schut (Ph.D. MIT); Van Zyl; unknown; Harry Tan; and Robert Langenberg.

As in every case with research with undergraduates in the Van Zyl era, stories abound about misadventures; Gene Heasley wrote that "Langenberg (above) was doing some reactions with sodium dispersions, preparing them with a high speed blender. The container disintegrated and he took the blast of sodium dispersion on his pants. He quickly yanked the pants off and went home in his lab coat."

William Maurice Dehn – a Hope chemist before the time.

Hope College was producing graduates that became Ph. D.s in chemistry early in its history. William Maurice Dehn was the first Ph. D. graduate (1902) of the University of Illinois' chemistry program, and was a Hope graduate ten years prior. Dehn, from Chesning, Michigan, spent his career at the University of Washington in Seattle. After graduating from Hope Academy, Hope College (1893); A.M. (1896). Dehn studied "during the summers" in William Rainey Harper's University of Chicago for six years and, in 1903 became an Instructor at Illinois. Dehn's early career was dominated by studies of halogenated hydrocarbons, as in bromoform, iodoform, di-iodoacetylene and tetraiodoethylene. One of his later papers, on the odor of urine, is a teaching paper even today. Dehn and Hartman maintained that the odor of urine, in some manner at least, was due to cyclohexene-3-one. Dehn was a prolific scientist and published at least seventy papers in American Chemical Society journals. His last publication was in 1950, published by a Ph. D. student who, after the war, could openly publish work that had been done during World War II.

Illinium – a Hope chemist's mis-adventure.

No story of Hope chemistry history would be complete without the story of Illinium, element #62. This element was putatively discovered at the University of Illinois immediately following World War I. The work was part of the Ph. D. dissertation of J. Allen Harris, University of Illinois, but the x-ray work on illinium was done at Yale by Leonard Yntema where he was a National Research Council Fellow. Yntema spent his early teaching career at St. Louis University, and left for industry during the depression. As all chemists know, the observations were wrong. Illinium did not exist at least in the form expected in 1924-1926, and the reports were over-interpreted.

"We base our claim to the discovery of a new element on three different premises:

1. The presence of lines in the arc spectrum of materials prepared in this Laboratory common to both samarium and neodymium and stronger in intermediate fractions. These consist of 130 lines in the red and infrared, and five lines toward the violet.

2. The presence in our intermediate fractions of absorption bands which become stronger as the characteristic bands of neodymium and samarium become weaker. The bands at 5816 A. and 5123 A. are especially prominent and their positions confirm the belief that there is a systematic drift in the absorption bands of the rare earth group.

3. The presence of lines in the X-ray emission spectrum corresponding closely to the theoretical positions for Element 61."

Edward Wichers (1913) was another Hope alumnus, frequently mentioned by Irwin Brink who spent at least one sabbatical with him. Wichers headed the chemistry division at the National Bureau of Standards from 1948-1958, was assigned to the Manhattan project in 1944-1945, and Hope College gave him an

Honorary Degree in 1941. His area of expertise was the atomic weights of the elements.[45]

Hope alumni establish a pattern.

Hope students chose chemistry if they were mechanically skilled, mathematically inclined, and intellectually able. For many it provided a way into the professions that combined innate talents. Chemistry was an experimental science, which meant manipulations and laboratory work. For those that had practical inclinations, it was a way to make inventions and practice their creativity. All of what Hope alumni did eventually had been dictated by the rise of the field from a country's young intellects – the young people of the evolving German nation. Germany was making the best for the Country because it had limited resources.

---

[45] L.F. Ynetema, "Observations on the Rare Earths. XIII. Studies in the Absorption Spectra," *Journal of the American Chemical Society, 45,* 1923; L.F. Ynetema, "Observations on the Rare Earths. XV. A Search for Element Sixty-One," *Journal of the American Chemical Society, 46,* 1924; J.A. Harris with B. Smith Hopkins, "Observations on the Rare Earths. XIII. Element No. 61, Part One. Concentration and Isolation in Impure State," *Journal of the American Chemical Society, 48,* 1926; and Harris and Hopkins, "Observations on the Rare Earths. XXIII. Element No. 61, Part Two, X-Ray Analysis," *Journal of the American Chemical Society, 48,* 1926.

Chapter 4

Hope at the Precipice

Hope College advertises itself to potential students today in words that I could have written in 1965. Hope says that it offers: "Liberal arts-based curriculum. Graduate school-style research. Hands-on opportunities. Challenging collaboration. The highest standards. These are the elements that help Hope students from every discipline succeed — in school and in life."

"Hope is consistently awarded more National Science Foundation grants for undergraduate research than any other liberal arts college and was ranked fourth for undergraduate research and creative activity by *US News & World Report*."

No individual represents Hope College, and the attention to undergraduate research it now offers college-wide, more than Gerrit Van Zyl. A graduate of Hope College in 1918 and a faculty member from 1924 to 1964, "Doc" was a physical chemist teaching organic chemistry courses. He was not a good classroom teacher by all accounts and students always wondered how he graded his examinations. The textbooks he chose both for the first course and the advanced course in organic chemistry were outdated, or at least unusual. He did not closely supervise the organic chemistry laboratories, though he was the instructor listed in the catalog for the courses. He knew little about instrumentation though he bought instruments for the department mostly because a salesman from one of the scientific supply houses sold them to him. He did not have much contact with research in organic chemistry, though he directed research. He really did not know how to write skilled research proposals, though he got them funded. When that funded research went on in the summer, the only time he came around was to hand out pay checks of $50, checks made out on a local bank. He was a mentor *in absentia*.

Yet, in the oft-mentioned North Central report dated April 1964, Van Zyl's chemistry department was singled out as being the most contemporarily academic in the College; and his brief statement for the chemistry department that was in that report, clearly demonstrated that he understood what was happening in his field. Though he knew he would not be part of it, Van Zyl was clear that Hope had to be with, or lead the effort into the mid-20th century research with undergraduates.

The reputation of the chemistry department was based, almost entirely, on Van Zyl's iconic insights into what was needed for students to succeed. Van Zyl's influences on "his boys" (most of his advanced students were male) was uncanny. He was a counselor *par excellence* and loved doing that. Van Zyl gave the students confidence that they could succeed in the broader world of higher education and that hard work in chemistry, like hard work on their father's farms, would pay dividends. Second, he made it a point to go to meetings and to connect with leading faculty members at least at Ohio State, Michigan and Illinois. He knew people. He gave good advice and placed his students in schools where he knew that they would not only succeed but stand out. He had an aversion to sending students to the University of Michigan, where they kept them for too many years (according to him) before they let them graduate with Ph. D. degrees. But aside from that, he treated the graduate schools fairly. This, he believed, was as important as the advice he gave to the students.

Gerrit Van Zyl had the benefit of working with J. Harvey Kleinheksel, himself a 1923 graduate of Hope. For all of Van Zyl's weaknesses, Harvey offered concomitant strengths. They were a successful pair: Harvey was the excellent teacher: a calm, coherent voice and colleague.

Doc Van Zyl came to Hope from Hospers, Iowa, and never taught anywhere else. He was a veteran of World War I and extremely active in Trinity Reformed Church where he served on the consistory for at least fifteen years. He never took a sabbatical, received an Honorary Doctorate of Science from Hamline University in 1952 and, when he retired, an honorary Sc. D. from Hope. He listed that he was a lecturer at the first college chemistry teachers institute funded by the National Science Foundation, in Laramie, Wyoming though the date is unknown.

Doc's doctoral work at the University of Michigan was done with Arthur Ferguson and he had two early papers on electrochemistry in a journal that can no longer be found. Van Zyl's Ph. D. thesis at the University of Michigan was titled "Measurement and Effect of Addition Agents on Decomposition Potentials and Transfer Resistance." This was published in the *Transactions of the American Electrochemical Society* in 1924. Van Zyl directed one graduate thesis at Hope, a Master's thesis authored by John. J. Mulder in 1931.

Cal VanderWerf returned to Hope because Van Zyl caused that. The appointment of VanderWerf was a recognition, at last, that chemistry was something more than a department to brag about when one was talking about Hope College. It was an academic discipline with high standards. As time would show, Van Zyl had standards that could have been too high for the rest of Hope College of his day that were translated through Cal VanderWerf when he became its president.

Van Zyl and Research

The Hope College of Van Zyl's early days struggled to stay in business because of the Great Depression. Van Zyl told me often that he and others gave up salaries during the 1930s at key moments; for instance, when the school needed to buy a new boiler for Van Raalte Hall. When Cal VanderWerf came through Hope College, graduating in 1937, there was nothing on the level of "undergraduate student research." The Science Building opened in 1942 but the only formal recognition of the event was a church service of dedication. If there was a scientific dedication, we could not find a program for it. With this building, and its labs in place, Van Zyl began doing experimental work with students during World War II.

American science lagged behind German science, and even French and English science was more advanced than was that in America. Our universities, save for some —Yale, Harvard, Johns Hopkins, Chicago, and the University of Illinois—were not strong or chemistry-oriented. The large California schools, save for Cal Tech (which was called Thorpe then) and UC Berkeley, were just beginning to make an imprint. Chicago provided some scientific leadership through Julius Stieglitz, and before that J. Ulrich Nef. But, though its faculty was excellent, its space for advanced student study was not large.

Research in the sciences is a combination of independent study and laboratory work. Van Zyl realized this and, beginning during the war, started students on work of their own while undergraduates. So, when a Hope student graduated, he had at least a little idea of what he would find when his graduate study started. A research project is first knowing what to do, then knowing how to go about it. Next, of course, comes doing all of the work. Once completed, the work must be written and published in a peer-reviewed journal. Research in chemistry is an individually developed skill. Nature is nature and does not behave because one wants it to, or asks for it to do so. Experiments must be set up correctly; more or less like cooking spaghetti, if one cooks it too long, one gets mush. But if one cooks it too short, one might as well eat sticks. The critical reasons for cooks are to make the pots produce good results, and Van Zyl became that director of cooks wherein his students were the learners.

The two students that really brought research onto the Hope College scene were the previously mentioned Gene Van Tamelen and physician George Zuidema. Gene led this. On the occasion of Van Zyl's retirement, Hope gave him an honorary degree: Van Zyl died of a heart attack while visiting family in Seattle, Washington.

Please read – 1st rough draft. (Van Zyl file)

(1) The many friends, throughout the nation, of Dr. Gerrit Van Zyl, modest and beloved former Professor of Chemistry at Hope College for 41 years will be pleased to know of the honors extended to him by his Alma Mater at Hope College's Centennial Year Commencement exercises Monday June 7 in the Civic Center in Holland, Michigan.

(2) Dr. Van Zyl was presented with a Doctor of Science degree by Dr. Calvin A. VanderWerf, Hope College President and former student of Dr. Van Zyl between 1933-1937.

(3) As head of the Chemistry Department at Hope from 1923-1964, Dr. Van Zyl' leadership resulted in the development and preparation of many students for graduate work in Chemistry in colleges and Universities throughout the country

(4) President VanderWerf said of his former Chemistry Professor, close associate and colleague, "The nation today is the benefactor of his leadership and dedication for a large proportion of his chemistry majors have distinguish themselves in medicine, research and college teaching".

As evidence of Dr. Van Zyl's outstanding capabilities and leadership, Hope College was rated in 1959 8th in the nation in the production of Ph.D. candidates in the sciences, according to the President's Scientific Research Board, "The Steelman Report".

In 1957 Dr. Van Zyl was named a Fellow of the New York Academy of Sciences, a signal, distinguished honor, conferred upon a limited number of Members, who in the estimation of the Council, have done outstanding work toward the advancement of science.

Along with five of his colleagues, Dr. Van Zyl was named in 1962, one of the six outstanding college chemistry teachers in the United States and Canada, by the Manufacturing Chemists' Association.

Cal VanderWerf's gave the eulogy for Van Zyl, delivered at the Chapel on December 14, 1967, "Doc Van Zyl was a many-splendored man," Cal regarded Van Zyl as having inspired more students to careers in chemistry than any teacher in the country and attributed this to his faith, a great faith, in each of his students. Van Zyl's faith was not misplaced; no decent Hope College student could let that be.

Van Zyl and the North Central Report

Van Zyl, even when seventy years old, and when American chemistry was entering a post-World War II fast track, had good insight for Hope College. His entry in the North Central Report highlighted grants for fifteen years from Research Corporation, resulting in over twenty papers from the chemical journals. He also highlighted the summer institutes held for science and math teachers. One of Van Zyl's major boasts was of the new Math and Physics building which would allow for Chemistry to expand in their present building and operate its programs better.

Van Zyl's suggestions included strengthening and broadening the scope of undergraduate instruction, because "new high school students were coming in better prepared and the American Chemical Society had recently set up new accreditation requirements." He was pleased that adding to the staff meant that more of the existing staff and students would be able to take time for more independent research.

The Incredible Gene Van Tamelen

It is unusual to feature a single individual among Hope's many distinguished alumni, but without great students, Van Zyl's efforts would have been a lot less than it was. The reputations Hope alumni established as graduate students at America's best schools made the Hope that we know today. Gene Van Tamelen was arguably Hope's most important chemistry graduate ever. He was bright, creative, hardworking, insightful, good critic and iconoclast.

Gene Van Tamelen was from Zeeland, Michigan, and enrolled at Hope during the war to study chemistry. When he graduated, he enrolled at Harvard taking his degree there with Professor Gilbert Stork. Stork, who later moved to Columbia University, was one of the world's most elegant chemist/research scientists. Gene, more than any individual, made Doc Van Zyl's research career. If there was a prototype of excellence, and the role a teacher played in that achievement, it was what Van Zyl did for Van Tamelen, and what Van Tamelen did, in retrospect, for Hope College.

Gene published six papers in the *Journal of the American Chemical Society* before he finished his Ph.D. degree. He maintained ties to the Hope community while at Harvard, coming back to Holland and Hope's chemistry department every summer to work in the lab. Harvard charged students, as in made them pay, for everything they used – chemicals, glassware, and equipment. There were also very few research grants at the time; no money at Harvard even for the best students. I never specifically asked Gene this, but think he returned to Hope during his summers at Harvard because of money and opportunity. At Hope, Gene knew where things were, and that he could work on projects without any cost to himself. The ideas he worked on were his but Van Zyl furnished both money and space. So Gene left the synthesis of cantharidin to Albert Burgstrahler at Harvard, while he went home for the summer. A small explanation of the canthardin synthesis is given below, because Stork gave Albert Burgstrahler most of the credit. "Gene went back to Hope for his summers." Gil hastened to also add that Gene had a girlfriend in Holland. Mary and Gene were married when he accepted his first job at the University of Wisconsin.

Before there was Viagra there was Spanish fly. Gene Van Tamelen and Al Burgstrahler synthesized Spanish fly at Harvard to partially fulfill their requirements for a Ph. D. degree. I quote from their paper on the subject below:

> "The active principle of cantharides was first obtained crystalline by Robiquet in 1810. The extensive structural investigations that followed culminated in 1942 when Ziegler, Schenck, Krockow, Siebert, Wenz and Weber succeeded in effecting the total synthesis of cantharidin (I) by a route which included, however, an unsatisfactory last step, resulting in a complex mixture from which cantharidin was eventually isolated in a 2% yield. Later efforts to render the synthesis stereospecific were unsuccessful. We have now achieved a stereospecific synthesis of, the principle constituent of Spanish fly.

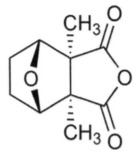

Cantharidin, a type of terpenoid, is a chemical compound secreted by many species of blister beetle, and most notably by the Spanish fly, Lytta vesicatoria. The false blister beetles, cardinal beetles and soldier beetles also produce cantharidin. It is a poisonous substance, acting as a blister agent, and can cause severe chemical burns, but these same properties make it effective as a topical medication." [46]

By that time, Gene had completed his Ph.D. at Harvard with Gilbert Stork, joined the faculty at Wisconsin, was promoted to Homer Adkins professor and, in 1961 moved to Stanford to be with fellow Wisconsin faculty Carl Djerassi and William S. Johnson. He was already recognized throughout the world for his chemically insightful elegance. Gene was classic synthetic organic chemist in its best sense. He would find a natural product, determine its chemical structure, and synthesize it in the laboratory by known chemical processes. On the way to a first total synthesis, he managed every detail in the construction of a natural product of known chemical structure. Van Tamelen was most concerned with the three-dimensionality of organic compounds, which chemists know as *stereochemistry—chemistry in 3-dimensional space*. His work, more the presentation of his work in formal lecture settings, had a certain cache – a cache that exuded brilliance and confidence. Van Tamelen and Woodward were the master chefs of organic chemistry. They specialized in dreaming up and cooking the finest molecular cuisines. They told the story with large pieces of colored chalk, exquisitely drawing each chemical structure. In left-handed Gene, the result appeared perfectly.

Gene Van Tamelen was renowned for "Tammy's brilliance." Sometime later, while I was a graduate student at Kansas in Lawrence, Gene, already at Stanford, gave one of the endowed lectures for the department.[447] He described his total synthesis of colchicine, which he said came from *autumn crocus* and was used

[46] Gilbert Stork, E. Van Tamelen, L.J. Friedman, and Albert Burtstrahler, "Canthardidin: A Stereospecific Total Synthesis," *Journal of the American Chemical Society*, 75, 1953.
[47]Van Tamelen, E. E.; Dewey, R. S.; Lease, M. F. Pirkle, W. H. Selectivity and Mechanism of Diimide Reductions. *J. Am. Chem. Soc.* 1961, 83, 4302-4303.

mainly to treat gout. However, Van Tamelen's synthesis, though it was not the first, paved the way for several others' syntheses. Van Tamelen published the full paper in *Tetrahedron* (Van Tamelen, E. E.; Spencer,T.A.; Allen,D. S.; Orvis, R. L. *14* **1961** p. 8-34).

Tom Spencer, emeritus Professor of Chemistry at Dartmouth College, was a Van Tamelen graduate student and worked on the colchicine synthesis in Madison. Eager to please his mentor Van Tamelen, Spencer stayed up all night to finish the preparation of ditrimethylcolchicinic acid. The next morning, he spilled about half of his compound. While Van Tamelen was not angry, it would take some time before the synthesis of the compound was perfected.

In 1967, as a brash assistant professor, I tried to entice Gene Van Tamelen, by then a member of the National Academy of Sciences, to visit Hope and lecture to our students. We had some excellent undergraduates at the time some of whom I thought were potential Stanford graduate students, and I wanted to share their skills with Hope's most famous chemistry alumnus. But Hope had never sent a chemistry student to Stanford, and as I understand it another Hope student had enrolled at Wisconsin intent on working in Gene's group, without success. So, as I found out years later, I was not generating a lot of interest from Professor Tamelen. Even though I thought Gene would be jumping for joy to recruit our excellent students, he had background information that I did not.

So what guile can one manage in such a circumstance?

The best way I could think of getting Gene to Hope was that we give Van Tamelen an honorary degree. If he accepted an honorary degree invitation, I could get him to Holland a day before the event for a seminar for our chemistry students who were staying for summer undergraduate research. I assembled a précis that President VanderWerf presented to the then Board of Trustees for their approval. Fritz Yonkman, already retired from Ciba Corporation in Summit, New Jersey, offered the motion for approval and read the précis for Van Tamelen at graduation, 1970, at the Civic Center. Gene received his degree with grace and though Hope did not have much civil unrest because of Vietnam protests, there was unrest in the air at every college. So it was not a good time for such formalities. Bucknell, was also giving Van Tamelen an honorary degree at that same time. Manning Smith, an organic chemist at Bucknell, had spent a sabbatical with Van Tamelen, and provided the précis for their nomination. My heart sunk when Gene told me that he could not arrive at Hope to give a seminar the day before our

80

graduation because he was at Bucknell receiving an honorary degree from that University.

The history of chemistry is replete with challenges to do that which none others could do. The coal tar product 'benzene', had captured the imagination of the chemists for many years. Its formula C_6H_6 was unusual given the other known compounds that contained just carbon and hydrogen. In benzene the ratio of carbon to hydrogen was low – as in one to one. How could that be?

Many thought up structures with one hydrogen per carbon atom. Sir John Dewar was one of those that worked on the problem. A few years before Gene died, I established contact with him again and asked him if he had ever read the papers of Sir John Dewar, on whose work another of his achievements – the synthesis of Dewar benzene - was based. He told me that he had not.

"Everybody knew about Dewar benzene! Why read those old papers?" was Gene's quick response. Dewar's formula, below, had one carbon and one hydrogen at the corners of the box like imagined structure shown on the right. That was wrong; but bright intellects like Gene tried to make it anyway, and establish its properties.

Dewar benzene

Gene Van Tamelen was a scientist whose work was characterized by care, delicacy and attention to fine structural detail, and he was an artist. His art was total synthesis as in making a complicated molecule from simple beginnings. He could also be quite disrespectful, particularly of ideas he did not think of value so I sincerely appreciated the time he spent with Hope students and me at a time when his career was at its apex. We at Hope were struggling with the future. Gerrit Van Zyl and Harvey Kleinheksel had retired and passed away, and we were in the process of developing what, for Hope, became its future mantra—research with undergraduate students. For us, Gene's visits in 1968-70 were real honors, and we were pleased that he did manage to include us in his schedule.

Gene's eventual visit to Hope for a seminar came in conjunction with a consulting visit to the UpJohn Company in Kalamazoo. I remember several things about his visit, but mostly I remember that it was my job to take him to Kalamazoo after his day with us was finished. We left rather late and I had a Mercedes 220 that I had bought in Germany but driven far too hard on US roads. This machine was the smallest Mercedes and had to be routinely serviced by a Mercedes dealer to continue functioning in the Mercedes way. When I got the car in Holland, the Mercedes dealer was located in Grand Rapids and getting my car to Grand Rapids for service took a whole day. So, I neglected my machine, and the car fought back. After a few trips from Holland to Urbana, Illinois, where I taught in the summer of 1970, it quit running smoothly first, and then at all. A diagnosis from a local mechanic defined to the problem to be "burned valves," which he fixed.

A few weeks after getting my dream car back from said repairer, the time came for me to deliver Gene Van Tamelen to Kalamazoo. We left in mid-evening, and got as far as Plainwell, where, when I stopped for a red light late at night, the car continued to run but my foot on the accelerator did nothing.

I jumped out, saw a big nut under the engine, opened the hood, found a rod hanging down, and – brilliantly if I do say so myself – connected said rod to another with said nut, jumped back in the car, and we drove on to Kalamazoo.

Gene left Stanford for an avocation that became almost a vocation for him – architectural design. Gene had a characteristic left-handed penmanship skill that was particularly impressive when he lectured on organic synthesis. His style was not that different from Woodward's. Woodward never used visuals – he wrote everything out with multicolored chalk. He would start at the top left of a lecture room's blackboards, and fill the boards from left to right with chemical formulae. Gene did more or less the same thing; though he was painstakingly left-handed (if there is such a thing.)

The impression the scientist Gene Van Tamelen gave was that he was meticulous to a fault. Those of us that knew him as a scientist, knew him there was an elegance about his writing, his papers, and most particularly his presentations. Like Woodward who, though a Professor at Harvard when Gene was a student was never directly associated with him, Gene could make even the synthesis of bathroom tile sound like a Nobel Prize winning accomplishment.

If undergraduate research in chemistry, and more, needed a first role model, none could have been better than Gene Van Tamelen. At Hope, because of Hope, and for Hope, he developed a career for himself and outlined a pathway to those that followed him in the labs of the old Science Hall showing how that could be done.

82

Chapter 5

Research with Undergraduates

Research is to an academic scientist as prayer to a professor of theology. Independently studying a specific area of chemical interest is the way a chemist practices his or her trade. I never thought about this when I returned to Hope to teach because I had thoroughly benefitted from my undergraduate research experience. It was the first thing I had done as a young student that I had been paid to do that I liked doing. Research was fun!

It was just natural that when I started teaching at Hope I paid a lot of attention to undergraduates that might benefit from the experience of independent study as I had benefitted. This was simple to do because it was my instinct. I was naïve but enthusiastic and hard-working. I kept on the lookout for students who wanted to do a special project. In the words of Arthur Amos Noyes, I was looking "for the seeds that needed planting." Jim Hardy ('68) was one of those; Paul Schaap was, too, but to a lesser degree.

How I operated is no secret. I always tried to find able students who could not sleep because they loved science so much. And I always tried to treat my research associates—undergraduate students, doctoral mentees, or post-doctoral affiliates—as collaborators. I made the point that "one plus one was always more than two." I wanted my colleagues thinking about their project at every stage; not just seeking direction. I operated this way even with undergraduate students. Though they had little background in the work, many learned quickly. I appreciated that they shared what they were thinking with me, and tried to use their ideas if they pertained. This is really hard for a young student to do. The students I remember most are those that told me I was wrong and set about proving why.

American university research was blossoming when I came into organic chemistry. The United States, driven by the initiatives of the Eisenhower administration, and the earlier leadership of Vannevar Bush, had managed to set in place the greatest growth in the intellectual United States in its history.

The National Science Foundation (N.S.F), though debated for a long time in Congress before Bush and his fellows (James Bryant Conant, President of Harvard; Frank Jewett, President of the National Academy of Sciences and Chairman of the Board of AT &T; Bethel Webster, a New York lawyer; and Irwin Stewart) prevailed, was authorized in 1950.[48] The idea had begun to emerge a decade earlier, over lunch at the Century Club in New York, in May 1940. That meeting seeded the National Defense Research Committee (N.D.R.C.).

Germany invaded Poland in September 1939. In spite of isolationists like Charles Lindbergh and Hitler devotees like Henry Ford, the majority of the leadership of the United States felt the country could not sit on the sidelines for much longer.

Vannevar Bush, head of the Carnegie Foundation in Washington, previously Dean at M.I.T., an operating chair of the five-person committee that began N.D.R.C., had managed to see President Roosevelt quickly after the May 1940 meeting in New York. With the core of the idea to mobilize the U.S. science establishment to protect America's best interests, the N.D.R.C. was approved and formally started in June 1940. The founders were joined by the President of MIT Karl Compton, a physicist, and graduate of the College of Wooster. Four individuals (Bush, chair, Compton, Conant, Jewett) thus became the face of civilian science directed toward military projects during World War II. Conant assumed the chairmanship of N.D.R.C. That committee became part of a wider organization the Office of Scientific Research and Development (O.D.R.C.) headed by Bush. In spite of meddling by the military, Bush's office was in the White House so he had direct access to the President. The O.D.R.C. had outstanding success during the war bringing science, engineering and medicine to bear on the United States' conduct of the war. After the war, the patterns set in managing civilian research were retained through a number of research organizations. N.S.F. was the most obvious of such benefactors.

By the time I came into science, the wartime N.D.R.C. had been subsumed into the peacetime National Science Foundation. By 1960, when I first started my research work, N.S.F. programs were starting to reach every university in the country. It would take a while for N.S.F. programs to reach the smaller colleges, and on that I played a significant role. But by 1960, N.S.F. graduate fellowships and N.S.F. cooperative graduate fellowships supported the best students.

[48] J. Merton England, *A Patron of Pure Science, The National Science Foundation's Formative Years, 1945-57*, National Science Foundation, Washington, DC 1983.

American Science Backgrounds

Chemistry, physics, biology, and geology were part of most liberal arts curricula from the beginning of young colleges and universities in the mid-19th century. Philip Phelps, Jr.'s first Hope academic curriculum included, as discussed in an earlier chapter, *Draper's Chemistry*. Chemistry did not start in the United States, but was new and exciting then (more or less like computer programming became a century later). So students were eager to find out what it was about. The best students knew that chemistry's formations were mostly in German laboratories, so beginning in the late-19th century, the best students in American colleges could often go to Germany for their advanced training in chemistry.

As U.S. entry into World War I approached, shipping blockades by the British and submarine warfare by the Germans challenged commerce. The US had to smuggle essential chemicals made by German companies in a German submarine into the port of Baltimore. Two submarines filled with dyestuffs to print the US paper money arrived that way in 1916.[49] The president of the American Chemical Society, Theodore Richards, asked President Woodrow Wilson if the government needed help from chemists. President Wilson answered his note with one of his own: "I'll get back to you." A few days later Wilson did get back to Richards saying he had checked, and the government already had a chemist.

Wilson, as we have pointed out, a professor of history and a devotee of the liberal arts, was tremendously short-sighted. At the outbreak of World War I, the United States was a technical desert. Britain and France were somewhat more advanced but not by a lot. The Germans in contrast, driven in part by their unification under Otto von Bismarck, had built research labs, and universities around them. From the mid-19th century, Germany led in research, with Bunsen at Heidelberg, Liebig at Giessen, and Wöhler at Gottingen. Bunsen was the teacher, Liebig was the businessman scientist, and Wöhler (who first converted ammonium cyanate to urea proving that organic compounds did not just come from living things) led in research. For a generation or more, the German chemistry personnel came from these three laboratories. Eventually, their students were the unquestioned leaders.

America was isolated behind two oceans, and Americans needed only to exploit their natural resources not develop

---

[49] Katherine Steen, *The American Synthetic Chemical Industry Between 1910-1930*, Chapel Hill: University of North Carolina Press, 2014.

universities. We were hunters and gatherers, not thinkers and tinkerers. To the extent there was any science at all taught in the United States, it was mostly astronomy or botany: putatively technical areas where one observes, one does not manipulate. In the documents that led to the National Science Foundation authorization in 1950 are statements, even from members of Congress, that the teaching of science in American schools was in disarray. Few high schools graduates studied chemistry and fewer still studied physics. The classic comparison is that of John D. Rockefeller (Standard Oil) with Karl Bosch (B.A.S.F.). Germany was a mostly landlocked, small nation and had to develop efficient agricultural methods in order survive without natural resources. One manifestation of this was the Haber-Bosch process, used to reduce nitrogen to ammonia. Haber and Bosch managed this out of necessity. Rockefeller on the other hand, looked for more gushers. He took an oil monopoly for Standard Oil and gave the chemical monopoly to B.A.S.F. (Bosch) during negotiations in the 1930s. By 1939, Germany was (or, at least, Hitler thought it was) independent of most of the natural resources other countries had provided it. It made rubber and gasoline/aviation fuel from coal and also fertilizer from the air. In the United States, we had just looked for more sources of oil and found them in Texas/Oklahoma. Rubber, however, was another matter, even for us. When World War II broke out, we had to move fast, accelerating a huge research program or our vehicles would have had no tires on which to ride. Natural rubber came from Southeast Asia and this was entirely under control of the Japanese. Led by scientists at the tire companies in Akron, Ohio, (with the assistance of the University of Akron), a crash program to make synthetic rubber was begun, and successfully completed in 1943.

I doubt the guns of August rang out because the Kaiser trusted his chemists, but he came to do that soon. It has been said that while World War I was the war of the chemists, World War II was the war of the physicists. What this means is that chemists penetrated battlefields first by producing, artificially from the air, the materials with which one could make nitroglycerin and T.N.T. They also penetrated battle fields chemical weapons, chlorine, and other battle gasses throughout the war.

After World War I, America, a victor, benefitted from reparations and the industries of our country built product lines based on reparation payments: German science products taken by the US at wholesale prices and resold by DuPont and other companies for profits. The classic case of Bayer aspirin has been described

86

elsewhere, but that story repeats itself over and over in the 1920s. America benefitted not from its own research and development, but by sending a million or two young men to France and Belgium, and picking up the spoils of their victory. So even after the "chemists' war," we did not do much to accelerate our teaching of science in our schools and universities.

Though some scientists in the United States were more insightful even during World War I (T. H. Richards, whose son-in-law, James Bryant Conant, must have had something to do with an Army chemical corps and laboratory at American University in Washington), we did not learn much from it. At the American University labs, organic chemists studied mustard gas and pushed Father Julian Nieuwlands' discovery of the reaction of acetylene with arsenic trichloride to a practical level. Conant, Ralph Lewis (Northwestern), and James Flack Norris (MIT) isolated the most toxic of the products, monochlorovinyl arsine, and outlined a commercial synthesis. This was put to use at Willoughby in Ohio in 1917-18 under Conant's direction. There on the site of an old automobile manufacturer's factory, tons of Lewisite were made, packaged and shipped. But when the war ended, we downgraded our research efforts. This lab went out of existence.

Training American Research Chemists

At the time of World War I, almost every organic research scientist in America had studied in Germany. Those chemistry professors who had not studied in Germany had taken their degrees with one who had: Ira Remsen (Johns Hopkins) took his Ph. D. in Giessen with Liebig; Jules Stieglitz and J. Ulrich Nef (Chicago) took theirs in Munich with Baeyer. William. A. Noyes, Sr. (Illinois) took his degree at Johns Hopkins with Remsen.

Few research departments in American universities or American industries could stand alone, because the scientists needed financial support to operate. Research was, and is, expensive. So various American universities developed product lines to support their research programs. Yale's chemistry department programs under Treat B. Johnson and others, looked to the chemistry of natural products and to medical solutions for its support. [50] There was also a close alliance with New Haven sewer projects. Johnson spent his

---

[50] Johnson T.B. and R.D. Coghill, "The discovery of 5-methyl-cytosine in tuberculinic acid, the nucleic acid of the *Tubercle bacillus,*" *Journal of the American Chemical Society,* **47**, 1924.

career studying vitamin deficiencies and common diseases. Rudolph Anderson, also a later life Yale chemist, spent his life studying plants.[51] After multiple frustrations in the United States and Sweden, Anderson finished his degree in Emil Fischer's labs in Berlin leaving Germany by train on the outbreak of war in 1914. Anderson's first job in the United States was with the U.S. Department of Agriculture in Geneva, New York, where he claims he "learned about cows."

Penicillin

Rudolph Anderson's Yale colleague and later instructor, Robert D. Coghill, later worked for the U.S. Department of Agriculture in their labs in Peoria, Illinois. Coghill, more than any American chemist, was responsible for the isolation of penicillin in 1943 in commercial quantities. Penicillin was a product of American research and development during the war. A team of scientists from Pfizer, Merck, Squibb, and other companies was assembled under the instigation of the N.D.R.C. Coghill told the story in *Chemical and Engineering News* in 1944.[52]

The British were so engaged in the war that they needed assistance in developing Fleming's discovery. Oxford scientists Howard Florey, and Edward Abraham came to the United States in 1941 to get the work started.[53] By the end of the war, every American soldier carried a vial with him in case of injury or disease.

Every miracle drug is a story in a story: penicillin is no exception. Belfer's *Fierce Radiance* is a historical novel, meaning that most of the incidents are historical and some of the facts are, too, but the main characters are fictional. As the *New York Times* review (June 18, 2010) of Belfer's historical novel pointed out: "Belfer combined medical and military history with commercial rivalry, espionage and thwarted love. She clearly knows her scientific material. She also knows how to turn esoteric information into an adventure story, and how to tell that story very well." *A Fierce Radiance* detailed the process of developing penicillin and other antibiotics during World War II. In her view, penicillin was seen as a "weapon" of the Allies, a marked advantage over the Germans they were fighting. According to my colleague at the Imperial War Museum Archives in England, Stephen Walton, there are

---

[51] Rudolph John Anderson, Biographical Memoir; by Huburt B. Vickrey Proceedings of the National Academy of Sciences, 1962, 1-50.
[52] Robert D. Coghill, "Penicillin Science's Cinderella," *Chemical & Engineering News*, 22 (8), 586-93, 1944.
[53] This has been detailed in Lauren Belfer's *A Fierce Radiance*.

fragmentary records of penicillin research in Germany: "Dr Schoeller at Schering AG in Berlin, Organon in the Netherlands, and a bit more detail on work at I.G. Farben, Elberfeld (this latter is mostly in CIOS Final Report XXV-54, 'Pharmaceuticals at the I.G. Farben plant Elberfeld')." Essentially though the German research establishment counted on sulfa drugs to fight the bacterial diseases of the Wehrmarkt. In Germany during World War II, research on penicillin was not a priority. Though the Germans led the way in the development of sulfa drugs and sulfanilamide after World War I, they must have felt sulfa sufficient for their antibiotic needs as they headed into World War II. That was certainly not the case. Penicillin use as an antibiotic miracle drug subsumed the market for sulfa drugs almost immediately after it became available in quantity.

The story is worth telling, briefly, here.

Penicillin was first produced for potential laboratory testing by cultivation of the fungus *Penicillium notatum* Westling on the surface of a liquid nutrient. That process is inefficient, costly, and laborious: in order to get enough penicillin from using it, huge numbers of pans and bottles must be washed and sterilized, small amounts of nutrients dispersed into the containers, and each container thus filled must be sterilized and inoculated. Then, following an incubation period of six to twelve days, penicillin activities at various levels can be detected.

Surface cultivation, as had been used for penicillin since Fleming's discovery, was contrary to other fermentation methods used in an industry where commercial preparations are carried out in vats into which a mold is submerged under the surface of the preparative broth. Though publication of the results of such tests followed after the war because of the strategic significance, excellent results came forth on submerged methods of culturing from the U.S. Department of Agriculture Laboratories in Peoria where submerged fermentation techniques had been previously developed for the synthesis of gluconic acid.

Gluconic acid as it is used in pharmaceutical industries is produced by the fermentation of glucose either by strains of Aspergillus niger, Penicillium sp., or selected bacteria. In the commercial process, a nutrient solution containing 24-38 per cent glucose, corn steep liquor, a nitrogen source and salts, with pH 4.5 is used to culture a selected strain of fungus in shallow pans or in submerged culture conditions to convert glucose into gluconic acid. The pH of the medium is controlled by the addition of a strong solution of sodium hydroxide.

89

The Peoria lab, also called the Northern Regional Research Laboratory of the U.S. Department of Agriculture, was founded for the specific purposes of finding other uses of corn and corn products. Some brilliant people worked there during and just after the Depression because the USDA provided jobs for scientists when others were unavailable.

Robert D. Coghill was a Kansan of prodigious scientific output. We found out about Coghill from the Ph. D. dissertation of Paul Block, Jr. an Ohio newspaper publisher who wrote a thesis in organic chemistry at Columbia. Block dedicated this thesis (*Problems Related to Thyroxine,* Paul Block, Jr. Ph. D. thesis, Columbia University, 1942) to Robert D. Coghill "for first inspiring the work." Coghill's own report completely outlines the development, more the commercialization, of this wonder drug. [54] By 1944, clinical tests had shown penicillin effective against numerous bacterial infections in humans including bacterimias, osteomyelitis, pneumonia, gas gangrene, gonorrhea, and possibly even syphilis.

Penicillin had been discovered by accident by Alexander Fleming, a professor of bacteriology at St. Mary's Hospital in London. Like Heinrich Hörlein and Gerhard Domagk, who discovered the sulfa drugs at B.A.S.F. while looking for bacteriastats, Fleming, too, came home from the war deeply impacted by the loss of many lives to bacterial infections and sepsis resulting from wounds incurred at battle. The antiseptics used by field hospitals were also more effective at killing soldiers than infections. Antiseptics worked well on the surface, but deep wounds tended to shelter anaerobic bacteria from the antiseptic agent, and antiseptics seemed to remove beneficial agents produced that protected the patients in these cases at least as well as they removed bacteria, and did nothing to remove the bacteria that were out of reach. Fleming was notorious for having a sloppy lab and left a number of molds in petri dishes at the start of his summer vacation. On returning, Fleming noticed that one culture was contaminated with a fungus, and that the colonies of staphylococci that had immediately surrounded it had been destroyed, whereas other colonies farther away were normal. Fleming showed the contaminated culture to his former assistant Merlin Price, who reminded him, "That's how you discovered lysozyme." Fleming next grew the mold in a pure culture and found that it produced a substance that killed a number of disease-causing bacteria. He identified the mold as being from the *Penicillium* genus,

---

[54] Robert D. Coghill, "Penicillin Science's Cinderella."

and, after some months of calling it "mold juice" named the substance it released penicillin on 7 March 1929.

Fleming published his discovery in 1929 in the British *Journal of Experimental Pathology* but little attention was paid to his article. He continued his investigations, found that cultivating *penicillium* was quite difficult, and that after having grown the mold, it was even more difficult to isolate the antibiotic agent. Fleming's impression was that, because of the problem of producing it in quantity and because its action appeared to be rather slow, penicillin would not be important in treating infection. Fleming also became convinced that penicillin would not last long enough in the human body (*in vivo*) to kill bacteria effectively. Many clinical tests were inconclusive, probably because it had been used as a surface antiseptic. In the 1930s, Fleming's trials occasionally showed more promise, and he continued, until 1940, to try to interest a chemist skilled enough to further refine usable penicillin.

Fleming finally abandoned penicillin, and not long after he did, Howard Florey and Ernst Boris Chain at the Radcliffe Infirmary in Oxford took up researching and mass-producing it, with funds from the U.S. and British governments. They started investigated the mass production after the bombing of Pearl Harbor. When D-Day arrived, they had made enough penicillin to treat all the wounded Allied forces.

There had been few, if any, clinical tests in England sufficient to justify additional research on the compound(s) though Harold Raistrick at the London School of Hygiene and Tropical Medicine had done enough testing to justify preparing some quantity. In 1939, a group directed by Florey at Oxford including Drs. Abraham, Chain, and Norman Heatley prepared material in quantity so that the drug could be tested on a few patients. The tests were somewhat positive so in 1941, funded by the Rockefeller Foundation, Florey and Chain flew to America with samples of their mold, *penicillium notatum* Fleming. Here, helped by the National Academy of Sciences and Charles Thom, principal mycologist at the US Department of Agriculture, the problem was referred to Coghill at the USDA's labs in Peoria. Chain and Florey arrived in Peoria on a very hot day in June 1941 and, as Coghill would say later, "Sold us on the problem." Adding A. J. Moyer, a fermentologist to the team, managed to increase the yields of penicillin per event ten-fold because Moyer added corn steep liquor to the broth, and developed submergible fermentation systems with new strains of mold grown at the USDA labs. Heatley came from Oxford bringing with him his penicillin assay. And, by transferring his knowledge to the American team, greatly improved the pace of research here. As Coghill went on to say in his 1944 report to the American Chemical Society, the discovery of the effect of corn liquor steep on the yield of penicillin was perhaps the greatest single contribution of this early, open to the public, research at the USDA. Soon five companies, Pfizer, Merck, Squibb, Abbott, and Winthrop were pounding on the door to manufacture penicillin. After patent rights were negotiated by the government, as "in the public domain" for the duration of the war effort, these companies began producing penicillin, too, selling it to the government at cost.

As I was researching this story, I came again in touch with Hans Wynberg, living still in Groningen, the Netherlands. Hans had been in my life when was I in my late twenties and as he aged, he became an internet conversant. Hans, as this story relates, came by Hope in the mid-1960s and from that visit came Paul Schaap's semester in Groningen in 1967. I struck up a correspondence with Hans again when I read his World War II biography, *They Dared Return* by Patrick O'Donnell.[55] Hans told me he had worked for Pfizer in Brooklyn between graduating from high school, and being drafted.

Wynberg said: "I had arrived at the age of 16 with my twin brother in Brooklyn (1939). The New York World's Fair opened at the place where they have the big tennis matches nowadays. But with just a nickel carfare and $100 per month from our parents, we were able to attend Brooklyn Technical High School (six thousand boys, all white, of course, and free of charge). Having had four years of Dutch high school, I breezed through in three years, placing eleventh out of 450 on the State Regents.

Although Pfizer at that time was a small company, making mostly citric acid derivatives, and hiring no Jews or blacks, I got a job as lab assistant in the analytical division (Fehling's titration, standardizing my own acids, bases and permanganate solutions and using the excellent (but old-fashioned) classical balances. After six months, I was promoted to research assistant with Dr. Peter Regna who had just gotten his Ph.D. at Brooklyn Poly while earning his living as musician in a jazz band. The research chief was Dr. Pasterneck, the plant superintendent was McKeen (later president of Pfizer), and C.E.O. and owner was Mr. Smith. My pay was $24 per

---

[55] Patrick K. O'Donnell, *They Dared Return: The True Story of Jewish Spies Behind the Lines in Nazi Germany* (Boston: Da Capo Press, 2010).

week with overtime if I worked on Saturday. The other project we worked on was the total synthesis of Vitamin C as well as of riboflavin. Yes, I have made Vitamin C all the way from sorbose! Regna was the one who devised the purification of the penicillin broth by passing it over a column of alumina and charcoal, I believe. I remember at that time, the structure was unknown. All we knew was that it was unstable towards heat, air, and acids of course. I was not called into the U.S. Army until August 1943 on account of this job! Pfizer gave people who went into the army a $ 20 war bond regularly!

While at Pfizer I had had only completed high school, of course, so I had enrolled in St. John's College in Brooklyn for night classes. I kept in touch with Regna after the war and was allowed to give the introductory speech when Regna got the Nichols Medal back somewhere in 1980s. He had worked with (Professor Robert Burns) Woodward on aureomycin when Woodward was consultant at Pfizer."

Arthur Amos Noyes and Chemistry on the West Coast

So what does all of this information have to do with research by undergraduates in Hope College laboratories in the 1960s? The answer is that research with undergraduate students took a lot of justifying – to have it supported financially in the first place, and to the non-science faculty at the college in the second place. Research was expensive because it had become so instrument-dependent. At the time, a typical organic chemistry laboratory needed hundreds of thousands of dollars of equipment. Without that, one was severely handicapped in the work. On the basis of equipment needs alone, many four year colleges fell out of the research business in the 1960s. Students that went to those colleges and majored in chemistry, if they did, were seriously handicapped if they started graduate or medical school.

Cal VanderWerf knew this. He determined, for Hope, that if science students, particularly males at that time, were to still find a Hope education of value, the science programs needed to be competitive with what a student could manage by studying at a major university.

The identification of the research for undergraduates as an important activity traces to Arthur Amos Noyes, first at M.I.T. and later Cal Tech Noyes, took his Ph. D. with Ostwald at Leipzig. He started the first university lab in the physical chemistry at M.I.T. after his association with Ostwald. He worked mainly with undergraduate students and probably stated, more clearly than anyone, what research is all about. He moved to Cal Tech (then Throop Institute of Technology) in the 1920s.

He wrote: "Though science, through daily experience, is recognized as vitally important, it does not just grow. It arises from *research*, and that research is a sensitive plant that will grow on successfully from carefully selected seeds—the best brains in the nation, and which must be protected against the frost of dogmatic intolerance, against the drought of administrative routine, against the flood of modern mass education, against over-forcing through the impatience of practical men and against the blight of poverty and social neglect. Research will come to its own in any community only when its members, in the words of Pasteur, regard their research laboratories as their temples."

Who Became Chemists?

Few children of Boston aristocrats became chemists. Most chemists had modest backgrounds, born into homes where education was as much from reading great books as anything. Science came into such lives when they undergraduates either by plan, or by accident. These students focused on chemistry because of the challenges presented by its do-it-yourself research environment *and* in many cases because advanced study in the sciences was supported by others. Tuition, such as it was, was handled by using the graduate student as a teaching assistant.

George Hammond had been undergraduate at Bates College in Auburn, Maine, and remarked that he found chemistry "easy," so he majored in it. He went to Harvard during the war to study, and managed with grants from industry and the government until he was awarded a National Research Fellowship in 1947. Nick Turro, in celebrating Hammond's 80th birthday, found an excellent quote:

"...Chemistry, that most splendid child of intellect and art. Chemistry provides not only a mental discipline, but an adventure and an aesthetic experience. Its followers seek to know the hidden causes which underlie the transformations of our changing world, to learn the essence of the rose's color, the lilac's fragrance and the oak's tenacity and to understand the secret paths by which the sunlight and the air create these wonders.

And to this knowledge they attach an absolute value, that of truth and beauty. The vision   of Nature yields the secret for power and wealth, and for this may be sought by many.   But it is most often revealed only to those who seek it for itself." -Sir Cyril Hinshelwood.

Hammond was the oldest of 7 children; his father died when he was fourteen, so he was in charge of the "farm" that he ran while going to public school and later Bates College. He lived in very cold Maine, about which he could never quit talking because his home had no central heat and had leaky windows into which paper was stuffed to keep out the wind. George became an extraordinary spark and an intellectual explosion could be expected upon most every conversation. From him, some of the greatest scientists in the late twentieth century grew.

Turro was one of his earliest Cal Tech students. According to Turro: "Everyone who was in George's group during the golden years at Caltech recalls with fondness the incredible highs that were had as George led discussions in which new ideas crackled like sparks. It seems that after an hour's discussion on the little blackboard near the entrance to the lab, within days new and exciting results were produced. George showed us all how to share his love and excitement with his muse, chemistry. We all owe many thanks and enormous gratitude to George for creating so much new knowledge and for changing the way we think."[56]

Perhaps this feeling for George is captured in still another quote from Sir Cyril Hinshelwood collected by Turro: "But of that most important kind of knowledge, that which does not seem to relate to any existing field, it is harder to speak on the basis of anything but faith. And yet in this knowledge lies the true seed of the future. It will come only from the least conforming of minds, and the discoveries of the greatest ultimate moment are the least likely to have been favored by official encouragement or support. They must be like the flowers of the poet ...daffodils, That come before the swallow dares, and take The winds of March with beauty."

Carefully selected seeds -→ an adventure in intellectual experience -→ that most splendid child of intellect and art.

Undergraduates are the seeds. It is a professor's responsibility to find them, plant them, nurture them, and then, let them grow.

All of this poetry

Poets writing about the beauty of scientific discovery make no contribution to the critical question. So who pays? Scientists need

---

[56] Nick Turro, "A Tribute to George S. Hammond in Celebration of his 80th Birthday," *The Spectrum* (The Center for Photochemical Sciences at Bowling Green State University, Bowling Green, OH) (Spring 2001), 11-12.

support, and that support comes because scientists in industry make products that can be sold at a profit. Without sales, there would be no more need for a chemist than there is for poets. So most Ph. D. graduates work in industry.

The Germans managed this interface to perfection, for better and for worse. Gerhard Schrader, the Bayer/Leverkeusen scientist that managed onto tabun and sarin while working to find rodenticides for ship holds was asked, at the Farben trial at Nuremberg, why he continued with this line of work after it was discovered that tabun was so poisonous. He said, "It was entirely logical that I do so. It was just the next experiment. Any chemist knows that."

Even in translation, Schrader's statements at Nuremberg rang completely true to an organic chemist. I knew exactly what he meant. In other words, it was not the direction of the work, it was the excitement of the chase.

B.C. Saunders, a British scientist/reader at Cambridge, continued looking for highly toxic substances, happening onto the extreme toxicity of halophosphinates. His son-in-law, Brian Ridgeway, said that, "His home had been bombed four times by the Germans and he would have done most anything to protect his family."

Schrader was a scientist once removed from the action. Saunders was in the middle of it because German bombs brought the war there. Though the motivations for their work differed, the outcomes were almost exactly the same.

It is not hard to feel when one, in one's personal life as a scientist, falls over the edge from student to scholar. In my case, though, I loved doing research as an undergraduate; it was only after I had worked for a year or so as a graduate student that I came to spot in the work that it was so exciting that I could not stay away...I just had to do the next experiment. I say, "I came to a spot" and that's a good description. But getting to that spot is another matter. To get to that spot as a student of chemistry one had to learn enough about how to do the work right first. When that part is done, one could feel the joy of discovery. Before that, it was mostly all the agony of defeat, the agony of failing, almost to the point that failure was to be expected.

So just how is it that one gets an undergraduate research student to the point of experiencing the joy of victory in an encounter with the unknown?

The Research Scholar Teacher – the National Science Foundation

Arthur Amos Noyes, who founded the chemistry school at the University of California at Berkeley, said that he believed students should start research as early as, or at as young an age, as possible, and I believe that, too. There is nothing like the excitement of discovery to trigger a zest for knowledge that will never go away.

But I have to add to that taking a bit of financial onus from work by students so that they might study science, and benefit it/from it, is a major part of the incentive. What sold research experience to the undergraduate students with whom I worked was a little bit of pay. The current Hope College seems to have addressed this with great success. Bill Anderson, who ran the budget office at Hope for many, many years, told me that summer time, which used to be vacuous on the campus, became much more exciting as the years went on, and more lucrative, too, because of all of this undergraduate research. In other words, undergraduate research was good business for Hope College. It was also excellent for the best students.

There is little evidence of organized scientific research in universities in the United States prior to World War II. The quality universities—Yale, Harvard, Columbia, Penn, Illinois, Berkeley, and, to a lesser extent, Michigan—had individual faculty members with strong programs and initiatives that were most supported by, or tied to, industry. Professor Moses Gomberg at the University of Michigan, whose library became Hope's after his death, actually co-published his first sixteen papers in German and in English. He had studied in Germany, so he published in German journals as a matter of course, but experimented with the *Journal of the American Chemical Society* at the beginning, too. Gomberg never married and was wedded, or so it was said, to his chemistry and the Michigan chemistry department.

What effort there was in research in America, in the words of Senator Harley Kilgore from West Virginia, to whom credit for the National Science Foundation is partly given, was "irrational and disorganized." So through O.S.R.D. and N.D.R.C., American scientists, heavily augmented by the work of Jewish refugees from Europe, built what became the American research establishment in the U.S. during the war. Kilgore had it that American science should be mobilized to solve national problems, and that the results of this science effort should be "public free to practice." Criticisms of Kilgore's efforts are not meant to belittle it: it was extremely important in the development of the National Science Foundation.

But he was obsessed with who would own the output of scientific research supported by the government.

Vannevaar Bush was committed to perpetuate the government's involvement in research when the war ended, too. But he opposed Kilgore's efforts because he felt the latter obsessed with patents/ownership and "making money." Kilgore's obsession is understandable at some level, given Farben's monopolization efforts broken up in the U.S. at the end of World War I. These had resulted in the Alien Property Act the property of whose ownership was eventually settled in a law suit in Wilmington, Delaware, in 1926.[57] Though one can argue with how the 'public property' was dispersed by the government, because the Chemical Foundation under Francis Garven, a former property act administrator, benefited too much, the result was the sale of confiscated property to the private sector.

During World War II, in contrast, the American science establishment was completely mobilized. As the war was coming toward its end, a letter from President Roosevelt to Bush indicated that there was significant concern that this scientific effort not be abandoned in peace time. The letter asked four questions of Bush and gave him a route to steer the Kilgore enthusiasm:

1. How can scientific knowledge developed during the war be released to the world quickly?
2. How can a program of medical research be organized to continue the attack on disease?
3. How can the government assist research by public and private organizations?
4. Can a program be developed of American youth to ensure high-quality research in the future?

Bush decided to stay out of the answers and asked committees to do the work.

According to the record, the first question was answered quickly: use the National Academy of Sciences and the military to screen, and then release for publication immediately. In practice, as the case with J. Robert Oppenheimer, head of the Manhattan Project, would later show, this became highly controversial. But at the outset,

[57] Kathryn Steen, "Patents, Patriotism, and 'Skilled in the Art:' USA v. The Chemical Foundation, Inc., 1923-1926." *Isis*, Vol. 92, No. 1 (Mar., 2001), pp. 91-122.

it seemed easy. Irwin Stewart, the O.S.R.D. administrator, headed this committee. Henry Allen Moe, Secretary of the Guggenheim Foundation, took on the education issue. The last question was answered in two months: it was estimated that it would take ten years to "rebuild" the nation's scientific talent pool: "To do this, award 6000 undergraduate four year scholarships and 300 post-graduate three-year fellowships, with the scholars chosen by state committees, and the fellows by a national competition." The committee offered several caveats that were specifically directed at returning veterans, but by the time I got into science these were long since forgotten and the programs for graduate students and undergraduates underway in some form.

The committee to study science support in the public interest, as in what eventually became the major role assumed by the National Science Foundation, was headed by President Isaiah Bowman of Johns Hopkins University. This committee took its "social responsibility" very seriously, and this colored many of the discussions and arguments. They realized that they had to replace the "intellectual bank" that European science had been before the war. But the academicians on the committee, and in the country feared government control. Industry, for its part, did not want its support going to scientific results that it could not own. Senator Kilgore, still a player, worried that public support would lead to private gain. The medical research community wanted its own authorization and this it eventually got with the authorization of the National Institutes of Health. But immediately after the war, that was anything but assured.

Vannevar Bush, though out of the direct discussions, had to balance all these divergent interests. The physician's community also took themselves out of discussions that include the basic sciences, opting to push for Institutes of Medicine that were self-standing. This happened and the National Institutes of Health became America's front face on the war on disease.

It took a while for the Bush committees to come to some level of agreement and when they did, Roosevelt was dead. So the Bush Report, sent as a single document authored by Carroll Wilson, Professor of Management and Contemporary Technology at M.I.T., and Oscar Reubhausen, general counsel of O.S.R.D., went to President Truman. It was entitled "Science: the Endless Frontier," and it outlined the blueprint for National Science Foundation.

Bush met with President Truman on June 14, 1945. Truman liked the report Bush recorded in his notes and agreed to its public release. This was coordinated by Samuel Rosenman, the ubiquitous

President's counsel, and occurred on July 19, 1945. Though the Report urged legislative action, that took five years. The original legislation was drafted by Ruebhausen and Carroll Wilson, and presented to Wilbur Mills from Arkansas, who agreed to introduce it in the House on the day the report was released, July 19, 1945. Warren Magnuson, from Washington, planned to introduce the bill in the Senate on the same day. The Magnuson introduction, which was for a National Research Foundation. angered Kilgore, and four days later he introduced similar legislation for a National Science Foundation. This is what it became.

The interesting story of how N.S.F. became law is best left for the reader to find in the England book. Its administration, though, was under a civilian board – the National Science Board. The first grant was to the Institute for Cancer Research for $10,300. Much of the National Science Board's job was to fend off the budget cutters in Congress.

National Science Foundation support for work done in small, four-year colleges was not unknown. But the entire N.S.F. funding for work in the smaller institutions for the period 1952-58 was but a little over $2 million. This was less than the funds given the University of California at Berkeley in just one year in the same period. So, when it came to N.S.F. support, small college scientists were of no consequence.

National Science Foundation operatives who interacted with faculty scientists were professional scientists serving at *program directors* at the Foundation. It was the program director's job to receive proposals, find reviewers for those proposals, collect the reviews, assemble the aggregate for a decision, make that decision, and forward that through channels. If the decision was for funding, it took a one course; if it were for declination, it took another. Categorically, when I first started teaching at Hope, the N.S.F. program directors were negative in their support of faculty proposals from smaller schools. They, themselves, were products of major universities and several aspired, and thought they should be, teaching at Harvard rather than Podunk U. Small state university faculty did not fare much better; and I can recall blasphemous remarks by Kent Wilson, an N.S.F. administrator, about the same issue at a meeting I attended at the now decayed Book Cadillac Hotel in Detroit in 1967.

But, faculty members who taught at small colleges, and smaller state universities, came from political districts, and it was arguably a lot easier for a scientist from Bowling Green, Ohio, to talk to his member of the House of Representatives (who also lived in

Bowling Green) than it was for a faculty member at Princeton, Harvard, or Stanford to do the same. Political pressure eventually bore fruit, and eventually too certain programs were started that were solely the domain of faculty from smaller schools.

Proposals came from institutions, not individuals. In the cases with which I was familiar, an office of "grants" would receive a faculty member's proposal, review its budget to make sure it was not committing the institution to something it could not afford or support, and then sign the proposal for the university. Thus a report of a successful grant application went to the individual scientist in the form of a copy of a letter to the person signing the application for the institution, while the arrangements for the money went to the chief financial officer.

Program officers at the National Science Foundation changed periodically. The chemistry program used a group called "rotators": chemists from the community, usually tenured professors, who would spend a time at the Foundation serving as junior assistant program directors. Other divisions relied on more permanent N.S.F. employees in the program director's position. That was the case with the division of materials science where almost all of my N.S.F. support came from over the years. Mostly I got along with the program directors assigned to my science and valued their insights, but some were just plain snotty. They wore the mantle of "N.S.F." as though it were a golden arrow conveying, to them, some degree of infallibility. The late Fred Findeis, under whom the program in analytical chemistry fell, was one of these. One would be insulted by his comments were one not to consider the source. Another was a program director in materials science, Andrew Lovinger, who came to N.S.F. from Bell Labs. Lovinger told me once, at a time when my group was as big and as successful as it ever was, that "his N.S.F." did not just support "productivity" as if one were productive, one was necessarily inferior. My sense is that the programs these two directors managed to support were no more, nor no less, successful in generating great scientific progress for the U.S. than were other program directors at the Foundation.

But most program directors were like John Showell, who gave me my first research grant. That grant was pretty successful; in fact, it generated two major businesses later. Though we did not have that many publications from it, those we did publish were important. When Showell sent my renewal proposal for review, it came back from reviewers with "too modest a productivity" to generate additional support. Showell was almost ill when he had to tell me this.

In other words, he had surmised that my program would be very successful in the long run, but the "old boy" university network was for keeping me out.

Eventually the community asked for, and received, all reviews from grant applications. This came about because the scientific communities were tired of anonymous peer reviews being used to decline their programs, without they, themselves, knowing what the reviewers had said. After N.S.F. became a serious player in American science, a scientist could see for himself what the community was thinking of his work. Every scientist that sought support from the Foundation knew the peer review process well. The most egregious case of improper review that I remember during my scientific lifetime was a review one of my colleagues at Bowling Green received. His proposal was reviewed by six reviewers. All graded his program excellent, save for one reviewer who wrote: "This is an excellent proposal and I would give it an excellent, save I do not believe in supporting research in this kind of institution."

I was on at least two N.S.F. standing committees, one in materials science, and the other in chemistry, at the time. So, I took this review to the Director of the Foundation with the question "What kind of tax payer funded organization is this?" Needless to say, my colleague's program was funded.

The Chemical Education Problem

I am a graduate of a small high school in rural New York state. My friends who had gone to college told me: "I was well prepared because our school had good teachers, and was a strong school academically." But, in the sciences and mathematics those teachers became the teacher. I had the same individual for every mathematics course my high school offered, for physics, and for chemistry. When I decided I needed trigonometry to be eligible for admission to one of the colleges of engineering to which I was applying, that teacher found a dusty old book and let me study it on my own. This was a good teacher. But his background was severely limited.

Standards in New York's small schools were controlled from Albany and the State Board of Regents. They did this through uniformly prepared and administered state examinations, called Regents examinations. Many subjects, mathematics, physics, and chemistry among them, required standardized tests made up and administered by the public education professionals from the State of New York. Little yellow books, targeted at the Regents exams, were made available to us for purchase at the beginning of the second semester. For most of that semester, we studied the course in point from these little yellow "preparing for the Regents' exams" books. If one paid attention to those books, it did not matter what one did in the class for the rest of the year because the entire grade was based on what one achieved on the Regents examination in the course.

Most of my school years, I did not pay a lot of attention during the year. I coasted through classes doing whatever a misguided teenager with lots of healthful opportunities to waste time did, counting on the Regents exams to be the leveler. Some would say I was not challenged; others would say I did not care. I would say I should have been kicked in the seat of the pants more often than I was. In some respects, it was probably for the best because I went to college with some terribly wrong information passed on in both my chemistry and physics courses. Freshman courses set that straight.

I was fortunate. I had those little yellow books and they served as levelers in New York. But other students from other states were less fortunate; if their teachers were unprepared to teach chemistry or physics, and most were, these students got off to terrible starts.

The chemical education problem that the nation faced after World War II had roots in secondary education. The National Science Foundation realized this, and began summer institutes for high school teachers in the sciences and mathematics in the late 1950s. At Hope, Jay Folkert in mathematics was the first to recognize the opportunity. He wrote a short, concise recollection of the summer institutes for teachers of math, physics, chemistry, advanced placement chemistry, and even geology. The advanced placement summer institute in chemistry lasted until the early 1990s. The others existed coming in and out, depending on the availability of faculty instructors from the late 1950s to the present time. I reproduce part of Folkert's report following.

1960

On June 25, 1959, Dr. Hollenbach wrote to Dr. Frissel who was doing summer work in Buffalo, NY and Dr. Folkert who was attending an NSF institute for college teachers in Washington, D.C., "I am sending along to Harry, all the material I have collected including the instructions and blank forms. You will note that there are two copies of the summary sheets and that the final proposal must be mimeographed and 15 copies sent on. My idea is that Harry should work through a proposal in rough stage and then send on the whole packet of materials, including his notes, to Jay. Then if Jay would revise it and add his ideas we should have a proposal in fair shape. I use this approach, Jay, in the hope that once you have reached that stage you might drop in on Ray Zwemer and talk it over with him. This cannot do any harm and might do some real good. These steps should be completed at the latest, by the time you return from Washington, July 24."

1961

When the grant of $38,900 was awarded later that year, the first phase of NSF institutes was underway. Voorhees Hall was selected for housing because a number of adjoining rooms could be used as suites for families. Meals were served in the Voorhees dining room. Classes were conducted in the Science Hall. Thirty-nine teachers participated including the following alumni: Norman G. Boeve, Richard W. Flaherty, Harvey R. Heerspink, Preston J. Petroelje, Ronald M. Schipper, James C. Schoeneich, and Willis W. Slocombe.

1962

Because the program was so enthusiastically received,
there was little doubt that a proposal should be submitted for
1962. The logical decision was to submit a proposal which
duplicated the one which had proven successful. But it takes
a team to carry out a program. The biology course presented
by Professor Crook constituted a vital part of the 1961 Institute.
He had other plans for the summer of 1962. Moreover, Professor
Jekel had plans for graduate work for that summer. With no
geology department at Hope, biology and chemistry courses were
essential to an institute which would appeal to junior high
school teachers of general science.

My own research – from whence did I come

I am originally an organic chemist – that is how and where I was prepared for my life in science. So my approach to science, and to the teaching of research as an experience for a young scientist, is that of an organic chemist. Organic chemists plan experiments, and carry them out, with the expectation that a result will be forthcoming relatively soon.

Among the most important advances in organic chemistry over my tenure in the field (1960 to the present) was the intersession of analytical instruments. Whereas in the days of the development of organic chemistry by German scientists before World War II, chemical reactions were carried to a logical stopping point, and the outcomes analyzed then, the advent of N.M.R., spectroscopies, and particularly chromatographic methods, let organic chemists follow chemical processes as they were going on.

One only needs to look at the synthesis of quinine by R.B. Woodward and William Doering in 1944 to see how much more difficult the synthesis of an organic natural product had to have been when one only had one's analytical information about the elements contained by the compound to confirm its existence. [58] Quinine had been of interest since Perkin and was said, by Perkin himself, to have given directed birth to the coal tar industry. And hence the organic chemical industry.[59] Pasteur had done work on quinine, too, or at least on the raw materials from which it is made. [60]The total synthesis was

---

[58] Woodward and Doering, *Journal of the American Chemical Society* **66**, 849 (1944).
[59] Hofmann Memorial Lecture, *Journal of the American Chemical Society* **69**, 603 (1896))
[60] Pasteur, *Compl. Rend.* **36** 110 (1853)

reported by Woodward and Doering in 1945 with the help of Edwin Land at Polaroid Corporation.[61] Scientists among the readership should enjoy the paper in the *Journal of the American Chemical Society*. But the synthesis was also reported in the *New York Times, Harpers,* and *Life* magazine. In the latter, Woodward walks the reader through some of the many processes required during the completion of the work. Until the advent of current analytical instrumentation, organic chemists were operating on intuition and best guesses. That is all changed.

At Hope, when I taught there, many students started working under my direction before they matriculated in college: Phil Van Lente was one and Michael Sponsler was another. Van Lente became an emergency room physician, one of the best in western Michigan. Perhaps all that lab impacted his career choice. Sponsler, after a degree at Cal Tech and post-doctoral work at Berkeley became, and remains, a professor at Syracuse University.

Undergraduate Research Participation was an afterthought at the National Science Foundation. U.R.P. programs were great for our colleges, and universities, but a modicum of what was justified in the fields, and actually needed. Research Corporation, and the Petroleum Research Fund, administered by the American Chemical Society, were two other granting agencies with a big impact on chemists in colleges, and universities.

Barbara Schwoen, professor emeritus of chemistry at the University of Kansas, shared the story of Kansas's Undergraduate Research Participation program. From the beginning, N.S.F. had a mandate to produce personnel in science and mathematics. But it did not do well with undergraduate scholarships so this mandate was not fulfilled. However, the nation became concerned after Sputnik with the pipeline problem. The U.S. had been on the top of all things scientific during the war but things had slowed down.

Barbara wrote, "N.S.F. approached Kansas to run a pilot project for undergraduate research. Kansas, though a major research university, had an extraordinary honors program for its best undergraduate students. The associate dean of arts and sciences was in charge, and funneled the N.S.F. Undergraduate Research Participation money to chemistry and other sciences and also (sub rosa) to political science and perhaps other disciplines. After a year or two of undergraduate research experimentation, U.R.P. was launched." Barbara moved to Kansas in 1977, and was immediately

---

[61] Woodward and Doering, *Journal of the American Chemical Society*, **67,** 860, (1945).

tapped to write the grants and run the chemistry U.R.P. at K.U. "I remember all U.R.P. advisors were required to attend an orientation in Washington, D.C., encouraging best practices and a level of conformity by member institutions. At one time, there were 4 URP sites at K.U.: Chemistry, Chemical Engineering, Biochemistry, and Medical Chemistry. After Undergraduate Research Participation was abandoned during the Reagan administration, Research Experiences for Undergraduates started up. I saw little difference except for the explicit mention of recruiting minorities and women. I received an R.E.U. grant in 1989 and chemistry was continuously funded for eighteen years. During these years, I helped organize and speak at a couple of N.S.F.-sponsored events at the National American Chemical Society Meeting and also at N.S.F. itself and became recognized for the quality of our programs. In 1998, I launched an ongoing annual University of Kansas Multidisciplinary Undergraduate Research Symposium, still going strong."

Hope's story is essentially the same: Undergraduate Research Participation began modestly with a successful proposal in chemistry in 1966. Save for the Reagan years, when the "undergraduates don't do research" crew took over, Hope College had support from U.R.P. and, now R.E.U., for many years.

Every quality teacher is proud of their students, so they refer to them as "my former student." Sam Adler, one of the world's most successful teachers of music composition, and orchestration, cannot talk about music but what he mentions are "my students." He is filled with stories about those that taught *him* at Harvard and elsewhere: Walter Piston, Paul Hindemith, Aaron Copland, and Randall Thompson. Sam must have been some sort of student devil: he and Thompson, who was his thesis advisor, came to an agreement to never talk again after he finished his dissertation. Charles Fried, the Beneficial Professor of Law at Harvard law school, and solicitor general in the latter Reagan administration, spoke fondly of Texas senator Ted Cruz as being his "student." When I asked him what meant, he said Cruz studied in two classes that he taught.

That is not what a scientist means when referring to his/her students. My students worked directly, one-on-one, under my supervision, and in most cases research grants I obtained supported their work, and their salaries. Day in, day out, I consulted with them, and they with me. It was a true collaborative arrangement between two persons. One was older and more experienced. The other younger, and eager to learn.

Out of my more than 400 research papers, only twenty-five or so were co-authored by undergraduates. And only in one case, that being the most recent, have undergraduate research associates done something I could not have done myself more quickly. In the most recent case, two video gaming brothers helped me with 3D-printing experiments. This is something I could not have done myself.

But literally in every case in which students worked in my labs, they learned, they had a chance to see their career develop, and they gained an understanding of whether or not a career in chemistry was for them. The earliest undergraduates in my labs benefitted from this philosophy the most. I cherish the interactions I had with them then as among my most significant career experiences. After Hope, my career become more directed toward research for its own sake. But in the beginning, it was research for the student's sake: that is what made it such a success.

I operated mostly on instinct. If one is not a people person, one will not do this well. The current president of Hope College, John Knapp, claimed when I listened to him at the alumni banquet in spring 2015, that one has to love people between the ages of 18 and 22 to be a strong undergraduate instructor. That is far too simple and is, frankly, superficial. Loving people, and loving to interact with people, is my take on it. Often, that means putting one's self in another person's position. The best mentors in the sciences are those that find their students as their partners realizing that one plus one can be, and often is, more than two.

To me, career advice is part of the job and I always really enjoyed talking with my students and my own children, about their futures. In part, I see some of myself in their planning, and only wish I had had better advice at the time I was considering next steps.

Choosing an Undergraduate Research Problem

Paul Bartlett said, when I was first assigned an undergraduate at Harvard, I said "Find her work you can do in an afternoon, and hope she'll finish it before the semester is over." Margaret Nothier was my first undergraduate, and went on to become a highly successful software engineer, probably because I drove her out of organic chemistry! Margaret (she preferred Peg) came regularly, one day a week during the spring semester of 1964 and I gave her an absolutely awful problem. So it was no wonder she left chemistry. I had been carrying out the alkylation of benzene with t-butyl chloride or alcohol. From this with oleum as the catalyst one got good yields of tri-t-butyl benzene. An excess of the alkylating agent caused multiple alkylations, and the result, in the case of t-butylation of benzene, was octamethyl dihydroanthracene.

I suggested to Peg that she try the alkylation with t-amyl alcohol. Aside from the reaction becoming overheated, and taking off the first time she tried it, the mixture of stuff she obtained was almost impossible to sort out. For my part, I figured she was a Harvard student (at that time a Radcliffe student) so she was really smart so she would be able to work through the products. Wrong! The semester ended, and I was left with several vials from Peg's work on my lab bench. She went off to the next assignment.

But one always learns. When I went to Hope and started my career there, I decided nothing is too simple. I literally gave myself, and the student(s) that worked for me at the beginning, targets that were impossible at which to fail. Physical organic studies are great that way. One takes a chemical reaction one knows works, and studies that reaction under a multitude of different conditions. I literally took an experiment I had thought about doing as a Ph. D. student, but did not have time, ran the experiment, and got the results I expected.

Program Support (As Opposed to Individual Investigator)

The National Science Foundation also used committees of peers on-site in Washington review proposals. Programs, as in collections of offerings, from a group of scientists, were reviewed that way. Unlike research proposals from individuals, program proposals had grant application deadlines. It was not unusual for the Foundation to receive hundreds of applications for individual programs for which they knew they could fund only a small number.

I was always a participant in National Science Foundation grants to support undergraduate students, and never a program director. I never had to carry this reviewer burden. But as my research program grew bigger and bigger, I was forever at N.S.F. reviewing program proposals from the large universities. N.S.F. was cautious about conflicts of interest, and interpreted such a thing to be the case for program proposals even if an individual came from the same university. For example, if the Biology Department at the University of Minnesota submitted a program proposal for a competition in small bug science, a ceramic engineer making glass beads prettier could not serve as a reviewer were he a faculty member at Minnesota, too. Bowling Green was so small, and so inconsequential in most of the sciences, that there were almost never any proposals from faculty at it; consequentially, I often ended up as a reviewer of major proposals. Materials science research labs were among the largest I reviewed. These were programs driven by groups of scientists all focusing on the chemistry of materials. Some great science came from them, and some efficient uses of scarce resources particularly for large equipment and instruments. I took particular joy in being on the review team that considered a proposal for Cornell University at some stage during these assignments. My high school was so small that there was no way to take all of the mathematics that the College of Engineering at Cornell required for admission. I was a good student, though, and played basketball at some level. So I had a few stones in my basket. I also was sure Cornell was too big for me, and I would not go there even if admitted. But they stuck by the requirement, and rejected my application.

Imagine my glee, thirty years later, when I shared the story at N.S.F. with the Dean of Engineering at Cornell. As I recall his proposal was $7 million per year for eleven years. As it turned out, Cornell got the money, or most of it. The persons with proposals in fields I knew well wrote good proposals so I supported the program in my vote. Two Bell Lab scientists did not, however, and the proposal was funded for a lot less than requested.

The National Science Foundation had obligations to students of science from the inception, though the universities that started to dominate decisions at the Foundation did not see this imperative as compelling. But high school teaching of science was another matter. There, because of political considerations, even the most jaded American research professor could be argued into agreeing it an important obligation.

As I look back on the processes I studied in my first year, I am struck by how simple they were, and how easy the results were to get. Some would say this impacted the value of the work. I disagree. Some of the work I did that I considered very simple has been among the more often quoted of my published experiments. So my first point of advice to the undergraduate research mentor is "nothing is too simple to study if it is studied well."

My second point is to learn from the students. From the New Testament Book of James;

"Not many of you should become teachers, my brothers, because you know that we who teach will be judged more severely than others. For we all stumble in many ways. If anyone does not stumble in what he says, he is a perfect man, able to bridle the whole body as well.... "

110

I have several good examples of where I was wrong, and the students taught me the errors.

My research group, which by the time I was done, had included hundreds who had worked at least one semester in my labs, stayed together for life. I still hear from colleagues that worked with me when I first started at Bowling Green. The more advanced in the program a person was, the more inclined that person was to stay engaged.

Teaching is instinct, personality, enthusiasm, and preparation. If one likes people, and is interested in their welfare they will, as the apostle James so clearly states, be taught even when they stumble.

Cal was that kind of teacher. Humble to a fault; interested beyond measure. He thought always of the student and never himself. He was a truly genuine, thoroughly professional academician. It was from his model that I learned.

Chapter 6

Hope 1960

The Hope College from which I graduated in the spring of 1960 was almost identical to the Hope that those who graduated in 1950 knew. Numerous members of the clergy dotted the administrative offices, placed in a church social structure that took care of its own. Some were former missionaries evicted, in certain cases, by their host countries. Irwin Lubbers was president, the chapel (both the building and the morning event) was still a force (though never on Sunday), and "Bible" was a series of courses and a department. Students, found an occasional beer at the Columbia Avenue One Stop, but were mostly well-behaved.

Hope was an all-encompassing environment and institution dominated by the Reformed Church in America. Undergraduates at the school would, as often as not, have gone to the same churches, the same church camps, the same church social events, and in some cases the same parochial schools while growing up. Many knew each other on arriving on campus; and if they did not know one another, they had relatives that knew their relatives; or ministers that knew them, their relatives and their college friends. It was a closed, though close and mostly helpful, collegiate society. Hope also happened to be chartered by the State of Michigan to give college degrees: chemistry had emerged as an alternative to being a minister for Hope undergraduates. Gerrit "Doc" Van Zyl and Harvey Kleinheksel managed well, so Hope's reputation in the field was widely known.

Some years later, when Sue and I first moved to Toledo, Ohio, we were looking for a church. On the Sunday we first went to Collingwood Presbyterian Church in the center city, high school graduates (one of whom was named "Snow") were given Bibles in preparation for life's next journeys. During coffee hour I overheard a tall, slender man say "I went to a college where one either became a chemist or a preacher." I turned around to introduce myself to Bob Snow.

So it was. The Hope College that survived the Depression, World War II, and the aftermath of both, was mostly a place where one went to be a preacher or a chemist.

Still, there were worms at work in Hope's apple at the time Cal was appointed President. One was the science vs. humanities conundrum. The scientist's mission in an institution that taught only undergraduates was the mission of pre-professional education.

Students with science inclinations, be they headed for graduate school in the sciences, or advanced study in the health professions, needed the basic instruction common nationwide. Though not recognized nationally as such until Van Zyl won the Scientific Apparatus Makers Award from the American Chemical Society in a nomination written by VanderWerf at the time he was Chair of the Division of Chemical Education, Hope was on its way to becoming one of, if not THE, premier educational institution for the pre-professional training of scientists in the United States.[62] This did not run counter to the Church; for the most part the chemistry part of Hope College was ignored by the Church.

By 1963, scientists had been out of administrative positions on Hope's campus, if they had ever been part of them, for a long time. The administration of the school, at all levels, had taken on a humanities/church professional mentality. Scientists were not ignored, and their value was not underappreciated, but Van Zyl was too busy doing what Van Zyl did, so he opted out of campus politics in any real way and the president, vice president, and dean were all humanists. Until VanderWerf showed up.

Cal stepped into an academic place that was the antithesis of what learning in the American community of scholars had become. Major universities like Princeton, Columbia, N.Y.U., and even by this time, Rutgers and Notre Dame, had divorced or at least distanced themselves from church roots long since, while at Hope the connection remained inviolable. The Church affiliation did not often get in the way, but it was clearly in the way of scholarship particularly in departments most proximate to issues of morals and ethics. The departments of philosophy, religion, at least in part English were severely impacted by this level of scholarship. At Notre Dame, by this time, a Dutch Calvinist from Calvin College could feel at home among the scholars, as the college concentrated on academics; at Hope, even a Presbyterian was out of place. Though Western Seminary might have been doing well in populating the pulpits of rural churches, the Reformed Church remained challenged in the greater America. Today, even its formerly flagship megachurch, the Crystal Cathedral, has faced debilitating internal conflicts that ruined it. Cal, in a sense, foresaw the irrelevance of the church of his Father.

---

[62] Gerrit Van Zyl, "Catalysts and Reaction Products," *Journal of Chemical Education*, **1955**, 463-8.

He came from a background of religious activism in Lawrence, and in the best of all possible ways – American church persons standing for the equality, in America, of all Americans.

President VanderWerf brought young faculty to the campus prepared in graduate schools with the terminal degrees in their fields. These were young scientists informed from their graduate school, and pre-professional educations, in established norms that were the norms in their fields of study. They were hired to teach at the Hope after Van Zyl's research start though not just in chemistry. In a real sense, the new faculty identified much more with their field of study than with other faculty members at the College. Scientists, in the emerging 1960s world of the National Science Foundation driven educations in their fields, were seeking and obtaining research grants; they were active on the campus year around; their grants paid them summer salaries; their grants paid for travel to professional meetings; and their work led to publications in peer reviewed journals. Some of them had the courage to write books including the text books Hope students would use in their classes. This attitude eventually permeated the entire campus, and is so prevalent there today that one wonders how it could have ever been any different.

Young faculty, being competitive in their fields managed research grants, and the result was that some were targets of envy and resentment on Hope's campus, most of which came about from misunderstandings of where the money came from. Research programs paid summer salaries for scientists, but these summer salaries came entirely from research grants. Scientists were forever on the road, peddling their skills to the next granting agency in search of the next dollars to support the next group of Hope College research students. Much, though not all, of the rest of the campus perceived this as unfair travel benefits for professional meetings for the scientists. I can remember snide comments made by a member of the religion faculty who held only a Master's degree, about a house purchased by one of the physicists; "Look at that guy; he can buy that banker's house."

However, a new day was on the horizon for Hope, and it had to be for the school to survive. Higher education was becoming a big business and former state teachers-colleges like Southern Illinois in Carbondale, Kent State, and Bowling Green in Ohio, as well as Western-Eastern-Central-Northern Michigan were to grow into universities as large, or nearly so, as the universities of Iowa, Kansas or Oklahoma. New places, like Grand Valley State in Grand Rapids, Michigan, were to grow out of blueberry patches. Agricultural schools like Michigan State, Texas A&M, and Colorado A&M were to turn themselves into mega research universities. The denominational schools, as Cal pointed out in his inaugural address, were to have significant competition from state universities set up, and designed, to offer college educations to every student in the specific states by being close to their homes. There were even predictions from higher educators around the country that the small, denominational liberal arts college was being rendered obsolete. Cal took on this challenge in his inaugural address.

Cal knew academic science a lot better than any of his faculty did. He had been one of America's most successful research entrepreneurs before he left the University of Kansas. He understood his young scientists precisely; but he never envisioned the intransience of campus humanists. He had marvelously interesting colleagues in the humanities at Kansas. At Hope, he ran into persons not currently in their field defending their every action as the way it had to be: (1) in a school that focused on undergraduate teaching, (2) in a school where undergraduates met full professors, and (3) in a school where the denomination of one's church attendance was marked as important criteria for faculty appointment.

Scientists, for Hope's part, were also more focused and directed than were their – "we take care of the College" counterparts. They were less often seen in the anteroom off the Kaffe Kletz set up as a faculty lounge where their humanist counterparts gossiped about them, and the President.

Undergraduate education – focusing on one to one instruction

Undergraduate students were important to research universities if one listened to the President or the admissions recruiters; they were the graduate students of the future. But when it came to federal money at the time, undergraduates were actually more or less incidental to the real objectives of the developing American research enterprise; grants, awards, publications, and dollars for the entrepreneurial faculty of American science. At least until we came on the scene at Hope. On several occasions I served on, or testified before, committees of the National Science Foundation, as to the value of undergraduate research, and the undergraduate research enterprise. Though mostly ambivalent, some politically connected university faculty also helped the cause, and overruled the staff at N.S.F. and N.I.H. on undergraduate matters. In one classic case, I served on a panel to judge research proposals in a specific initiative N.S.F. Though a long time ago, I recall the top rated proposal, of the several hundred the competition received, was one on single molecule spectroscopy from the then beginning Professor at U.C. La Jolla, W. E. Moerner. Moerner was the 2014 Nobel Prize winner in chemistry for his work single molecule spectroscopy. I did no read Moerner's proposal, but I did read several including one from a biology faculty member at California's premier science liberal arts college, Harvey Mudd. As I recall the incident, this proposal and another from a faculty member at Case Western Reserve University targeted virtually the same scientific principle, so only one could be funded. I took the position – as did another younger colleague serving on the panel – that funding the faculty member in the undergraduate institution would do much more for the science than funding the program at C.W.R.U. But another member of the panel, one who it was said had a major role in the invention of the laser, took the opposite point of view. He said a faculty member in a research university will accomplish more science with about the same amount of money. He won; the money went to C.W.R.U. – that time.

But many long hard arguments later were won; and research experiences for undergraduates became part of the N.S.F. lexicon in the 1980s, never to leave.

With that came money for individual instruction, and indirect charges that went to the college or university. A student's tuition could not cover, even then, the costs to operate a Michigan or an Ohio State. So the states made up more and more of it, but much of the campus budget was soon to come from indirect costs incurred by research agencies when faculty at the universities used university laboratories for funded research. Scientists, engineers, research physician professors, plus others were to become the cash cows of the research university to be.

That only-in-America word, "competition," was also on the horizon in higher education as in competition for students. For a long time, Hope had a lock on a goodly group of potential students, the sons and daughters of Dutch immigrants who remained members of the Reformed Church. For instance, I grew up in rural New York State and New York had no state university system until 1968. So, because my family had emigrated from Gelderland in the 1850s, and my parents and grandparents were pillars in the local Reformed Church, and my grandfather, mother, and father were all alumni of Hope College, it was a natural choice for me. But with state universities growing, and their growth being enabled financially as states tried to handle the crush of students now wanting a college education, the higher educational system of the state of New York took over and no one from my town would again go to Hope after about 1975. There were too many cheaper, and often better, opportunities in state universities in New York.

Cal inherited a school facing the competition of state universities and rigorous recruiting by other private colleges in the 1960s. The attention of a much larger percent of the population of the new educated generation of American college age students was up for grabs. Hope had to get with the program or be crushed. And another also important matter – a majority of male students at that time, because of national priorities, preferred to study science and engineering. So his presidency had to almost completely rebuild the school. Though his presidency changed the institution forever, he was not around to see just important those changes were to become. The incest due to the local Reformed Church carried the day in the end and he was forced out.

Chapter 7

Young Cal VanderWerf

Cal VanderWerf was born in Friesland, Wisconsin, on January 2, 1917, where his father, Anthony VanderWerf (born in Niezijl ((near Grijpskerk)), the Netherlands, March 23, 1873) was a Dutch Reformed Church minister.

Cal as a little boy, with his family; all family photographs courtesy of Klasina and Gretchen VanderWerf.

Cal's father died when he was nine years old, and he spoke often of the lessons he learned from him. Cal's mother, Antje Schaaftsma (born August 12, 1873, Dantumadeel, the Netherlands), met and married his father during his first pastorate. Cal's heritage was not just Dutch, it was decidedly Friesian. The family moved to Holland, Michigan, when Cal was a boy, where his father served the First Reformed Church.

Cal was literally raised by three sisters, though, who were much older than he: Lucille (later Veneklassen), Anne (Wabeke), and Joan (Brieve), and they doted on him. Cal worked his way through Hope during the Depression as a bellhop at the Warm Friend Tavern, a local hotel popular with tourists and business persons. He was blond with dimples, and so was dressed up by the owners as a little Dutch boy.

118

His wife, Rachel, told me that a wealthy Chicago family by the name of George tried to adopt him which, of course, could never have been done because his mother was still living and gave him a good home. But Mrs. George sent Cal some silver and linen table cloth said to have been used to serve the Governor of Illinois when they were married.

Cal was an inveterate sports fan. As a boy, he would visit his sister Lucille in Chicago, and she would let him go to Chicago Cubs games. During the years when he was young, the Cubs won the pennant every three years – 1929, 1932, 1935, 1938. Following one of those years of victory, Cal managed to get his baseball signed by every member of the victorious Cubs' team, a treasured possession.

Cal graduated *summa cum laude* from Hope College in 1937. The Reformed Church (it was the Dutch Reformed Church then) took good care of its own so Cal's mother and sisters were supported by the Church after his father died. He always felt a strong affection and sense of obligation to Hope because he paid no fees for his years there attending on a scholarship as he was a preacher's son. Following graduation, Cal enrolled in the Ph.D. program at The Ohio State University, turning down a later offer from Harvard, because Holland was much closer to Columbus than Cambridge, graduating in 1941. Rachel often wondered, she told me years later, just how Cal's life would have been changed had he chosen Harvard over Ohio State. For sure, from the author's perspective, had Cal been at Harvard in 1941, he would never have been allowed west because via Conant, Louis Fieser, Paul Bartlett, George Kistiakowski and Woodward, courtesy of Polaroid Corporation, were all engaged in secret war research. Ohio State chemistry was thriving at this time, too, and he managed pretty well. During his four years in Columbus, he met his lifelong partner, Rachel Good, an economics major working for the local Y.M.C.A., and finished his degree with, according to Mel Newman, his Ph. D. mentor, only one catastrophic accident (not his fault) slowing his progress. Cal was appointed assistant professor of chemistry at the University of Kansas in Lawrence in 1941, spent a year there, and in June hitchhiked from Lawrence back to Columbus to get married.

Cal's Ph.D. mentor at Ohio State was the Regents Professor of Chemistry Melvin Spencer Newman. I knew Mel Newman well, because in my first year of teaching at Hope, Cal farmed me out to Ohio State, where I taught organic chemistry to pre-dental students for the summer. Mel and Bea Newman were our hosts. Like all of Mel's students, I sincerely appreciated his good nature, his humor, and his love for life and chemistry. Newman was promoted directly to Professor from Assistant Professor at Ohio State in an incident about which his colleagues still marvel. Newman was surrounded by university leadership, William McPhearson, a chemist, was President of the University and Willie Evans was chair of the Chemistry Department, so it was a vibrant place to study.

During the course of his career, lasting forty-seven years, Cal rose from Instructor to Professor at the University of Kansas (1941-63), eventually becoming head of the department of chemistry. He served as President of Hope College (1963-70), spent a sabbatical year at Colorado State University in Fort Collins, Colorado, was appointed and served as Dean of Arts and Sciences at the University of Florida in Gainesville, Florida (1972-78), and spent the remainder of his career teaching freshman chemistry in Gainesville (1978-88).

In fact, Rachel told me that on the day before he was stricken by the aneurism that took his life in 1988, he finished grading his classes for that spring semester. He literally taught chemistry until the day he died.

Cal VanderWerf with Governor George Romney.

Cal had different reputations depending on who was giving the evaluation. To his earliest students, he was patient, careful, kind and thoughtful. Hope College faculty, some of them, saw him as impatient, hard on others, hard on himself, and expecting to turn the school which, they thought, was pretty good as it was, into a university that they did not want. "His expectations were too high." Still, some of his colleagues saw him as one of the academy's most insightful persons. He was an individual that saw the schools at which he served as arms of that great university in the sky—places and programs embodied with mankind's most significant, and awesome

120

responsibility, the education of its young. Some administrators that worked for him, and with him, saw a man whose expectations were way beyond their abilities. And there were still others who knew him (his colleagues at K.U. for example) and thought him to be disinterested in his day job, chemistry, and much more interested in other things like athletics.

In fact, he was a complex person, and one who, just when one thought one knew him well, would surprise with still another insight and still another battle to fight. He was proud, but not vain, committed so as not to be dissuaded. He was committed to Rachel, the love of his life, and to his six children. Rachel was his intellectual partner - his foil - his collaborator. There were a tremendously effective team.

Growing up, Cal was a prize-winning orator, and when he was a Hope College student editor of the student newspaper, the *Anchor*. He wrote clearly and well, mostly because he spent a lot of time trying to understand what he was to write before he wrote it. If he told me once, he told me several times, "No one is a born writer; writing is 99% perspiration; 1% inspiration. And before you pen a word, you have to be sure you have a plan for what you want to say." As a writer of chemistry, he always considered a topic from the perspective of it not having been presented before. He struck clear, logical lines in his writing as well as in his speaking. From the beginning of his academic career to the end, he would work calmly, efficiently, and in a direct, friendly way with students and others of all ages. One of Cal's classmates at Ohio State, Edward J. Cline, wrote that it was apparent early on that Cal would contribute substantially during his lifetime, and "this was exactly what happened."

Mel Newman, Cal's Ph.D. mentor, kept a copy of Cal's Ph.D. dissertation. It is typical for the time, following the German model. Newman's influence on the young Cal VanderWerf was profound.

Melvin Spencer Newman, whose own work was seminal for organic chemistry graduate students from the late 1940s onward, was an icon.[63] Cal was Mel's third doctoral graduate, and they grew up together in the academy. Mel, according to Cal, did not have a political bone in his body. When the President of Ohio State called to

[63] L.A. Paquette, "Melvin Spencer Newman," *National Academy of Sciences Biographical Memoirs* (National Academy Press, 1998).

ask Mel to serve on a committee, Mel's first response was "Why should I do that?" Secondly, he would ask "Just what do you think that might accomplish?" Most of the time, according to Cal, the answers would not satisfy Mel, and he would say "No thanks!" He enjoyed his beloved organic chemistry, his golf game, and Louis Armstrong record collection too much to be bothered with academic committee service.

Gerrit Van Zyl, Hope's distinguished professor of organic chemistry, proudly displayed a device Newman had made for him that illustrated the actions of the nucleophile, as in a second order nucleophilic displacement reaction, in the SN2 reaction. The thing was worth more as an antique teaching tool even then, because no nucleophilic displacement reaction ever went to the trouble ascribed to it in that chemical machine. But nonetheless Mel used it, and Doc was seriously proud that Mel had had one made for him. The Newman projection, outlined originally to help students and scholars alike visualize groups placed along a carbon-carbon single bond, as in butane, remains important today in the teaching of stereochemistry. So I was thrilled when Cal called me to his office one day in the spring of 1965 to tell me that Mel Newman had called and asked if he (Cal) had anyone who could teach organic chemistry in the summer quarter at Ohio State. Cal had more than one motive; he had promised me one-third of my academic year salary "for the first summer" when I took the Hope job in 1964 so he was saving money by sending me to Columbus for the summer. And I am sure he wanted Hope's scientific reputation in the VanderWerf era to be spread to his graduate alma mater.

After dinner one evening (we stayed with them one night or two), Mel retired to his study to read a Ph.D. dissertation draft, but first he shared his extensive collection of Louis Armstrong's jazz with my wife, Sue. Mel was an Armstrong devotee and Sue was duly impressed that at least one chemist knew something about a favorite musical icon. Mel told us the story that night that is also related in his National Academy of Sciences biography. That story was about Robert Burns Woodward, the Harvard future Nobel Laureate. Woodward was almost a scientific recluse by the time I studied at Harvard (1963-1964), but this was before that. Woodward rarely left Cambridge, and more rarely left his office on the first floor of Converse Hall. He would arrive at 8:30 AM, park his blue Oldsmobile in a parking space near located between Converse Hall, and the Harvard Museum next door. His parking space was also painted blue the deed having been done reportedly by Bartlett graduate students with J. Michael McBride, recently retired Professor of chemistry at Yale, confessing to me within the year that he and Kathy Schueller "bought the paint." Woodward, on the first day his spot was painted, did not notice it. As he always he moved silently toward his office dressed, as always, in a dark blue suit and light blue tie. Woodward rarely traveled except to Switzerland where he consulted for CIBA and collaborated with Albert Eschenmoser at the ETH in Zurich. As I remember Mel's story, Woodward had emerged from his den on the first floor of Converse Hall at Harvard long enough to venture to Columbus to give some lectures at Ohio State. Mel's colleagues did most of the work in preparing for Woodward's visit, and Mel, when he was to serve as Woodward's dinner host, was conflicted because Louis Armstrong was giving a concert on the very night he was supposed to be doing chemistry with Woodward, too. So Mel took R. B. Woodward to the Armstrong concert. Mel, on introducing Armstrong to Woodward, called R.B. the "Louis Armstrong of organic chemistry." Whereupon Armstrong laughed and said "He must be really good then." Each time re-telling the story, Mel would break into an absolutely infectious and almost childlike laugh.

Mel was a Jew. Though we think it should have made no difference to one whose career was to be determined by his skills in chemistry, it did in the 1930s in America. Mel finally got a teaching job in the chemistry department at Ohio State University in 1936. Being an easterner, Mel said that he and Bea looked for Ohio on the west side of the Mississippi River first they were so geographically challenged. He told me many years later that there were several negative votes both on the question of his employment, and on the question of his tenure a few years later, because he was Jewish.

Mel was a simple, brilliant, happy professor of organic chemistry. He drove Corvettes because he liked to drive, and would play golf with anybody that would play with him - which most students who played the game wanted to do. His religion was a personal matter and not worn on sleeves. Only years later when he came to speak at a Symposium in Paul Block's honor at Bowling Green did he tell me the story about his appointment and his tenure. By that time, he was old enough to be reflective. There was no bitterness in his telling the story, and Paul Block Jr. was Jewish, and a Yalee, too. He just related what had happened.

Mel was obviously a good research director, and one-on-one teacher of chemistry. Students likely chose his mentorship easily particularly when he first started. His first student was Milton Orchin, a gentleman scholar who spent his entire teaching career at the University of Cincinnati. Mel remarked that Cal VanderWerf was perfection personified (and those of us that knew Cal knew exactly what he was talking about), but a flying stopper from someone else's explosion broke a glass flask filled with product on Cal's lab bench toward the end of his graduate study. "That was the only lab mistake Cal ever made."

Cal's scientific career was graced with strong industrial, government and private agency interactions, it is important that we talk first about the scientific projects he undertook as a young research scientist.

Mel Newman shared a copy of Cal VanderWerf's dissertation, "Preparation and Properties of Cyclohexane-2-acetic Acid," with me toward the end of his career as Regents Professor of Chemistry at Ohio State. Cal took personally that young scientists could do so "much more with the new experimental techniques" than he was able to do as a student in the 1930s and 1940s. Cal was Mel's third doctoral student and they stayed in touch regularly after he graduated.

Chapter 8

Hope College Makes Its Mark

As their current advertising says, Hope College assumed the university educational model in the sciences, except at a reduced pace. Students were exposed to chemistry, or physics, as though they were professional chemists or physicists in training. This started as early as it could be justified with that student. Now Hope advertises "From Day One". Then we said we were teaching students to do chemistry as 'chemists did it'. In other words, we were outlining a pattern of study for each student that had no pattern. Each student was to be mentored in a program that was uniquely his/hers and their mentor's. Hope understood, and made a campus practice of individual self-study experiences.

Science in Universities in 1941

Being a teacher of chemistry is a lot more than standing in front of a classroom or a lab, showing young students the difference between a pH meter and pH. Chemists also have to be entrepreneurs or they will not succeed in the academy. This became the norm with research support for the sciences demanded by the events in Hawaii on December 7, 1941 and the subsequent declaration of war by Adolf Hitler on the United States on December 13, 1941. Scientists were among the first to be mobilized on behalf of the United States even before the beginning of the war. They remain mobilized in the national interest to this day

So, what do I mean "scientists were mobilized"? The majority certainly did not pick up an M-16, learn how not to drop it, and head to the South Pacific or North Africa. They were sent into their labs to work on government projects – at least the most able of them were. Specific projects of interest to the U.S. military (for example the synthesis of the most violent nitrogen containing explosive, or the preparation of synthetic rubber from a petroleum feed stock) were assigned to specific scientists, called principal investigators (P.I.s). Some P.I.s were just picked by the managers of the US research effort. Others were identified from their submissions of proposals that indicated their capabilities. Before the war, except for significant exceptions and people, chemistry professors in colleges and universities taught classes of chemistry students, and laboratories filled with those same students did experiments that were reasonably well planned. There were standard laboratory manuals, for example, that set out a prescription for what one might do in a particular experiment. The role of the laboratory in organic chemistry, for example, was to teach people to do preparations and make compounds by prescribed procedures – the students were learning to manipulate compounds. The Ph. D. experimenter was supposed to set out his (for in that day most scientists were men) new preparations and procedures or he was to study a process to provide new details about it. Exceptional people were persons like Melvin Spencer Newman or Paul Bartlett, or George Simpson Hammond, who won National Research Council Fellowships to pursue advanced study beyond the completion of their Ph. D. degrees. These persons could work on research for longer than the Ph. D. required. But for most, the student completed the degree and found a job either teaching chemistry or working for a chemical industry.

During and after the war, those with faculty positions in universities became known by their published works. The scientist led the university in the "publish or they would perish" idea of why one was a university professor. Some places carried this to extremes. During the campus uprisings of the early 1960s, most notably at the University of California at Berkeley, publish or perish became a *cause celebre*. "We're paying tuition, so why aren't our instructors the best? The full professors?" cried the students. Most schools achieved a balance in chemistry when they made tenure decisions. But from the day that the US government began supporting research work in universities in June 1940, under the direction of the National Defense Research Committee (N.D.R.C.), teaching chemistry or any science at a university would never be the same. It became William Rainey Harper's ideal for the job. One's research was important; what one's undergraduate students thought or did not think made little or no difference in a university.

That is a world where a Hope College could fit in. There, its faculty could do both, publish good work with undergraduate students, and set the standard for what excellent teaching could be. Hope's currency was its students, the alumni.

Today, scientists are judged by the number of papers they publish, the number of citations those papers receive from their peers, the research money (and attendant overhead funds for a university) the scientists accumulate, and the awards they garner along the way. Teaching, as in how well one's students do (and how well the students think the professor is doing), still counts on a local, university level, but save for the students that graduate from one's program, classroom teaching has almost no impact on a person's scientific reputation.

VanderWerf's Science

Cal developed his career as an organic chemist. He published about 100 papers. His Ph. D. thesis, with Newman, was published in 1945 and was based on work he finished at Kansas.[64] Given he was to become Hope's president, and that presidency would be controversial, it's fair that we understand the science he published in his academic scientific career.

The Germans were first to develop collaborations between government, universities and research laboratories. As in most cases, it was politics that drove their directions of work. In their case, they wanted their country to be as natural products independent as much as possible. But when this came to America, Cal managed in it as a master.

Cal's research program, as it developed at Kansas, was driven by his sources of funding for it. His career was impacted early on by programs funded because they suited a national need. One source of his funding came from the American Petroleum Institute. The American sources of gasoline waxed and waned in importance to our supply even before the Arab oil embargo of the early 1970s. American cars were riding on other countries' resources long before O.P.E.C. (Organization of Petroleum Exporting Counties). American companies also toyed with American's health. For example, this humorous *New York Times* story from December 31, 1932 that announces that no colored gasoline can be sold in Nebraska. All gas must have the anti-knock tetraethyl lead.[65]

---

[64] Melvin Newman and Calvin VanderWerf, "Isomeric Lactone Pairs Related to Cyclohexanone-2-acetic Acid," *Journal of the American Chemical Society, 67*, 1945.
[65] "Curb on Colored Gasoline," *The New York Times* (December 31, 1932).

CURB ON COLORED GASOLINE

Nebraska Bars Sale of All but That With Tetraethyl Lead.

LINCOLN, Neb., Dec. 31 (*P*).—A proclamation prohibiting the sale in Nebraska after Feb. 15 of all colored gasoline except that containing tetraethyl lead, was signed today by Governor C. W. Bryan.

The Governor is empowered to limit the sale of gasoline to that which is "water-white and clear."

The Governor said that many colored gasoline mixtures were commanding an additional price on the pretext they contained the qualities of the tetraethyl fluid, whereas they were merely colored.

Shortly after that, a *New York Times* story worried that America was running out of oil. A year later (May 22, 1927), gushers in Texas and Oklahoma had changed the picture entirely.[66]

A DELUGE OF OIL NOW PLAGUES ITS PRODUCERS

Lucky "Wildcatting", Artificial Renewal of Old Wells, and Chemistry's Tripling of the Gasoline Yield Produce Surplus That Leads to Demand to Check Production—A New Field in Oklahoma

Oil, too much or too little, meant there were jobs for scientists. The American Petroleum Institute (A.P.I.) represents all aspects of the American Oil and Gas industry. VanderWerf developed a program funded for research on a project #52 which was for the study of nitrogen-containing compounds in petroleum. Each summer, he and several graduate students would drive in the VanderWerf's car to Laramie, Wyoming, for A.P.I. meetings where they reported on their latest results, and shared insights with colleagues.

Vic Heasley ('59), almost blew up a lab on the 5th floor of Malott Hall at Kansas with an explosion of 1,4-diazidobutane. It was early morning and there were just a few of us in the labs when we heard this horrific explosion from Vic's lab. Vic had been distilling his product and left to talk with the stockroom keeper one floor above. Charlie Matusak and I were on the 5th floor in other labs at the time, heard the explosion and went across the hall to see if anyone was hurt. I can still remember Charlie, who was a post-doc and older

---

[66] "Deluge of Oil Now Plagues Its Producers," *The New York Times* (May 22, 1927).

than I, sticking his head gingerly in the lab and calling, meekly, "Is anyone in here?" Fortunately, no one was.

Vic Heasley then went to VanderWerf and asked if he could begin to write his dissertation. In a fit of self-protection, Cal said yes. The thesis was written, Vic graduated, and the saga came to a happy conclusion. Save for the windows in the lab, the bottles broken by the explosion, a magnetic stirrer that was under the distillation flask, and a few other odds and ends, the lab was none the worse for wear so we all went on with our work. Vic was appointed at Pasadena College, which later became Point Loma College in San Diego, CA. While there, he developed one of the more distinguished careers of any organic chemist teaching in an American liberal arts college. As I was the first to observe, he was an extraordinary teacher that loved what he was doing, and took the best interests of his students as his highest calling. Point Loma chemistry graduates were sought by industries and universities around the country.

VanderWerf also held consultancies with Smith, Kline and French as well as Phillips Petroleum. At the time he was considering leaving Kansas, he was being actively courted by DuPont as research director. According to his wife, Rachel, he turned DuPont down at least 4 times. He was also a good collaborator and published with both William E. (Bill) McEwen and Jacob Kleinberg, colleagues in the chemistry department. Cal was a great humanitarian, and as previously mentioned, a conscientious objector during World War II. One of the first three Japanese students of organic chemistry to come to the United States to study after World War II was Shigaro Oae. Following additional post-doctoral work with Charles Price at either Notre Dame or Pennsylvania, Oae who Cal called "Sig" returned to Osaka City University. Professor Oae had a distinguished, though rocky, career in Osaka. Student unrest hit that University hard in the 1960s and he was targeted. Eventually he moved to Tsukuba University where he finished out his career.

Chemical weapons – what did Cal know, and when did he know it?

Cal's last major scientific contribution, with Bill McEwen and student Karl Kumli, was the first successful resolution of a tetracovalent phosphonium salt.[67] This work, originally supported by the Army Chemical Corps, then by the National Institutes of Health

---

[67] Kumli, K. K..; McEwen, W. E.; VanderWerf, C. A.; "Resolution of Non-Heterocyclic Quarternary Phosphonium Iodide" J. Amer. Chem. Soc. **1959**, *81*, 248.

(N.I.H.), was sufficiently significant that it was published initially as a Communication to the Editor of J.A.C.S. But in retrospect it was a lot more significant than that. I believe Cal VanderWerf was the first organic chemist in the world to recognize that sarin, the nerve gas, had more than one three-dimensional form.

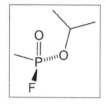

One has to wonder what Cal and Bill McEwen knew, and when they found it out? Sarin, to the left, is optically active at phosphorous. One has to wonder, after unfortunately all the principles are no longer living, just when they undertook their work on phosphorous chemistry? Neither had published anything on phosphorous before; but both of them were very well connected. They were strongly entrepreneurial in the newly developing world of research grant and contract support for their work. McEwan and VanderWerf must have known, or at least could have postulated that the chiral phosphorous in sarin and soman could be important in their reactivity.

The VanderWerf/McEwen work on phosphorous in the early 1950s triggered careful consideration on my part as more and more frequent exposes of chemical weapons stores in Syria and Iraq became news items in recent years. What follows is a relatively complex trail through the organic chemistry of "phosphorous containing" organic compounds, some of which were, and are, horrible toxins. My information was found in the now available papers and much was garnered through now open, though not easily accessible archives.

Chemists are citizens; and vice versa, citizens may be chemists. When wars are engaged by governments, citizens at all levels become fair game for the right. Scientists often have value off the battlefield and are employed in chemical corps. That was the case in America, in Britain, in France, and in Germany before, during, and after both World War I and World War II.

Invariably, this work is secret, hidden, embargoed, and classified. There are still environmental issues arising from World War I chemical experiments, done in Washington, D.C., for example, continuing to surface. There are also details of that work which would be useful for us to know—when the population is larger. The information is either lost or so well hidden that it might as well be. One specific question many have is what happened to the eighty tons of Lewisite supposedly manufactured and shipped from Willoughby,

Ohio, to Baltimore in the summer of 1918. It was said to have been on a ship scuttled at sea after the war ended.

The sense of the German chemistry during World War II, however, is something that we know much more about. What was discovered about German organic chemistry in 1945 was that scientists in Leverkusen and Elberfeld had made sarin developed— an entire series of acetyl cholinesterase inhibitors now known as nerve gasses.

As indicated in the initial publication on the resolution of tetracovalent phosphorus, others had tried to resolve tetracovalent phosphonium salts in which the phosphorous atom was not part of a heterocyclic ring, but with no success. The failures of others had been rationalized by assuming tetravalent phosphonium salts ionized to trivalent, planar phosphines destroying the original molecular asymmetry. The publication basically said that this was bunk.

After scientists in the U.S. knew the structures of the phosphates, GB (sarin), tabun, and later somin, and learned the Russians had begun programs studying these, Morris Kharasch at the University of Chicago convinced the Army Chemical Corps that American university scientists needed to learn more about the organic chemistry of phosphorus-carbon compounds. Kharasch generated the funds to seed research among young scientists in American universities. Kharasch was in a succession of Chicago organic chemists central to the war efforts both in 1914-1918; after that war; and later as the Nation prepared for, fought, and worked after World War II. Julius Steiglitz was a central figure in the theft of German property as executed because of the Alien Property Act in 1917. This was as much a theft of intellectual property as real. Much of this has been recorded in excellent articles by Steen.[68] These young scientists were expected to be on the scene for a while, so the country put them to studying organic phosphorus chemistry. Programs, in addition to that managed by Kharasch, were begun at the University of Maryland, Brooklyn Polytechnic Institute, Columbia University, and the University of Kansas as wells as others. At the time, the

---

[68] "Patents, Patroitism and 'Skilled in the Art'; USA v. The Chemical Foundation, Inc. 1923-26", *Isis 92*, March 2001; also "Politics and Property: The United States and German Chemical Assets, 1919-31." John Lesch, ed., *The German Chemical Industry in the Twentieth Century,* Dordrecht: Kluwer Academic Publishers, 2000.

phosphonate (P=O, O) group of sarin was unique; not well known in the literature.

The records on the Kharasch program are not easily accessible. So much of what I report is from 1) conversations with the scientists that are still living who worked on the project (Earl S. Huyser and Sheldon Buckler); and 2) from a search of the literature of the day on phosphorus. I looked at the acknowledgements.

There were also new ways to construct the P-C linkage, or at least chemists including Earl Huyser thought that there were. Here's what Huyser told me recently about his recollections, regarding the structure of the nerve gasses:

"The only one that I knew the structure of was sarin, which we, at that time, referred to as GB. I remember it to this day since I memorized the name and avoided writing out the structure anywhere (Isopropyl methanephonoflourotdate), for fear of leaving it around somewhere. As for the others, I had no "need to know." The whole business of secrecy was supreme. We were in a cold war and quite paranoid about a lot of things. The one thing that was made clear to me was that the carbon-phosphorus bond was a significant aspect of GB. This probably is why Khrasch had me working on the addition of diethyl phosphite to ethylene."

The Germans, for their part, had developed a pretty good route to sarin. Well after the report that the Germans had developed and manufactured tons of sarin, soman and tabun—some of which was brought to the United States—the U.S. had brought a convicted Farben defendant, Otto Ambros, and several others, to the U.S. to assist in our scaled up synthesis of sarin. Isolated facts:

1) According to Annie Jacobsen this was a solo effort: the result of leadership and decisions of General Charles Loucks, then head of the Army Chemical Corps.[69]
   Dow Chemical Co. among others, was also strongly committed to producing compounds of phosphorus commercially.
2) Huyser, VanderWerf, and McEwen were organic chemists in the same division of a relatively small midwestern university's chemistry department. But according to Huyser who is the only one still alive they never talked about phosphorus chemistry.
3) Huyser did little phosphorus chemistry after Columbia; nothing similar to what he did as an Army private on a post-doctoral.

---

[69] Annie Jacobsen, *Operation Paperclip* (New York: Little Brown, 2014).

4) I never had a conversation of consequence with Cal VanderWerf about this that I remember either. But he mentioned it in his inaugural address at Hope (November 8, 1963). He also talked about the work at least once at an American Chemical Society meeting in western Michigan during his Hope presidency. During that talk, VanderWerf went out his way to complement Marvin Meltzer, a chemistry graduate student known for his horrible laboratory technique. Cal pointed out that it was Meltzer's insights that led to the eventual successful reactions.

5) Bill McEwen was working hard to get additional students to work on his phosphorus project when I was student.

One critical question that continues is when highly toxic fluorine compounds such as the phosphonate esters became known in the organic chemical community. The British found shells filled with the toxic nerve gasses in ammunition dumps in the west while liberating the conquered Germany and there is some indication that a German defector had told the British about these weapons are early as 1943 – but the information was ignored. Those working on organophosphorus chemistry in universities in the United States after the war surely knew the structures of GB (sarin) and its analogs as early as 1950 or 1951. Industries had major programs developing the chemistry of phosphoric and phosphorus acid even before the war. Gerhard Schrader, at Bayer, was working on pesticides – and I think that a legitimate analysis – he was working on new pesticides – when he discovered the excessive toxicity of the halogenated phosphonates. Research in Britain on similar systems was started at Porton Down, near Stonehenge/Salisbury in 1935 in collaboration with both Universities – Cambridge and Oxford as well as with Imperial College, London. [70] The wartime work of B. C. Saunders is overviewed in a book summarizing Saunders' worldwide lectures on the subject after the war. [71] Saunders' son-in-law, Brian Ridgeway, explained this work as being in self-defense; Saunders' students at Cambridge, Ian Fleming in particular, noted that Saunders was extremely introspective about it; though after the war he was more

---

[70]A special edition of *Chemistry in Britain* was devoted to the history of Porton Down in *Chemistry in Britain, 24,* 1988.
[71] B.C. Saunders, *Some Aspects of the Chemistry and Toxic Action of Organic Compounds Containing Phosphorus and Fluorine (*Cambridge University Press, 1957).

forth-coming. A defector to England (1941) had directed British attention to methyl fluoroacetate, $FCH_2CO_2CH_3$.

Saunders' group at Cambridge was asked to study the fluoroacetates, in secret. A commercial preparation (published) was developed and this compound's toxicity tested in detail. Saunders' work continued throughout the war, though, according to his former students at Cambridge, not without ethical concern. He had long conversations from the laboratory phone each evening with his family priest; he was married, had two children, but was a fellow of one of the Colleges so rarely at home – at least during the term. The lectures summarized in the book were given all over the world, even in behind the then Iron Curtain countries.

The Germans, well experienced with "sledge hammer" organic chemistry since the development of the Haber-Bosch reduction of nitrogen to ammonia, had developed a high yield sledge hammer synthesis of sarin before its commercial manufacture was begun. The synthesis started with dimethyl phosphite, which was methylated by the Wurtz reaction, chlorinated with phosphorus trichloride, and esterified/metathesized with isopropyl alcohol/NaF. All the reactions were conducted at high temperatures, no doubt in pressurized vessels, and they occurred in high yields. Schrader reported that GB/sarin's toxic action was, in comparison to hitherto-known substances, astonishingly high.[72] As one can see from his own report, B.C. Saunders came close, in structure, to the German compounds. Schrader visited Porton Down in 1946-1947 and, it is said, was offered a job there. But, he said he would no longer work on toxic compounds. Given that was what the British wanted him for, no relationship was consummated, and Schrader stayed at Bayer until he retired. He died in old age though this chemist can barely see how he managed that. He wrote his thesis at Braunschweig in 1930 on reactions of OsO_4 and RuO_4 with liquid HCN!

Cal VanderWerf and Bill McEwen, with their students, attempted the resolution of tetravalent phosphorus – so called

---

[72] Volumes have been written about this and it's not the purpose of this interlude in VanderWerf's phosphorus chemistry to detail the chemistry of the nerve gasses; the question is who knew what and when did they know it? Schrader told British Intelligence Officers all that he could remember on interviews in Elberfeld, May 1945. These are summarized in BIOS reports (British Intelligence Objectives Sub-Committee) – in this case, Final Report 714 (1945). Reported by S. A. Mumford and E. A. Perren of the Chemical Defence Experimental Station, Porton Down.

phosphonium salts – in the early 1950s. It was known, or at least it was reported in 1955, that one enantiomer of the sarin which was chiral at phosphorus was more reactive (i.e., toxic) than the other. VanderWerf and McEwen succeeded when they reported, with student Karl F. Kumli, the first 'completely successful resolution of a quarternary phosphonium salt in which the phosphorus was not part of ring'.[73] Methyl ethyl phenyl benzyl phosphonium iodide enantiomers were separated as the l (+) dibenzoylhydrogen tartrate salts, and were optically stable.

$RR_1R_2R_3P^-$, X^- where R= Me, R_1= Et, R_2= Ph; R_3= $PhCH_2$

The McEwen/VanderWerf work was supported by both N.I.H. and the Army Chemical Corps and, in the last paragraph the authors said the compounds were being tested for their toxicological properties in the School of Pharmacy at the University of Kansas by Dr. Duane G. Wenzel.[74] Thus they must have known of Schrader's work;

 Sarin is a O-isopropyl methylphosphonofluoridate. It is optically active at phosphorus, as are all of the nerve gasses in the series. One of sarin's two stereoisomers is much more reactive toward the amino acid serine, in acetylcholinesterase as the other.[75] As early as 1954 it had been reported by Michel that when injected in rats, sarin P- enantiomer is reactive; the P+ enantiomer accumulates; though it is also toxic. Thus, the toxicity of sarin is biphasic. Even if an antidote counteracts its actions at the beginning, later reactivities are lethal.

America needs more research on organophosphorus compounds

The United States Army made a conscious investment in research on compounds containing carbon phosphorous bonds in the early 1950s. On the east coast, at Columbia, Earl Huyser ('50) was

---

[73] Karl Kumli, William McEwen, and Calvin VanderWerf, "Resolution of a Non-Heterocyclic Quarternary Phosphonium Iodide," *Journal of the American Chemical Society, 81,* 1959.

[74] D.M. Coyne, William McEwen, and Calvin VanderWerf, "The Synthesis and Resolution of Compunds of Tetracovalent Phosphorus," *Journal of the American Chemical Society, 78,* 1956.

[75] The reactions of stereogenic phosphorus nerve gasses has been reviewed in Benschof and DeJong's "Nerve Agent Stereoisomers – Analysis, Isolation and Toxicology," in *Accounts of Chemical Research, 21,* 1988.

also working on phosphorus in an effort also supported by the Army Chemical Corps.[76] He was following work based earlier experiments by Robert Boyle (1681) and, curiously, Richard Willstätter, who reported that his coauthor Sonnenfeld was taken into the military service so the work was left unfinished. Earl was studying the reaction of phosphorus with oxygen in a hydrocarbon solvent, cyclohexene. The observation was that hydrocarbons present the glow seen when white phosphorus reactions with air, in air. Curiously, Willstätter later won the Nobel prize for his work on chlorophyll, but later left his position at Munich in 1924 because he felt Germany too racially prejudiced, but not before he directed the Ph. D. dissertations of Richard Kuhn (1924) (soman, and later to win the Nobel Prize for his work on Vitamin A) and Otto Ambros (1926)!

Sheldon Buckler, while also serving in the Army was working with polymer chemist at the University of Maryland, William J. Bailey. Buckler became Vice President of Research at Polaroid Corporation, remaining that position until the company's demise, in the early 1990s. For several years during this period, Buckler worked on, and published on, organic phosphorus derivatives.

The Army Chemical Corps decided that it needed to know more about phosphorus compounds after the war, probably in concert with U.S. attempts to produce commercial quantities of sarin, soman and tabun. Because of the connection between the Army, and the University of Chicago, Professor Morris Kharasch was assigned the task of finding a number of research programs in universities where phosphorus chemistry could be studied. Huyser, himself, became part of one of these programs, as a post-doctoral fellow with Cheves Walling at Columbia. But, before he was assigned to Columbia he was assigned to Edgewood Arsenal. There, he was asked about a study he had done with Professor Kharasch; the study was of the addition of diethyl phosphite to ethylene. He told me had written the report, when he had a *secret* security clearance. But, upon arrival at Edgewood Arsenal that clearance had not been reapproved for him;

---

[76] C. Walling, F.W. Stacey, S.E. Jamison, and Earl Huyser, "The Reaction of Olefins with Oxygen and Phosphorus," *Journal of the American Chemical Society, 80,* 1958; and C. Walling, F.W. Stacey, S.E. Jamison, and Earl Huyser, "Chemical Properties of the Reaction of Cyclohexene with Phosphorus and Oxygen," *Journal of the American Chemical Society, 80,* 1958.

so, when a report that he had written was being studied by two Army scientists at the Arsenal. There was a question about chemical detail, and as addressed in the report by the Army scientists, Huyser could not look at a report he had written. He did not have the clearance to read his own written material. I looked for historical reports on this initiative at the National Archives, but finding none decided to see what I could find in the literature. The three programs I have found all reported the results of their studies. I understand that there were other supported programs.

Sarin and the other Schrader-like nerve gasses, soman and tabun, have been the subject of much discussion in the popular press since their discovery in Syria. Banks' review stated in so many words that 'the development of systemic insecticides, arising from the independent wartime work by Saunders in England and Schrader in Germany is an excited story that is not as well-known as it deserves to be. The secrecy of some of the early work at Porton, in collaborations with institutions has previously prevented its disclosure; the present review seeks to remedy this.

This topic brought on a discussion between the author and a Washington friend about how much was known about these compounds – particularly their preparation – before the current outbreak of their use. My comment was that anyone skilled in the art of organic synthesis could make sarin. My colleague's comment was that the synthesis of sarin should be embargoed, as in classified. To the extent that situation exists now, I think it should remain. But the fact remains that anyone even marginally skilled in the art could make these toxic gasses – that is if they were stupid enough to do so.

The idea is to prevent governments and terrorists from wanting or needing to make the compounds... and, as our wartime friends are bold to say – has anyone ever questioned the use of the explosive R.D.X.? It was far more directly deadly than any chemical weapon; but guns, and the use of high explosives seem to remain.

The VanderWerf/McEwen collaboration succeeded with a number of additional publications on the subject and, as indicated above, it was Cal's last scientific effort of consequence. Later work, in which McEwen postulated a hexavalent phosphorous atom, was

137

supported by the Petroleum Research Fund, administered by the American Chemical Society.[77]

[77] C.L Parisek, William McEwen, and Calvin VanderWerf, "Decomposition of Optically Active Methylethylphenylbenzylphosphonium n-Butoxide," *Journal of the American Chemical Society, 82*, 1960.

Chapter 9

The Kansas Chemistry Department

I took my Ph. D. degree at Kansas. I went to Kansas less for the chemistry than for the quality of education I might receive, or Cal VanderWerf's personal magnetism. I was driven by the romance of life on the prairie, and by Vic Heasley's persuasive personality. Vic convinced me that I would like the University of Kansas and its department of chemistry. I was an incurable sports fan and knew of Wes Santee, Ray Evans, Glenn Cunningham, and Wilt Chamberlain from reading Sport Magazine. Kansas sounded like fun to me. So I dragged Sue there, convinced Cal I could only come if he found a job for her, and drove her to the plains of Kansas during about the hottest week on record in June 1960.

I identified immediately with the University of Kansas (K.U.): the history of the university, the history of the chemistry department, its sports programs/teams, and everything in which I was interested as it went on there. Sue and I did not have a lot of money, but we were comfortable because Sue had a teaching job at Lawrence Junior High School. Her salary was $4000/year + $125 for mentoring the pep club! My work had its ups and downs, but so does everyone's. Few get through graduate study in a physical science in the United States unscathed. As I aged, I learned of Nobel laureates that flunked quantum mechanics!

Cal VanderWerf's chemical career, by the time I enrolled at Kansas in the fall of 1960, was clearly in remission. Cal was honest at his core, and was not interested in presenting a story that was not true. By the time I met him, he had lost interest in doing research in chemistry. The Kansas chemistry department at that time had several who were excelling in their field. Along with VanderWerf there was F. Sherwood Rowland (later the first head of the chemistry department at the University of California and winner of the Nobel Prize in Chemistry in 2000) and William E. McEwen (later the chair of chemistry at the University of Massachusetts, Amherst. Bill's tenure as a department chair was fraught with tension, but while he as a professor at K.U., he helped keep chemistry at Kansas at the cutting edge.

Earl Huyser and Kansas Chemistry

Travel for a successful research scientist went with the territory. When I interviewed at Kansas in June 1960, Cal was

139

consulting at Phillips Petroleum in Bartlesville. Earl Huyser, another Hope alumnus, was assigned to be my scientific host and he showed me around the chemistry department housed by then in Malott Hall. I liked Earl instantly and knew that if I did not work for Cal, I could work for Earl. In a few months, when I was allowed to select a research advisor, I selected Earl's group.

At the graduate level in chemistry, students required a thesis advisor or mentor as in, as we have said, the pattern long established by the Germans. The mentor serves as chair of student's Ph. D. advisory committee. Long before that, though, he/she has guided, cajoled, taught, encouraged, baby sat the student through the ups and downs of the competitive graduate educational enterprise. In most cases in chemistry at that time, after the first year, the advisor's research support paid the student's stipend as salary. The student is a general partner of the graduate advisor – a young colleague in training in a program that bears some resemblance to being an apprentice.

I took my Ph. D. in Huyser's lab. I never really considered working with anyone else, although I talked somewhat with Sherry Rowland before making the decision. Sherry's lab had some interesting scientists working there with him, but some of the work going on appeared more about the technique than about the results. I wondered if one might get stuck making things, and miss the opportunity to do experiments on new ideas. I also interviewed Cal, or at least tried to. But he was too heavily involved with the K.U. athletic department at the time: they were funding raising for the addition of a new press box to Memorial Stadium, and Cal was seriously active in that effort. In fairness to him, it was clear that his career was transitioning from research and teaching to other things at the time, so he was no longer interested in assembling a new group of graduate students.

Earl was a Hope graduate and I had a strong inclination to work with him for that reason. He had spent at least one summer working with Van Zyl because he lived in Holland. Earl, Paul Cook and Van Zyl published one paper on the reactions of 2-thienylsodium and 2-thienylmagnesium bromide with epoxides.[78] This paper represented a combination of VanderWerf's ideas given Van Zyl (epoxides) with Howard Hartough's (thiophenes).

---

[78] Gordon Van Zyl, J.F. Zack, Earl Huyser, and P.L. Cook, *Journal of the American Chemical Society*, 76, 1954.

Earl graduated from Hope in 1951 and enrolled immediately at Chicago. He told me that he went to Chicago because Hope chemistry chairman Gerrit Van Zyl got a letter from the chair at Chicago asking "Why doesn't Hope ever send us a student?" So he sent Chicago an application, and a few days later received a telegram from Chicago informing him that he had been accepted, with a teaching assistantship. Earl ended up working in William Urry's group more or less by accident.

Earl's career at Chicago started with difficulty. First, he almost failed his first semester course in organic chemistry taught by Frank Westheimer. Second, his background in physical chemistry was marginal so he waited eighteen months to take his preliminary exams because he was way behind. He told me that he pulled himself from the destruction surely in store in physical chemistry by a strong performance on his preliminary oral exam in December 1952. Nobel Laureate H. C. Urey was on his committee, with Professor Weldon Brown. The question was asked by one of the examiners, what is the crystal structure of NaCl? Earl expected the question, he told me, and answered that its crystal structure was face centered cubic. Professor Urey thought otherwise, insisting it to be simple cubic. So, off the entire group marched to the library to find the answer in one of the dusty books there. It was, as Earl maintained, face centered cubic and he later said that he passed the exam because he knew that answer. Urey, for his part, insisted the books were all wrong! Earl's friend, the late Jim Wilt later a Professor at Loyola University in Chicago, was in the library and saw this action up close and personal. Together we had a good laugh about it, but that was many years later. I'm sure it was anything but funny to Earl at the time.

Earl finished his thesis on free radical reactions with Professor Urry in the summer of 1954. Expecting to be drafted, he stayed at Chicago and worked on chemical weapons projects, as a post-doctoral fellow, in the laboratories of Professor Morris Kharasch. Earl's work with Kharasch was to study the free radical addition of phosphines to olefins. Years later, I understood why Kharasch undertook this project, but I can imagine it was a pretty smelly activity! Earl told me many years later that when he was drafted and sent to Edgewood Arsenal, his director there asked him about his report on his work. He said "I didn't remember the details exactly. So, I asked the Army guy that was seeking the information, if I could look at my report."

Kharasch was a polished industrial consultant, and had been intimately involved in the development of synthetic rubber as part of a U.S. rubber project during the early years of World War II. The chemical warfare service used the Kharasch laboratory, and him, to bring chemical weaponry to the attention of the many assistant professors occupying faculty positions in universities following the war. Kharasch directed an Army project in which he was asked to find twelve young professors who could manage research programs around this subject, and he selected, among others, his former students Cheves Walling at Columbia, William Urry at Chicago, as well as Charles Overberger at the Brooklyn Polytechnic Institute, and Henry Rappaport at Berkeley. Cal VanderWerf was intimately involved in this work too, but came to it from the Navy side.

When Huyser's draft notice did not come soon in early 1955, he took a position at Dow Chemical Co. in Midland, Michigan. Almost immediately after arriving at Dow his draft notice came! Because he was a chemist, he was again assigned to the chemical warfare corps. This required a classified security clearance which took a while to obtain and he was assigned to Fort Leonard Wood in Missouri for basic training while it was being arranged. When his security clearance was finally arranged he was asked to look at the synthesis of LSD, because the Army had aspirations of using it as a chemical weapon. Unfortunately, before he could get a plan, Woodward synthesized lysergic acid, and a budding hippie movement, under the direction of Timothy Leary at Harvard, was beginning to experiment with it.[79] Eventually, the Army assigned Huyser to Edgewood Arsenal in Maryland, and then to Walling's lab at Columbia, where he spent two years as a post-doctoral/army private. While at Edgewood, and before his clearance was approved, he worked in the director's office on a library project. While working in this capacity, the director and an Army scientist came to talk to him about a report he had written for the Army project he worked on at Chicago on radical reactions of diethyl phosphite and ethylene. When questions got detailed, Huyser asked to look back at his own report for the answers. Unfortunately, his superiors said, "You can't do that because you don't have security clearance." In other words, Army regulations were such that one could not see one's own written materials once it managed its way into the Army's classified system! Such was the protection we were getting from the Army even back then!

---

[79] E.D. Kornfeld, J.G. Fornefeld, B. Kline, M.J. Mann, D.T. Morrison, R.G. Jones, and Robert Woodward, "The Total Synthesis of Lysergic Acid," *Journal of the American Chemical Society, 78,* 1956.

Huyser introduced me, in the fall of 1960, to a problem in photochemistry. Little did I know that the research question would occupy me for the rest of my scientific life: "Why is benzophenone photoreduced in alcohols to benzpinacol in 100% yield, while acetophenone photoreduction leads to some pinacol, but also some disproportionation product - ☐ ☐phenethyl alcohol?"

To work on the project, Earl, being a typical Dutchman, had bought the least expensive mercury resonance lamp one could find. A 100-watt medium pressure bulb was set it up in one of the hoods on the 5th floor labs in Malott Hall. I used it, plus a bevy of available sun lamps purchased at a local drug store, to expose everything I could find in the labs, or laying around the organic store, to putatively *ultraviolet rays*. I say putatively because except for the low wattage mercury bulb, we had no spectral outputs for the drug store lamps and they gave off more heat than light of the proper wavelength. But we plowed forward. The 100-watt bulbs and our irradiation apparatus took on a life of its own. A few months after I started my work, Albert Burgstrahler and his students tried to use it to cyclize 1,2-di-tert-butylbenzene to its Dewar isomer. The lamp Earl had bought was not intense enough. Around the same time Gene Van Tamelen and S. Peter Pappas reported a similar successful experiment.[80]

I made two new observations that became the nucleus of my Ph. D. thesis. First, one could assemble a reducing media for an aromatic ketone in a hydrogen donating solvent by adding an oxidizing agent – a peroxide - to the solution! Think on that for a minute – to reduce an aromatic ketone to a pinacol or a carbinol one uses a hydrogen donor and an oxidizing agent![81] The second new observation I made had to do with alpha-keto esters, so called phenyl glyoxylates and needs more explanation. I will deal with the oxidizing agent reduction of aromatic ketones first.[82]

It is often said that education is wasted on the young and if ever there was a good case illustrative of that, it's me. To be fair, young persons are so busy assembling a data bank of knowledge to carry them forth to work in their chosen fields, that they have all than can handle doing just that. If it means missing the forest for the tree

---

[80] Eugene van Tamelan and S.P. Pappas, "Bicyclo[2.2.0]hexa-2,5-diene, Dewar Benzene," *Journal of the American Chemical Society*, *85*, 1963.

[81] Earl Huyser and Douglas C. Neckers, "Dialkyl Peroxide-Induced Reductions of Aromatic Ketones," *Journal of the American Chemical Society, 85,* 1963.

[82] Earl Huyser and Douglas C. Neckers, "The Photochemical Reactions of Alkyl Phenylglyoxylates in Alcohols," *Journal of Organic Chemistry, 29,* 1964.

– so be it. And, though I soaked up a lot more or less like the proverbial sponge, I missed a lot too.

My first discovery, above, is a case in point.

Earl Huyser wrote the paper; he must have. My job was to provide Professor Huyser with the data to support the verbiage he was penning for the Journal of the American Chemical Society. I recall him chuckling when he came up with its title: "Dialkyl Peroxide-Induced Reductions of Aromatic Ketones." But the observation, which he made, is significant. The mechanism had precedent – in the photoreduction of aromatic ketones in the benzophenone series as first reported by Ciamician and Silber, and later expanded upon by Cohen, and then by Weismann, Bergman and Herschberg. In the photoreaction, the oxidant was the excited triplet state of benzophenone. In our case, the oxidant were radicals generated from the thermal decomposition of peroxides. It was this paper, written by a Kansas professor based on the experiments gathered by an enthusiastic, still learning young graduate student that confirmed the chemical mechanism of benzophenone photoreduction. The spectroscopic mechanism would await the development of transient spectroscopy, and femtosecond experimental capabilities.[83] But the chemical mechanism came out of my Ph. D. dissertation.

Earl Huyser was a young assistant professor when I chose to work for him. One day he came into his office – we used the best balance in his possession for weighing things that were of special value and this was kept in his office. He carried a letter announcing to him, and the Dean, that the Department had recommended that he be promoted to Associate Professor with tenure. He was self-effacing about it but proud.

Earl said "How's that! From assistant professor to associate professor with tenure in three years. That's pretty good!" His students, of whom I was probably the youngest at the time, did not have a clue what this was all about.

F. Sherwood Rowland, Ph.D. 1952, Nobel Prize, Chemistry 1995

Sherry Rowland was also an Associate Professor in the Kansas department when I began my studies. F. Sherwood Rowland (he never used his first name) obtained his undergraduate degree from Ohio Wesleyan, and he had little choice in his family but to attend

---

[83] B.K. Shah, M.A. J. Rodgers, and Douglas C. Neckers, "The $S_2 \square S_1$ Internal Conversion of Benzophenone and *p*-Iodobenzophenone," *Journal of Physical Chemistry, 108,* 2004.

University of Chicago and was admitted in autumn 1948. At that time, the chemistry department of the University of Chicago had a policy of immediately assigning each new graduate student to a temporary faculty adviser prior to the choice of an individual research topic. Rowland's randomly assigned mentor was Willard F. Libby, who had just finished developing the Carbon-14 dating technique for which he received the 1960 Nobel Prize. Libby was a charismatic, brusque dynamo (at their first meeting, he said "I see you made all A's in undergraduate school. We're here to find out if you are any damn good!"), with a very wide range of fertile ideas for scientific research. Rowland settled automatically and happily into his research group, and became a radiochemist working on the chemistry of radioactive atoms, "Almost everything I learned about how to be a research scientist came from listening to and observing Bill Libby."

The first nuclear reactor had been built by Enrico Fermi in 1942 under the football stands at the University of Chicago, and the post-war university had managed to capture many of the leading scientists from the Manhattan Project into the physics and chemistry departments. This was an unbelievably exciting time in the physical sciences at the University of Chicago. Rowland's physical chemistry course was taught by Harold Urey for two quarters and in the third quarter by Edward Teller; inorganic chemistry was given by Henry Taube; radiochemistry by Libby. Rowland also attended courses on nuclear physics given by Maria Goeppert Mayer and by Fermi. Urey and Fermi already had been awarded Nobel Prizes, and Libby, Mayer and Taube were to receive theirs in the future.

Rowland was an associate professor at Kansas while VanderWerf was there, and a Nobel Prize was not that clearly in the offing when I knew him. As it turned out it would not be for radiochemistry, but for the discovery of the decomposition of the Freons in the upper atmosphere with Mario Molina. Rowland had just begun studies on the gas phase photochemistry of ketones, and I thought about that for a time while considering with whom I should work in my first year of graduate school. But he had at least two students making a fetish building glass vacuum lines priding themselves in blowing glass, which I feared, would not be my forte.

I also had a more or less irrational loyalty to Hope College at the time, and believed that if I did not work for Cal VanderWerf, it would still keep it in the club if I worked for Huyser. So given the Hope connections, the fear of glassblowing, and the fact I felt more comfortable mixing, stirring, smelling, recrystallizing and distilling

145

than working with gasses I could not see, I found the Huyser labs more to my liking.

Rowland left Kansas for U.C. Irvine in 1965, where he built the department of chemistry and where the chemistry building is now named after him. I visited Irvine a couple of times and one time, Sherry was my host. Ralph "Buzz" Adams was also in town at the time, and joined us. Adams and Rowland were two of the exciting young chemists on the faculty at Kansas at the time I studied there, and I learned as much as I could from each of them as a teaching assistant in their freshman laboratory courses. Adams was approaching late middle age at the time I saw him in Irvine, and he had completely redirected the course of his study to that of the chemistry of brain chemicals.

Rowland's Chicago was also Huyser's Chicago though Earl was several years younger, and worked on the organic side of the chemistry department as opposed to the physics side where Rowland studied. Earl's Chicago created an indelible image for me, because it had such an important role to play in his scientific career.

The chemical warfare corps, as has been mentioned more than once, had decided there were not enough young chemistry professors working on projects involving chemical warfare, so twelve assistant professors' laboratories were delineated, given money by the Army, and directed to work on "something of significance in chemical warfare."

Cheves Walling had a strong Chicago connection, and was one of the young scientist's labs identified to do this work. Walling decided to work on white phosphorus because it was used in incendiaries, and because he knew that the German war gasses were phosphorus containing. This was almost exactly at the time the Army was building its arsenals in Muscle Shoals, Alabama, and Rocky Mountain Arsenal near Denver, so there had to be some rudimentary knowledge of sarin and tabun in the interested scientific community.

At Columbia, Earl worked on the chemistry of phosphorous following with a project begun by his fellow Chicago collaborator, Frank W. Stacey. Stacey, was drafted, despite being Canadian; he was later given American citizenship. He used this toward a successful career at E. I. DuPont de Nemours in Wilmington, DE.

Huyser told me that he had to wear his private's uniform in the lab in Havemeyer Hall, the chemistry building where he worked at Columbia, a building that was directly adjacent to the library where the President's office was. Grayson Kirk was held captive in this office by student protestors about ten years later.

146

But back to phosphorus. In 1681, Robert Boyle reported that the glow accompanying the oxidation of phosphorus in air was not observed with solutions of phosphorus in turpentine, an observation which may be considered the first on the effect of olefins on the oxidation of phosphorus.[84] The similar effect of ethylene on the slow combustion of solid white phosphorus was studied by Graham, in 1829, and the subject continued to excite interest well into this century, providing one of the important bases for the theory of chain reactions developed by Semenoff. In 1914, Willstatter and Sonnenfeld published a short paper on the products obtained on exposure of benzene or cyclohexane solutions of phosphorus and olefins (chiefly cyclohexene) to oxygen, but the only subsequent work we know of is the report by Montingnie of the formation of an organophosphorus derivative from phosphorus and oxygen in the presence of an unsaturated steroid.

Willstatter reported that phosphorus-cyclohexene mixtures in benzene rather rapidly took up oxygen on shaking to yield initially a product $C_6H_{10}P_2O_3$. Further oxygen absorption was "slower," giving finally a product $C_6H_{10}P_2O_4$. This "phosphorate" was a white or pale yellow, benzene-insoluble, hygroscopic solid that reacts vigorously with water, dissolving with evolution of heat. The only product identification involved treating the phosphorate with 40% HNO_3. Under these conditions the phosphorus was further oxidized and approximately half was liberated as phosphoric acid, precipitated by magnesium ion."[85]

Huyser suffered a major medical issue in October 1956, when a cancerous testicular tumor was discovered and this had to be removed in surgery in October. He was out of the lab until mid-December. Fortunately, the tumor was encapsulated and he managed a complete recovery to finish at Columbia in May 1957. Huyser was at Columbia until mid-year 1957, where he made a connection with B.C. Saunders, the British chemist at Cambridge who independently discovered chemical analogs to sarin and somin. Though they were not as active Saunders, by reading German patents from Bayer had figured out what the Nazis knew, and the Allies did not; that fluorophoninates were exceedingly toxic gasses useful, potentially, in battle.

---

[84] Robert Boyle, "New Experiments and Observations Made upon the Icy Noctiluca," London: Printed by R.E. for B. Tooke, 1681/2.
[85] R. Willstatter and E. Sonnenfeld, *Chemiste Berichte.,* 47, 1914.

Earl returned to Dow and remained there until fall 1959, when he was appointed Assistant Professor at Kansas. His first project with a graduate student was that with Corwin Bredeweg, also a Hope graduate '59), who studied reactions of alcohols with peroxides.[86] In this paper, Huyser addressed a process significant particularly in the uncontrollable (explosive) decompositions of peroxides in alcohols. The explosives used in the London British Musuem bombings in 2008 were peroxides; and there's more than a little suspicion that those used to down the Russian airliner in the Sinai, and in the Paris events, were too.

Following the assembly of a large group of aggressive graduate students and the completion of the older members of that group's Ph. D. dissertations, Huyser applied for and received an N.S.F. senior faculty fellowship. He used this fellowship for study in Groningen, the Netherlands, in the laboratories of Hans Wynberg. There he worked with a Dutch laboratory technician, Hans Sinnige, on addition reactions of radicals to highly hindered, or crowded, olefins.[87] Continuing the work studying the effects of size of groups on radical addition reactions in his Kansas laboratories, Huyser studied bromine additions to highly-hindered styrenes.[88] In one of these papers, he reported that the reaction of bromine with 2,4,6-trimethylphenyl-2-propene yielded a yellow solid at -78° that decomposed at room temperature with the evolution of HBr. The yellow solid was likely the bromonium ion, bromide complex (equation).

Huyser's yellow solid

The Huyser group was the first in the world to isolate such a complex, though they did not address it in the publication as such,

[86] Earl Huyser and C.J. Bredeweg, "Induced Decompositions of Di-t-butyl Peroxide in Primary and Secondary Alcohols," *Journal of the American Chemical Society*, 86 1964.
[87] Earl Huyser, F.W. Sieberg, H.J.M. Sinnege, and H. Wynberg, H., "Free-Radical Reactions of 2-t-Butyl-1,3-butadiene and 2,3-Di-t-butyl-l,3-butadiene," *Journal of Organic Chemi*stry, 31, 1966; and Earl Huyser, H. Benson, and H.J.M. Sinnege, "Stereochemical course of free-radical additions of mercaptans to trans-DELTA.2-octalin," *Journal of Organic Chemistry, 32*, 1967.
[88] Earl Huyser and L.Kim, "Reaction of Bromide with Hindered Olefinic Bonds," *Journal of Organic Chemistry, 33*, 1968.

and they had no direct evidence; their evidence was only circumstantial. More direct evidence was obtained a year later by Strating, Wynberg, and colleagues.

For some time, graduate enrollments in chemistry stayed high. But our national and international situation was too good to be true, and the aggressive anti-Communists took leadership roles in government. A small activity in Vietnam, became a war by the mid-1960s. The draft, which had been furloughed in late 1950s, was reintroduced, and by 1967, college graduates could look at careers that would be put on hold while they served in some military capacity mostly in Southeast Asia.

The Vietnam War altered Earl Huyser's approach to university teaching. In 1970, there were two killings on the K.U. campus, protestors burned the student union, there were street curfews, and no gasoline could be sold by the can. Larry Chalmers, the Chancellor of the University at the time, held a rally at the football stadium to try and quiet the community. These events were devastating for some and deeply discouraging for others. It would take years for the campuses and universities to recover.

Enter Adamantane: Schaap, Wynberg and the Development of Singlet Oxygen Chemistry

Every so often, compounds come along that capture chemist's imaginations like few others have. This capture is often a fad; but occasionally a compound and its derivatives develop uses, and they then have much longer lives, and chemist's attention spans. Adamantane was one of these. With three fused cyclohexane rings, admantane took on a carbon assembly in the solid that had the charisma of diamond. Adamantane chemistry was a carbon chemist's prelude to the C-60 buckyball chemistry that would capture the 1990s.

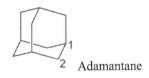

2 Adamantane

149

Adamantane, isolated from petroleum in 1933, was synthesized initially by the Croat/Swiss scientist Vladimir Prelog, and conveniently synthesized by Paul Schleyer at Princeton.[89]

As is often the case, new compounds with unusual structures get research attention, and uses are found for them that one would never expect. For a time, E. I. DuPont de Nemours tried to market amantadine, 1-aminoadamantane as a treatment of viral flu.

Amantadine

Scientists interested in structure reactivity relationships in small molecules found adamantane intriguing because it had three tertiary carbon atoms fused at three bridgeheads. They C-H centers were almost completely unreactive and followed the then strongly advised theories about reactivity at saturated carbon centers.

Photochemists later found adamantane interesting because stabilized dioxetanes could be made from olefins with adamantane as at least one of the terminae. Medicinal chemists were interested in its biological properties. Hans Wynberg's students took him into the adamantane arena. Wynberg was a professor of organic chemistry at the University of Groningen at the time.

Wynberg

The province of Groningen in the Netherlands borders Germany and Groningen, its largest city, is by far the largest city north of the Amsterdam megalopolis to the southwest. Groningen is mostly a farm province of which crops like sugar beets form a mainstay. Much of the unrest in the Dutch Reformed Church that so impacted the transposition of Albertus Van Raalte and his gang to the United States in 1846, came from the small town of Ulrum. Ulrum was located a few kilometers from the Lauwersee, near the slightly larger towns of Niekerk and Zoutkamp. Curiously, when we lived in

---

[89] P. Schleyer, "A Simple Preparation of Adamantane,"*Journal of the American Chemical Society, 79*, 1957.

Groningen, 1968-1969, we joined a students' walking party through the wad (muck) to the off shore island, Schiermonikoog. The walk was about 15 km, and took half a day under the careful attention of a well-trained guide. Today, the Lauwersee (Lauwersmeer in Dutch) had been enclosed with a dam, and the *wadlopen* from the site, is still possible, but more difficult.

The Rijkuniversiteit Groningen has been in existence since Shakespearian times. The University has had strong programs in the sciences for many, many years and (even though its science buildings were at the time near the city center) survived World War II relatively unscathed. Not so for the city. A battle near the Rijkuniversiteit took place in the spring of 1945 and many buildings on the town square were damaged or destroyed.

The trail of a few Hope College students and alumni from the United States to the north of the Netherlands traces to Earl Huyser.

Hans Wynberg, became Professor at Groningen in 1961. Wynberg was born in the Netherlands in 1922, and sent by his Father along with his twin brother to the United States in 1939 to live with a business partner, Maurits Dekker, in anticipation that German policies of anti-semitism. The Wynbergs were Jewish and, save for the twins sent to the U.S., none survived the Holocaust. After graduating from high school in New York, Hans joined the U.S. Army where he was selected, because of his facility in English, Dutch and German, to join an OSS operation scheduled to parachute behind German lines in Austria. Wynberg became the radioman for the trio and was involved in the infamous "Operation Greenup," as later featured in Patrick O'Donnell's *They Dared Return.*[90]

Jan Strating, Hans Wynberg's lector colleague at Groningen took a particular fancy to adamantane chemistry probably because some of his former students took at an interest in it. For a number of years, he and his students reported chemistry that can best be described as "ordinary chemistry on an extraordinary compound." He found Hans Wynberg a ready collaborator when Hans arrived in 1961 and through the hard work of some extraordinary students at Groningen, the two professors (by the time he retired Strating had been promoted to Professor) managed some important, and interesting new observations. A year after Huyser and Kim's work on possible stable bromonium ion, Strating, Wierenga and Wynberg reported isolating the first bromonium ion, bromide ion salt from the

[90] Patrick O'Donnell, *They Dared Return: The True Story of Jewish Spies Behind the Lines in Nazi Germany* (Boston: Da Capo Press, 2010).

reaction of bromine with adamantylidineadamantane.[91] Adamantylidine adamantane had been synthesized at Phillips Duphar Research Laboratories in Weesp, for whom Wynberg was a consultant.

3   Adamantylidine adamantane 3

When treated with Br_2 in chloroform or carbon tetrachloride, adamantylidineadamantane gave a yellow product after six hours that analyzed, for the addition of two molecules of bromine. Elemental analysis is the only evidence given for the structure of the bromonium ion product in the first paper because, given its instability and insolubility, it was not possible then to record its N.M.R. spectrum.

Subsequently, the chloronium ion complex was also reported.[92]

Chloronium ion

Adamantylidineadamantane synthesis

Adamantane smells like camphor, has a high melting point for a hydrocarbon (mp 270°), sublimes even at room temperature and has most unusual chemical reactivity due to the presence of so many C-C-C bridgehead bonds. Wierenga, Strating, Wynberg and Waldemar Adam first reported adamantylidineadamantane dioxtane in 1972 and Joop Wierenga, in his thesis in Groningen, November 1973, included the preparation of it by methylene blue sensitized singlet oxidation.[93] Wierenga also reported a blue green

---

[91] J. Strating, J. Wierenga, and H. Wynberg, "The Isolation of a Stable Bromonium Ion," *Chemical Communications, 908.*

[92] J. Strating, J. Wierenga, and H. Wynberg, "The Reaction of Chlorine with Adamantylidineadamantane," *Tetrahedron Letters,* 1970.

[93] J. Strating, J. Wierenga, W. Adam, and H. Wynberg, *Tetrahedron Letters,* 1972.

chemiluminescence upon heating to 180.° Such luminescence, given the appropriate substitution on the vinylidine of the adamantane, gave stable dioxetanes the luminescence of which could be triggered by enzymes leading directly to the field of chemiluminescent diagnostics.

(III)   (IV)

The patent literature is extensive on this subject, and fights over who owns what traversed an entire area of the photosciences for nearly a decade.[94] In other words, in this application, and a subsequent patent the writer draws a line in the sand that says we plan to pursue water soluble dioxetanes based on adamantane derivatives that are water soluble in order to… "make them eminently suitable for use as reporter molecules in many types of analytical techniques carried out in aqueous media, and especially in bioassays."

Such fights among scientists, and the companies they start are not unusual. Though they are expensive, in some cases the only way to figure out exactly who owns what in the progress of a research in science, is for the courts to hear both sides of the argument(s). We have been lucky as a society that one of the most significant scientific advances of the last generation, the production of penicillin, did not get so encumbered. Vannevar Bush and James Bryant Conant, among others, convinced companies that would eventually play a role to forgo their patent right and work for the national defense. Several companies actually worked together on and claimed no patent rights. Penicillin was, because of America's science, made available to American soldiers in 1943 and at least one author, Lauren Belfer in *Fierce Radiance,* suggested that this sped the end of the war. Ironically, Hans Wynberg who figures significantly in the Schaap story, worked for one of these companies in Brooklyn – Pfizer –as a young high school graduate.

It is hard to determine where the Wynberg lab fit in this at the end because a major law suit developed between A. Paul Schaap (Lumigen) and Irene Bronstein (Tropix). But the Wynberg/Strating labs were part of the beginning of adamantane chemistry, and

---

[94] I. Bronstein and J. Voyta, U.S. Patent No. 4952707, August 28, 1990.

153

specifically pioneered the development of, in fact they were the first to report, chemiluminescent dioxetanes, containing adamantane nuclei. A later paper from the Groningen group, of which there were still others, deals with the crystal structure of one of the dioxetanes.[95]

Waals H · · · H distances.‡

The X-ray structure of (I) cannot of course by itself prove its decomposition mechanism. Two decomposition modes are feasible. A planar, non-torsional mode[3] is pictured in the Scheme in which a stretching of the O–O bond is coincident with a compression of the two adamantyl substituents. A second mode[9] involves a concerted torsional decomposition as the two adamantane units appear to 'unscrew.'

(I)

Paul Schaap

Jerry Mohrig and I started teaching at Hope in the same fall, and we agreed that Jerry would teach all of the students in the lecture course in the fall semester and I would manage the laboratories. In the spring semester we would switch. When I first starting teaching at Hope in the fall of 1964, three young, smiling students occupied the western lab stations in the organic lab on the second floor in the old Science Building: David Anderson, Paul Schaap and Fred Van Lente.

We used a relatively new text book by Robert T. Morrison and Robert N. Boyd *Organic Chemistry*. This book, in its second edition by then, is still published and remains one of the most important textbooks ever written by chemists. We probably used Brewster, McEwen, and VanderWerf's *Laboratory Experiments in Organic Chemistry* as the laboratory manual because it was already being used at Hope –not for the obvious reason that Hope's new president was one of the authors! Soon, Mohrig proposed that we write our own laboratory book, which we did later, but in our first semester of teaching we used books that were already well broken in by generations of students.

---

[95] H. Numan, J.H.Wierenga,, H. Wynberg, and J. Hess, J., "X-Ray Structure of the 1,2-DioxetanDispiro(adamantane-2,3'- [1,2]dioxetan-4'''‴-adamantane)," *Journal of the American Chemical Society, 591, 1977.*

The chemistry department at Hope because of its aggressive research program for undergraduates required seniors to sign up for a year of research with a specific teacher. John Stam, David Stehower, and Jim Fargher were assigned to work in my lab but none of them had a lot of time. So I looked for younger students to work in the lab as well. Schaap was a student that I contacted then, as was one of freshman advisees, Jim Hardy.

Hardy was extremely interested in working in the lab and would spend Friday afternoons and Saturdays working with me. I could not have gotten my lab started without him.

I emerged from graduate school at just about the time commercial instruments were becoming available for laboratory work in chemistry. I saw my first two infrared spectrometers in the fall of 1960 in the 5th floor labs of Malott Hall at Kansas, and we had at least one introductory lecture to U.V./visible spectroscopy that involved at least a demonstration of the Department's Beckman DB spectrometer. Each of these instruments had an associated strip chart recorder with an ink pen/stylis arrangement on each to record absorptions or transmissions to whichever version of spectral reporting the recorder was set. When the new chemistry faculty (Mohrig, Dave Klein and I) arrived at Hope in September 1964, we found boxes containing an infrared spectrometer, a U.V./visible spectrometer and a monstrosity set up in the office I was to occupy that someone called a gas chromatograph. It was ancient even by the then standards, and instead of having a variable temperature device to control the column temperature, and another to control the temperature of the detector it a set of six mercury thermometers customized to control temperatures at intervals of 50° C. One day, I braved to turn the contraption on only to find after a few minutes another small thermometer broken in the detector oven. A pool of mercury lay on the bottom of the oven which was now being heated, because there was no thermal regulator, to whatever temperature it pleased to pursue all the while emitting mercury vapor into my office.

The next day, I went to Irwin Brink, who was chairman and told him that I had to have a new chromatograph. I could not function with the ancient of days sold to the retiring Gerrit Van Zyl by some salesman charged with the duty of getting the thing "off his company's hands."

Irwin Brink used department money to buy a gas chromatograph. After a tank of helium arrived from who knows where (laboratory gasses were tough to come by in the Holland of those days), Jim Hardy and I went to a local plumber's shop to get a gas regulator, the appropriate copper tubing, and gas fittings to hook the tank to the chromatograph. It was not a difficult job, and I had had experience in doing this while a graduate student, but it was very exciting for Jim and for me to see it happen when we were completely in control. In a few days, we had a functioning gas chromatograph, attached to a relatively new Sargent strip chart recording, just awaiting to produce new results from my new labs at Hope.

The research projects I undertook in those early days at Hope were simple and derivative of my Ph. D. studies at Kansas. One could easily criticize me for doing it this way, but faculty in liberal arts colleges were handicapped (I thought) by a lack of graduate students (or so I thought at the time) so I needed to work on a project that I could do on my own. If a student or two could contribute, that was all the better. I managed to buy the starting materials from Aldrich Chemical Company, and get my research started using them.

Paul Schaap was the first student I paid from a research grant to work in my labs. He was paid for the summer 1965. As it turned out, in order to manage a summer salary, I had been farmed out to teach at Ohio State.

At some time in the spring of 1966, Hans Wynberg had appeared in the Hope laboratories. It was a Friday afternoon, relatively late and I was about the only one around. Hans introduced himself, said that he was in Holland visiting a student who was spending a semester abroad at Hope and whose Dutch parents were friends of his. This student was Johannes Huber. He was a Prins Bernhard scholar and had come to Hope specifically because of a visit the Princes had paid the College a year earlier.

From left: Phil Van Lente, Ted Oegema, Linda Kozel,
Doug Neckers, Paul Schaap, and Jim Hardy.

I chatted with Hans for a time because I knew of him from Earl Huyser's experiences in Groningen, asked him if he wanted to give a seminar, which he did not. We probably talked a little bit about possible future relationships between Hope College and Groningen, and then he went on his way.

I had a number of students working in the labs at the time, including Jim Hardy, Linda Kozel, and Ted Oegema each of whom is pictured in the 1966 picture. But Schaap and Hardy were the most enthusiastic. Oegema was married and more mature, or so it seemed, than the rest. Linda Kozel was enthusiastic but a year younger than Schaap. Phil Van Lente, a high school sophomore at the time the picture was taken, worked off and on in my labs for several years. He was exceedingly diligent, an excellent worker, a nice person, and just plain fun to have around.

At some point in the fall of 1966, I wrote Hans Wynberg a letter and inquired if he would be interested in having an undergraduate from Hope spend a time in his lab. This was a shot in the dark because I was not sure I had a student who could do this if it came to pass. But Hope had a strong Dutch connection from its history, so it seemed logical to me. Hope was at a distinct disadvantage in dealing with the Dutch university mainly because the Dutch did not understand what a college was. To the Dutch, a college was a course or two and certainly nothing of consequence could come from such things with regard to research. But Wynberg had an American higher education, and understood what a college was because he had taught at Grinnell, Iowa for three years immediately after his finishing his Ph. D. with W. S Johnson at Wisconsin. That does not mean he thought highly of the American colleges of which Hope was, at that time at least, a relatively lesser member. But he knew what we did, and who we were. So he eventually wrote back and said he thought an American in his labs would be a plus. What did I have in mind?

What I had in mind was first Schaap and then Hardy would have the benefit of a unique experience – spending a semester doing research in a Dutch laboratory. Holland, Michigan, was settled by the Dutch in 1847, so what could be more logical?

Schaap going to Groningen did not take long to arrange and, eventually, I brought up to my colleagues in chemistry that I was working on this, and that it would mean Schaap would miss the second semester of his senior year. Only Dwight Smith was concerned about because he would miss an advanced instrumentation course and, perhaps, be disadvantaged when he started his graduate work. But Dwight did not protest mightily, and it was agreed upon that Paul could spend the spring semester in Groningen and still graduate with his class in June.

Paul traveled to Groningen and began work with Dick Kellogg in January. Paul won a Woodrow Wilson Fellowship while he was away. I remember being a little miffed when the publication that resulted did not identify Hope College.[96] But Paul returned to the U.S. in time for the fall A.C.S. meeting in New York, where he gave a paper on this work, having graduated from Hope in absentia.

I was personally more preoccupied with a proposal to the Alfred P. Sloan Foundation that, when funded the subsequent December, became the largest grant Hope College had ever received. I remember going to New York for the meeting and telling my colleagues I had research responsibilities and a paper to give at the meeting. I left the proposal to Brockmeier to finish. Cal had to rewrite the proposal and make English from Dick's physics-write.

During the time Paul was in Groningen, he communicated on a couple of occasions. He had been accepted for graduate study at both Harvard and Cal Tech so one long letter dealt with his upcoming decision about which graduate school to choose. In the end, he chose Harvard, and probably because he was intrigued by my stories about working for Paul Bartlett, ended up in Paul's laboratories, too.

The next I remember hearing from him or about him was in communication from Nick Turro, my friend from Harvard, now teaching at Columbia telling me about Schaap interviewing for a job there. Later, I found out that there were only three jobs in organic chemistry at that time in the entire country. One of those jobs was at Wayne State.

At some point, I went to Detroit to do some mass spectrometry experiments in Don DeJongh's lab. Don had established a start-up program at Wayne as the mass spectroscopist in the organic division. He and I were Hope colleagues, if not exactly classmates, and we had kept in touch over the several years since we had both graduated. Don was one of the gang of three from Burnips' in the Hope class of 1959, all chemistry majors. During that visit to Wayne, Carl Johnson came to Don's laboratories and introduced himself. A few months later at an A.C.S. meeting, I ran into Carl again – Carl was also a Wayne State organic chemist – and he asked me if there was any way he could spend some time at Hope, in the Hope chemistry department.

I was flattered that one from a major university with a Ph. D. program would think that spending some time at Hope was a good

[96] R. Kellogg, Paul Schaap, and H. Wynberg, "Bromination, Deuteration, and Lithiation of Dithienyls." *Journal of Organic Chemistry, 34,* 1969.

idea, so I contacted Henry Boersma, who was in charge of operations at Hope at the time, and he arranged for Carl and his wife to live near campus. During the summer of 1967, Carl used the house to write what became Pasto and Johnson, "Organic Structure Determination."[97]

Carl, Mary, Sue and I became good friends. Both the Johnsons liked to cook, as did Suzanne, so we would have occasional cooking meets in which one of the gourmet cooks tried to out-cook the other. The treats were absolutely delectable and were particularly nice for me, because all I had to do was eat them and put a few dishes in the dishwasher when the evening was over. Carl spent the following summer in Holland, as well. Sue and I were preparing to go to Groningen ourselves at the end of that summer. The Johnson's accommodations during the second summer were much less than the first but we enjoyed that time, too. I remember as I was planning our flight to the Netherlands, I had to arrange the timing so that I could meet John Hollenbach, a Hope professor of English spending the summer in Llubiana on September 7. John had rented a Mercedes I had purchased from the factory in Stuttgart for my year in Groningen, and John picked the car up at the factory, driven it to Yugoslavia, and then returned with it to Munich so I could claim it for my year in Groningen.

We had time to kill on the way and Carl suggested that we go through Zurich, Switzerland, which is what we did. We stayed in the Hotel Central Zurich in September 1968, in sight of the train station, for 35 Swiss francs. After two nights in Zurich, we took a train to Munich and met John Hollenbach at the airport, and drove the autobahn, eventually managing Groningen two days later.

I can hardly imagine what Hans Wynberg thought – he had lost his entire family in the Holocaust - when he saw me driving into the Grote Markt in Groningen with a Mercedes, but such it was and so it became. That I had a Mercedes at all was due to 1) naivete and 2) they could be purchased for less than a comparable American car at the time because of the exchange rate.

In the spring of 1970, after I returned to Hope from the Netherlands, Carl and I were sitting by his swimming pool in Detroit, talking about department affairs. Carl was bemoaning the fact that the organic division at Wayne was having trouble signing up a distinguished industrial chemist then at Proctor and Gamble, to be a faculty member, Paul Kropp. They wanted to hire a photochemist and

---

[97] D. R. Pasto and C. R. Johnson, *Organic Structure Determination* (Prentice Hall, 1969).

were having a hard time making a deal with Kropp (who eventually took a position at North Carolina, where he remained until retirement.)

I encouraged Carl to drop Paul Kropp and hire Paul Schaap instead and that is precisely what happened. That fall, Paul was appointed assistant professor at Wayne State. Don DeJongh had left Wayne State for the University of Montreal, meaning that Paul Schaap was Don DeJongh's replacement on the faculty. In fact, he moved into the office that Don vacated.

In the fall of 1971, I had been appointed associate professor of chemistry at the University of New Mexico. This was an interesting move for me professionally for a number of reasons, but two things were apparent almost immediately. First, the New Mexico chemistry department was not used to putting up with aggressive guys like me. Guido Daub, the chairman, referred to me as malignant: he spreads everywhere. Of course, I would add that nature abhors a vacuum, and there was nothing in most of the corners and quarters of the chemistry building at New Mexico at the time except lab benches and empty spaces, so no harm was done when my group spilled out of its labs a bit. The second was that the students at New Mexico, even though older, were, as a rule, much, much less able than the students at Hope College. The faculty at Hope was more productive, too. I counted that Hope's seven faculty in chemistry in 1971 published 42 papers! That was more than the much larger chemistry department at New Mexico published in the same time. I learned a strong lesson from this: it was not the age of the student that counts, nor the degree that they seek but their ability and work ethic.

Albuquerque was a good place to live, but not productive for me professionally. Within two years, I was on the move again, this time to Bowling Green, where I became head of the chemistry department in 1973. Schaap and I stayed in touch by seeing one another at meetings and more particularly at Gordon Research Conferences. His work with Paul Bartlett had begun to appear and, though this was not the only paper reporting the formation of a stable dioxetane from singlet oxygen addition to an activated olefin, certainly was sufficiently noticed to make a quick difference in the visibility of his career.[98] Schaap's first paper on the dioxetane of

---

[98] Paul Bartlett and Paul Schaap, "Stereospecific Formation of 1, 2-Dioxetanes from cis- and trans-Diethoxyethylenes by Singlet Oxygen," *Journal of the American Chemical Society, 92,* 1970; and C.S. Foote and S. Wexler, "Chemistry of Singlet Oxygen: Isolation of a Stable Dioxetane from Photooxygenation of Tetramethoxyethylene," *Journal of the American Chemical Society, 92,* 1970.

adamantilidineadamantane was a collaborative study published in 1975.[99]

It is not clear whether it was this paper that caused misunderstanding, and some mistrust with Wynberg, or a prior paper in the *Journal of Organic Chemistry*.[100] This reaction was precisely that used in Groningen by A. C. Udding and about which, at least Wynberg surmised, Schaap must have known.[101]

Regardless of the details and validity, I recognized a major inconvenience in singlet oxygen syntheses involved highly colored, visible light absorbing, photosensitizers that were soluble only in polar solvents, or water, and messy dyes, too. In spring 1968, I had given a short course on photochemistry at the UpJohn Company in Kalmazoo, and one of their scientists, William Schneider, showed me a series of five or six large reactors filled with a bright red solution that Schneider said was Rose Bengal. Schneider pointed out that the oxidation step was one of a series in the synthesis of a steroidal birth control pill then marketed by UpJohn and I wondered, though not out loud, just how much Rose Bengal remained in the birth control pills when they were actually sold? The red dye was very difficult to completely get rid of. In fact, some years later we participated in some studies with a cardiologist at one of the hospitals in London in which he was using nanomolar solutions of Rose Bengal to assess recovery from rat perfusion injury. These solutions were active even though the dye concentration was so low, that one could not see its red color.

But, my Upjohn observation was one of several that caused a research proposal to N.S.F. that was funded, to study the photosensitivity of polymer based photosensitizers. Out of this relatively modest amount of funding came three important concepts: polymer protected reagents,[102] polymer Rose Bengal,[103] and polymer-based benzophenone triplet sensitizers.[104]

---

[99] G. B. Schuster, Nick Turro, H.C. Steinmetzer, Paul Schaap, G. Faler, G. Adam, and J.C. Liu, "Adamantylideneadamantane-1,2-dioxetane. An Investigation of the Chemiluminescence and Decomposition Kinetics of an Unusually Stable 1,2-Dioxetane," *Journal of the American Chemical Society, 97,* 1975.

[100] Paul Schaap and G. Faler, "A Convenient Synthesis of Adamantylidineadamantane,"*Journal of Organic Chemistry, 38,* 1973.

[101] A.C. Udding, Ph.D. Thesis, Groningen, September 28, 1968.

[102] Douglas C. Neckers, D.A. Kooistra, and G.W. Green, "Polymer-Protected Reagents: Polystyrene-Aluminum Chloride," *Journal of the American Chemical Society, 94,* 1972.

[103] E.C. Blossey, Douglas C. Neckers, A.L. Thayer, and Paul Schaap, "Polymer-Based Sensitizers for Photooxidations," *Journal of the American Chemical Society, 95,* 1973.

[104] E.C. Blossey and Douglas C. Neckers, "Concerning the Use of Polymer Based Photosensitizers," *Tetrahedron Letters, 4,* 1974.

I had been appointed chair of chemistry at Bowling Green State University in the summer of 1973, and was in the process of moving my labs from New Mexico there, I asked Paul if he would like to collaborate by studying photoxidations using polymer Rose Bengal and several other immobilized sensitizers for singlet oxygen. He said he would, so I packed up a box of dyes immobilized on poly(styrene) beads and sent them to him. The system that worked the best, and which became the staple—the only sensitizer to make dioxetanes he used going forward—was polymer Rose Bengal.

I recognized the potential of the product and used the only process available to me at New Mexico, because polymer Rose Bengal was made at New Mexico by Erich Blossey and myself, Research Corporation, to apply for a patent on polymer Rose Bengal. Patenting began in the summer/fall 1973, the patent was finally applied for in February 1979 and completed with the issuance of a U.S. Patent.[105] Shortly after the patenting work was begun, Research Corporation began trying to license the patent, and managed to do so with a company called Hydron Laboratories which mostly made pH paper. A scientist at Hydron, Peter Valenti, was assigned to work on the patented product soon after the license had been signed. Not long thereafter, a paper appeared in the *Journal of Organic Chemistry* co-authored by Schaap on which I was neither a co-author or acknowledged and which listed as references to our earlier work: "This sensitizer is available under the tradename SENSITOX I from Hydron Laboratories, Inc., Chemical Sales Dept., New Brunswick, NJ 08902."[106]

I probably should have made more of this at the time than I did, but I chose to do what I always have done when seemingly wronged – I withdrew. I spoke to Schaap years later, after my father had called and said that he had been at a Hope alumni banquet at which Paul Schaap received an award, and Schaap had mentioned the work I had done with him when he was an undergraduate. I called Paul and thanked him for doing so.

By this time Paul, had been through a major patent infringement litigation. The details of this are described in great length elsewhere, and have no currency here. This was awkward for me because sides were drawn among my photoscientist colleagues at

---

[105] Douglas C. Neckers, E.C. Blossey, and Paul Schaap, "Polymer Based Photosensitizers," U.S. Patent No. 4,315,998, February 16, 1982.

[106] Paul Schaap and Steven D. Gagnon, "Chemiluminescence from a Phenoxide-Substituted 1,2-Dioxetane: A Model for Firefly Bioluminescen," *Journal of the American Chemical Society, 104,* 1982.

a time when I was trying to get the Center for Photochemical Sciences started. Irene Bronstein, an Edwin Lamb protégé and former Polaroid employee, thought one from a small college like Hope or an inconsequential state university like a Bowling Green of no consequence in her most rarified of intellectual worlds and only made contact when she wanted a favor or had a complaint.

I understand from my former professional colleagues at Wayne State that Paul first repaid the institution for the support he received in the patent infringement litigation. Later, he was responsible financially for a new wing on the chemistry building at Wayne. His substantial contributions to Hope College are the direct result of the work much of which involves adamantane derivatives, polymer Rose Bengal, and much else directly related to his undergraduate experience at Hope College. He has done very well, and Hope College is rightfully proud of his contributions.

Chapter 10

The VanderWerfs and Race Relations in Lawrence, Kansas

Kansas was admitted to the Union on January 29, 1861, after a decade or more of guerrilla battles over whether the Kansas territory would enter the Union as a slave or free state. As the more than three hundred victims of Captain William Quantrill's raid on Lawrence in 1863 proved, the Civil War touched the small town in a serious way, because it had been regarded a good target because many free state abolitionists that had moved there in the 1850s.[107] African-Americans who moved into Kansas after the Civil War were tolerated in some communities, though not in most. Topeka, Lawrence, and Wyandott had the largest black populations, though the races really did not mix much even there. Tensions, mostly subliminal, existed in Lawrence when Cal VanderWerf moved there in 1941. African-Americans lived largely to the North of the Kaw River in Lawrence, a trickle usually that nominally flowed into the Missouri near Kansas City but that offered an occasional flood.

The VanderWerfs were married in the summer of 1942, and moved to a house on Alabama Street, close to the University. Klasina and Gretchen, the VanderWerf daughters, both remember the neighborhood being somewhat integrated, and they had friends across the street who were African American that treated the girls like grandchildren. By the fall of 1960, some but not all of the effects of segregation in the community had modified.

There was no unofficially sanctioned segregation, *per se*, but racial prejudice was part of the Lawrence community and a standoffishness lay just under the surface in most everything affecting the people living there. Margery Argersinger, wife of chemistry professor Bill Argersinger and a community activist, said that when she led the campaign that eventually succeeded in bringing an integrated community swimming pool to Lawrence, phone calls using racial slurs to her home from callers were a norm.[108] The

---

[107] Chronicled at these sites:

www.civilwaralbum.com/misc/kansas_lawrence1.htm;
www.kansastravel.org/quantrillslawrenceraid.htm; and
www.sonofthesouth.net/leefoundation/civil-war/1863.

[108] Interview with Margery Argersing and the author, May 3, 2011.

community finally found funds for a community pool on the third try but only after she left the committee. Kansas summers were exceedingly hot, and there was essentially no air conditioning at the time, so a pleasant late afternoon swim could really cut the heat. The VanderWerfs and others were members of a private swimming pool located near Route 40 west going out of town toward Topeka called Jayhawk Plunge. When, in 1960, the owners banned blacks from swimming in the pool, members of the community boycotted the pool too and the VanderWerfs, even though they had little children, pulled their membership. Cal managed, by encouraging business in the direction of a new Holiday Inn at the corner of 23rd Street and Iowa Street, the edge of Lawrence at the time, to convince the manager to let his children swim in the motel pool—and this they did for a time. Though that pool did not come up to Jayhawk Plunge in size, it took the heat off on many a summer day.

Racial tensions were the norm in Lawrence before World War II, though, and some of the younger university faculty tried to implement change. Cal and Rachel VanderWerf were founding members of the Lawrence League for the Practice of Democracy (L.L.P.D.). The L.L.P.D. was made up mostly of university, but also black and white racial activists from the community. It was founded in the VanderWerf's living room on Alabama Street. According to Margery Argersinger, the L.L.P.D. never grew much larger than forty or fifty members because many blacks were afraid to join it, thinking their jobs would be in jeopardy if they did. Until *Brown v. Board*, and even after, elementary school students went to segregated schools, theaters were segregated, blood for transfusions was segregated into black blood and white blood, the Eldridge Hotel (the only hotel in town and the specific focus of Quantrill's raid almost 100 years earlier) would not serve blacks, and black and white kids could not swim together.

A group of K.U. faculty members that included Arthur Davidson and his wife, and others from the community like Russ Carter, a minister at Haskell Institute, Mazone Wallace, a young black mother, and Ethyl Molton, a cook who had served dinners in most of the best Lawrence homes, met with the VanderWerfs to found an organization created to encourage exchange between the races and foster change in the community. The L.L.P.D. took the first steps in the slow integration of local businesses and schools. When Weslie Sims, a returning World War II veteran, and his wife sat in the wrong seats at the Varsity Theater on Massachusetts Street, the theater owner first advised them they were seated in the white section and they needed to move, and then called the police who escorted them from the theater. Blacks were supposed to sit in seats at the rear of the theater that were delineated by three inch white strips across the top of the backs.[109] This incident caused a visit by the Reverend G.E.E. Lindquist on November 26, 1945, to remind the Lawrence City Commission of the Kansas state law that made segregation illegal in theaters and other service businesses that required licenses to operate in a municipality. The Lawrence city commission, according to Kris McCusker,[110] rescinded the Lawrence statute that required theaters to be licensed, thus protecting the theater owners to make their own rules about seating in the future.

According to Rachel VanderWerf, while the L.L.P.D. could be "more talk than action," it also provided a necessary forum where black and white civic leaders could associate and try to assert a moral presence in the community.[111] Cal VanderWerf was the "light people followed," the organization's instigator and "idealist." But, according to Argersinger, Cal was also a politician "backing away when ideas failed" while taking credit for those ideas that worked. Most, though not all, L.L.P.D. members belonged to Protestant churches in Lawrence. The VanderWerfs were members of Plymouth Congregational Church in Lawrence which was a "very liberal congregation," according to Bill Hillegonds.[112] To Hillegoods, who encountered the VanderWerfs when they moved to Holland, Michigan, that could mean congregants might meet for a picnic on Sunday, or go for a swim on the Lord's Day. But, it may have had something to do with the fact that black and white congregants worshipped together at Plymouth Congregational church, too. Nonetheless, Plymouth was a Congregational church which meant it was organized by the congregation having little connection with other congregations in the same denomination. But, to the rigid Dutch Reformed Church of Cal's upbringing (still extant in Holland when he moved back there), it had to be suspiciously full of sin and sinners. Arthur Davidson was a distinguished alumnus of the City College of New York who, after World War II, encouraged numerous of the new

---

[109] *Lawrence Journal World*, November 1, 1945. A1.

[110] K. McCusker, *The Forgotten Years of America's Civil Rights Movement: The University of Kansas - 1939-61*. M.A. Dissertation, 1992; and *Kansas History, 17*, 1994, pg. 17.

[111] Interview with Rachel Vanderwerf, June 5, 1989.

[112] J.C. Kennedy and C.J. Simon, C. J., *Can Hope Endure? A "Historical" Case Study in Christian Higher Education* (Grand Rapids, MI: Erdman's, 2005), 129.

C.C.N.Y. alumni, many of whom had come to the U.S. as European refugees and survivors of the Holocaust, to study at Kansas. Several VanderWerf students of the postwar era were refugees: the Davidsons, Kleinbergs, Paul and Helen Gilles and their families, as well as the Argersingers, were all members of the chemistry department. All of them started and ended their careers in Lawrence at the University of Kansas. All of them continue to evidence the passion for equality that was struck in the 1940s when they were founding, or early joiners, of L.L.P.D.   Idealists or no; politicians or no; talkers or no; the L.L.P.D. was a *mostly* university organization wishing to make changes in the town, the state, and the country. The community of Lawrence itself, in a classic town-versus-gown confrontation, viewed the L.L.P.D. as a bunch of "communists," referring to Cal VanderWerf as the "king of the Communists."[113] If the town was conservative and the university faculty, in general, liberal, members of the L.L.P.D. were viewed sometimes as radicals. They were even too radical occasionally for that otherwise liberal Plymouth Congregational Church congregation. When the L.L.P.D. organized an interracial conference at his church in 1947, and the congregation of his church objected, the pastor of Plymouth Church cancelled the conference.

Cal and Rachel believed in education to the depth of their respective beings. Cal's personal life story—raised by a single mother of four during the depression, and working his way through grit and determination—evidenced what one individual could do given the chance. The L.L.P.D.'s plan was to demonstrate an educational plan for the community symbolizing interracial cooperation as an example of democracy at work. They connected occasionally with activist organizations for racial equality, as for example the Congress of Racial Equality (C.O.R.E.) headed by James Farmer, and invited Farmer to meet with them in 1949. But the L.L.P.D.'s actions were rooted in democratic ideals, not confrontation and they disagreed substantially with Farmer's approach to achieving racial equality. The L.L.P.D. did not back from involvement in local politics, though, as when Cal VanderWerf ran for a city council seat in 1946. His opponent was the iconic Phog Allen, the basketball coach, and Cal lost by six votes. Rachel suggested that Cal captured black votes and lost white votes because of his reputation as a racial radical. The former was not enough to outweigh the latter. But standing back from the then fray, any time a

---

[113] Interview with Rachel VanderWerf, June 5, 1989.

basketball star/coach runs for political office against a professor of organic chemistry, the name recognition of the coach, be he an angel or scoundrel, will likely get him elected.

The L.L.P.D., as a missionary for democracy, also started a cooperative interracial nursery school. The group had a hard time finding a church to house the integrated day care center until the First Methodist Church came forward. The idea was that children would learn equality of the races before they learned prejudice. Each mother had to spend one day a week working at the school, a requirement that eventually caused the plan to fail. Black mothers had to work outside the home to support themselves and their families, so working at the nursery school could not be managed. When the school itself would not allow students in the school whose mothers did not also come there one day a week, the school could no longer function.

For some reason, the movie theaters in downtown Lawrence became a major target of racial issues. The Jayhawk Theater, the largest of the four theaters in the town, proved difficult to integrate. An L.L.P.D. group went to a movie there in 1949, and sat throughout the theater in small groups of blacks and whites until, eventually the theater owner stopped the film to tell those who were there that they were disobeying theater policy sitting as they were. Strong words followed with eventually the theater owner agreeing to meet with a group of those who were there the next morning. When he had agreed to do this, he turned the movie back on. The next morning, a group of organizers went to see owner Stan Schwanther, demanding that blacks should be allowed to sit anywhere in the theaters.

Two pressures, one from the University and another brought on the community by Wilt's arrival, did more to change the policy than the work of the L.L.P.D. University of Kansas Chancellor Franklin Murphy, who arrived in Lawrence in 1950, gave the theaters a push by saying that if they did not accede to the group's demands, he would start showing movies on weekend nights at reduced prices on the campus. Theater owners had their minds changed when Wilt Chamberlain arrived, and theaters quickly were integrated. But the hardest of the actions were those required before there was any call for action and Rachel and Cal VanderWerf were there at the outset, and visibly in the community. Kansas was not a large community after the war, with only around 4,000 students, so the work of the VanderWerfs was well known throughout Lawrence.

One incident has been singled out in Cal VanderWerf's almost-immediate involvement in racial equity matters in the community, even as a young assistant professor. This happened in

1942, and it also caused the beginnings of the campus community's involvement, specifically the members of the chemistry department, in racial matters in the town. As part of the build-up to World War II, the government decided to move a Hercules Powder manufacturing facility from Wilmington, Delaware, to Lawrence to get it away from the coast, and away from the nucleus of chemical manufacturing facilities around DuPont. When Hercules Powder, as part of the war effort, was to open a manufacturing plant called the Sunflower Powder Works east of town, a number of workers and their families had to be moved into Lawrence from the east to work in this powder manufacturing facility. With so many new workers moving into the community, school space was needed for their children. In order to accommodate the war workers' children and the new students arriving in the Lawrence schools, it was proposed by the school board and the town council that all black students be bussed to Lincoln School, across the Kaw River, and north of Lawrence. Lincoln School was an all-black school segregated legally according to Kansas law.

The black citizens in Lawrence, though probably reticent to protest, thought this a terrible idea and a meeting was held at the Ninth Street Baptist Church to decide what to do. Cal VanderWerf had heard about the meeting from two stockroom keepers in the chemistry department, and went to the meeting. There he stood up and made a speech in opposition to the bussing. Cal became a black community ally and helped nucleate the black community's protest. The bussing idea was soon dropped as a bad idea.

Plessy v. Ferguson was the benchmark Supreme Court decision that justified "separate but equal" as the rule for school segregation in the South. Though presumably a "free state," segregated schools were legal in Kansas since 1879, having been the subject of legislation of the Kansas Legislature.[114] In that 19th century legislation, cities with populations larger than 10,000 inhabitants could segregate black elementary students from whites in a separate but equal school. The history of *Brown v. Board* is legend in the annals of Supreme Court decisions in the 20th century leading, as much as anything, to the subsequent defense state's rights that pervades judicial appointments in the U.S. Senate to this day. While activist judges are anathema in the South and to conservatives everywhere, the *Brown v. Board* decision demonstrated how

---

[114] *Beyond John Brown: The Story of Race Relations in Lawrence.* larryville.com (accessed Feb 12, 2009).

progressive the judges really were. In just four thousand words, the decision changed the history of education in the United States and forced a confrontation of civil society in America.

Brown v. Board struck down an 1896 decision by the Court, *Plessy v. Ferguson*. That case, first heard in the Supreme Court in December of 1892, concerned an 1890 Louisiana law dictating that the races could be served by the railroads in separate but equal railway cars. It served as the genesis of segregation laws and policies throughout the South, and was important in Kansas for the role it played in school segregation. In the case, Louisiana plaintiffs argued that a long line of prior rulings had shown that separate but equal accommodations for "coloreds" in railway cars and in schools did not violate the 13th or 14th Amendments because there was no enslavement, nor were the separate cars unequal. Post-*Plessy,* "separate but equal" was subrogated to include restaurants, theaters, swimming pools and virtually any other area of commerce or recreation potentially enjoyed by both blacks and whites. It also applied to schools, though in the case of Kansas, only to elementary schools and only then, to schools located in cities with populations greater than 10,000.

In 1879, the Kansas Legislature gave cities with populations above 10,000 the authority to establish racially separate grade schools, and Leavenworth and Topeka promptly did so. But Lawrence had succumbed to racism five years earlier in 1874, when a school for black students only was built in North Lawrence. Until 1954, a dual-school system functioned in North Lawrence, and two grade schools were maintained: the Woodlawn School for white children and, down the block, the Lincoln School for African American children. The school existed substantially beyond that and remained mostly, if not entirely, segregated for many more years.

Access to high-quality schools was a primary motivation for southern migration to Kansas, and the illiteracy rate for African Americans in the state dropped dramatically from 32.8 percent in 1890 to only 8.8 percent in 1920. African American children generally received good educations in Kansas whether in integrated or in all-black schools.[115]

But black families felt their children were being deprived socially by being isolated from the general communities in which they were growing up. In Lawrence, though a small group of blacks lived on the south side of the river, most did not. So even had the

---

[115] Interview with William Tuttle, May 1, 2012.

schools been neighborhood schools, the result would have been the same—mostly black elementary school students studied together.

Lawrence itself, and the University of Kansas, were practically, if not by law, segregated entities. The University of Kansas, founded in 1865 as a symbol of the liberties retained by the fight to keep the Union, enrolled blacks from the late 1870s onward. The first black student enrolled at K.U. in 1876 and the first graduate, B. K. Bruce, received his degree in 1885. However, it took another approximately sixty years before the first teaching assistantship was awarded to a black student (1947-1952). That student, I. Wesley Elliott, told me years later that he was "a black student with little self-confidence studying at K.U. during the war;" adding, "I could not serve in the military because I had very poor eyesight."

"Professor VanderWerf discovered me in his organic chemistry class," Elliott continued, "and encouraged me to seek an advanced degree in chemistry at K.U. When the University would not provide a black student a teaching assistantship, Professor VanderWerf went to the Chancellor of the University (Malott) and pushed him to make the award he eventually did."[116] Elliott finished his degree with Professor William E. McEwen, returning to K.U. after a disastrous misadventure at Southern University in Baton Rouge, LA. He then taught for a number of years at Fisk University in Nashville, Tennessee. Cal VanderWerf, Elliott noted, was a real role model for him.

Though Topeka, and Lawrence, were only nominally segregated rather than blatantly so as remained the case in the South, the Topeka school's cases were, arguably, the most important civil litigation of the 20th century. In *Brown v. Board*, the Rev. Oliver Brown sued the school board in Topeka seeking admission for his daughter to a white school that was near her home. Kansas law (1879) was cited against his suit so the Kansas Supreme Court voted 3 to 0 to uphold that law; in effect, voting to uphold segregation in elementary schools in cities with populations larger than 10,000 persons in Kansas. The case was carried forth to the Supreme Court, where it was argued by Thurgood Marshall and the Court considered *Brown* and four other school segregation cases as one in the 1954 term. The Court ruled unanimously that "Separate but Equal" was not constitutional under the 14th Amendment.

---

[116] Communication between Rachel VanderWerf and Wesley Elliott; see I. Wesley Elliott, *Hydrogenation Studies on Reissert Compounds: Synthesis of Some Hexadecanes and Octadecanes*, Ph.D. Dissertation, University of Kansas, 1952.

Black undergraduate students at the time, according to Professor Tuttle, were mostly self-supporting and a number came from surrounding states to study at K.U. because their own state universities were entirely segregated. Black students rarely had time for extra-curricular activities because most had to work to stay in school. Wesley Elliott, for example, and two of his friends lived in the basement of the Chi Omega sorority house where they worked as cooks and cleaners to afford their room and board. They were not wanted on campus either. In fact, Phog Allen, K.U.'s basketball coaching icon, was quoted as saying in 1927 "that no colored man will ever have a chance (on the K.U. basketball team) as long as (Allen) is there."[117] The K.U. Chancellor at the time, Lindley, also strongly supported a segregationist stance and, according to Miller, "the University essentially ignored its black students." A new dormitory built for women in 1928 carefully excluded black women, so most lived with families in north and east Lawrence. Coach Allen, when asked a year later about his statement above denied saying it, but also said he "didn't think it appropriate for blacks and whites to play contact sports together." What is more, several of the universities on K.U.'s schedule—Oklahoma, Oklahoma A&M, and Missouri for example—were segregated so if K.U.'s basketball team carried black players, the team would have to either play two teams— one white and one mixed—or two schedules.

Allen obviously was forced to change his stance thirty years later when Wilt Chamberlain came on the scene. The Big Six— Kansas, Oklahoma, Missouri, Nebraska, Iowa State and Kansas State College—also officially excluded blacks from all athletic events. However, track and field was an exception and in 1943, W. W. Davis, who was a K.U. history professor and the faculty representative to the conference, took a motion drafted by students to a Big Six meeting. The motion boldly stated: Since "Negro men are good enough to pay taxes and to serve in the armed forces, it was only fair . . . that they be allowed to compete in intercollegiate sports." Davis's motion died for lack of a second.

It took a while, but integrated athletic teams and integrated events eventually happened at K.U. in the 1950s. Though even when I was a student (1960-1963), there were not many black football players in the then Big 8 Conference; Oklahoma had few, if any. The

---

[117] Loren Miller, "College," *Crisis,* 34 (January 1927), 138. For more information, please see also Miller's "The Unrest Among Negro Students at a White College: The University of Kansas," *Crisis,* 34 (January 1927), 184.

Hornfrogs from Texas Christian University (T.C.U.), not a member of the Big 8 but from a sister conference and located in Fort Worth, Texas, came to Lawrence to play the Jayhawks in the fall of 1960. On that team included Bob Lilly, the eventual Mr. Everything lineman in the N.F.L., but no blacks played for T.C.U. Syracuse, came from the east with a team led by Ernie Davis and Art Baker, both black, to play against K.U. in 1962. Missouri, by 1963, had recruited the running back, Johnny Rowland, who made a major difference in their team but Missouri, as a state, remained decidedly segregated even in the 1960s. On at least one occasion when I was a student, a fight broke out in a Kansas/Missouri basketball game at Allen Fieldhouse, allegedly triggered by a racial remark made by a Missouri guard named Doughty to one of the K.U. guards. That guard, Nolan Ellison, eventually became president of Cuyahoga Community College in Cleveland.

Deane Waldo Malott was chancellor at K.U. from 1939-1950, eventually leaving to become President of Cornell University. According to things written about K.U. during and just after World War II, his presidency was one of continued segregation and inequality mainly because Malott was an old fuddy-duddy, according to the students. Rachel VanderWerf excused him a bit in correspondence with Kris McCusker, though pointing out that during the war, the University lost a substantial amount of revenue because the undergraduate populations were decreased by the military effort. So, Malott was tasked with the issue of paying for the University's programs which may have had something to do with his racial stances. The administration's attitudes toward racial equality changed when Franklin Murphy became Chancellor in 1950. And the entire country changed, or at least started to, with the *Brown v. Board* decision in 1954.

George Axelrad was a Czech Jew whose entire family survived the war in Europe. George came to the U.S. in 1945, finished high school, and then studied chemistry at the former City College of New York, C.C.N.Y. Through Dr. Arthur Davidson's efforts—he was also a graduate of City College—a number of young chemistry students, mostly refugees like George, came to Lawrence to study for their Ph. D. degrees.

"Soon after I arrived at K.U. in 1954, I was recruited by Cal to participate in the Brotherhood Program in Lawrence. Cal and Rachel, I believe, were intimately involved in the Brotherhood Program (L.L.P.D.). The highlight of the Brotherhood Program came on April 30, 1954, shortly before the *Brown v. Board* decision came

173

down, as the L.L.P.D. welcomed Thurgood Marshall, then gaining notoriety for his delivery of the argument. He remembered how important the Brotherhood dinners had become, and "in addition to the graduate students in chemistry, many chemistry spouses also had big jobs to do to prepare for them. Margery Argersinger told me that Edna May Brown was the "slave driver." Edna May, who had worked for years as a cook in the fraternity houses and dormitories at K.U. knew how to cook for, and serve, a crowd. So she ordered Margery, Helen Gilles (whose husband, Paul, was also a K.U. chemistry faculty member), and others to get the tables set just so. Margery made Jello for more than 100 people for one of these dinners. Yet Rachel and Jane Kleinberg both told Kris McCusker in 1992 that the Brotherhood Banquets, for which 559 tickets were sold in 1954, really did not work that well. Though racially mixed, most attendees had little in common, and conversations were said to be stilted and uncomfortable."[118]

George Axelrad knew the Wilt story, too, and told me (as many others have) that "Cal went to Philadelphia and spoke to Wilt's mother and family, persuaded them that, if Wilt came to K.U., he Cal, would make sure that he got a 'good education' and that he would be his faculty advisor."

Cal followed through and became Wilt's faculty advisor. According to George, "Cal came one day into the laboratory where I worked and said 'I want to introduce you to our new student who is a great basketball player, from Philadelphia.'"

As he reached out to shake a very large hand and looked up, George said "Wilt towered over me, I may have reached about his midsection. Because my lab was next to Cal's office I saw Wilt visit him several times."

Cal managed also that Wilt's attendance at K.U. would have an important effect in improving racial relations in Lawrence and beyond, which it certainly did. Barbershops in Lawrence would not cut the hair for black people. They had to go to barbershops "on the other side of the tracks," in North Lawrence. When Wilt walked into the most popular barbershop on Main Street and sat in the barber chair, what barber could refuse to cut Wilt's hair? The barbershops became eager to have Wilt and other black athletes come to their shops. Some businesses actually advertised that "Wilt Chamberlain shopped there."

---

[118] Correspondence between George Axelrad and Doug Neckers, 2011-2013.

George shared other anecdotes, including that Cal VanderWerf was a vigorous stamp collector. "He used stamps to teach his own children. I had some stamps from Europe, and especially from Germany of the Hitler period, which I gave him."

Gretchen VanderWerf remembered Axelrad clearly even though she was a small girl when he was her father's student. "George had been before a firing squad."

George had, as the war was being lost by the Germans been rounded up (along with several other boys from the Budapest ghettos), taken to a ditch, and summarily shot. George, though severely injured, managed to survive. He was nursed back to health by medical servicers in the ghetto and lived to tell the story.

George said "I tell this, because I was present when Cal was sorting out some of my stamps in the presence of his children. He took this opportunity to tell them about the terrible times during World War II and the Holocaust. As a Holocaust survivor, this made a very strong impression on me."

The girls clearly remembered the story, too, as Gretchen pointed out when I asked her about George. I asked Cal's older daughters, Gretchen and Klasina, who remember living in Kansas, if they understood why their parents were so passionate about civil rights.

Klasina explained: "Mom's parents were Mennonites who had moved out of the community and who didn't really practice the culture. But they came from a tradition of peace and service. Dad was the youngest of four; his siblings were all sisters. His father died when he was nine years old. The sisters helped raise him and doted on him, and, because he was so bright, they really expected him to succeed. I believe he felt he owed a tremendous debt to those who had helped him along the way."

"I don't know if Lawrence was a typical mid-west college/university town or not. But what I remember in terms of 'race' is this: Most black families lived in North Lawrence across the river. Many had jobs working for white families as 'domestics,' a word Mrs. Wilburn, the VanderWerf's live-in colleague, later used to describe herself. This meant they cleaned, did child care, cooked, etc.; the men did yard work. I believe some walked to work, which would have been a long distance. Sometimes they got rides home. I remember Mom often picked up and dropped off Mrs. Wilburn. Domestics also took the bus to work, so riding the bus must have been possible.

"There were sort of integrated neighborhoods in town, though I believe there had to be a certain number of black families already living there for another black family to purchase a property. Our first house was on Alabama Street. White families lived on our side of the street, and black families lived on the other. I remember two black homes in particular that we often visited as children. One

was that of Mrs. Suttles. She was already fairly old at the time, or at least seemed so to us. She would roll her nylons down to just above her knee, and you could see this roll that always struck us children as particularly funny. The other family was Mr. and Mrs. White. All of these people doted on us as children, and we loved to go visit them."

Rachel and Mrs. Wilburn.

"My folks raised chickens at one time on Alabama Street, I believe something many people did to supplement their meat supply which may have been a hold-over from World War II. The house had a closed in back porch that was where Mom sewed. Years later, I visited that house (we knew the people who bought it and used it as a rental), and there in a build-in cabinet, was a saying Mom had painted across the back, which was very apropos: 'A place for everything, and everything in its place.'"

"I went to kindergarten at Cordley School. Both Gretchen and I believe that we had black classmates. I remember one black boy who lived not far from school who always wore a long-sleeved white

shirt, even in summer. It took Mrs. Wilburn to explain to me that he did this to keep cool. I think several black families lived in that area."

The VanderWerf children at Klasina's first birthday party.

"When we went to junior high school, we went to Central Jr. High that was the old high school. Central was integrated but the black kids from North Lawrence walked to school. It would have been pretty cold to walk across the bridge on a windy winter day. Blacks and whites didn't mix a whole lot as friends, as I remember. I don't remember much outright racism, kids just seemed to gravitate that way. The real division in junior high school was between regular kids and 'hoods' which were both black and white. Hoods wore black leather jackets, often did a peroxided streak in their hair (girl and boyfriends to match), and the boys wore their hair in ducktails."

By far, Cal's most important contribution at least in terms of its long-term consequences to racial equality was in his work with Wilt Chamberlain: and I do not just mean in recruiting Chamberlain, but in encouraging him to help change the Lawrence community forever.

Competition for "Wilt the Stilt," as he was later called, was fierce. Phog Allen pulled out all the stops so, when Cal was to visit Smith, Kline and French in Philadelphia where he consulted on the development of new anesthetics, he made an appointment with the Chamberlains at their home at 401 North Salford Street and talked with the family. A May 12, 1955, letter written by Allen to Mr. and Mrs. Chamberlain after the visit sets the tone.

The sister to whom Coach Allen refers is Barbara Chamberlain Lewis. She and Wilt were very close at home as Cal's letter seems to indicate. And, as those of us who knew Cal well also can comment, he was not just recruiting Wilt Chamberlain, the basketball player, when he visited the Chamberlain family. He was the consummate teacher/scholar and believed in education to the

---

May 12, 1955

Mr. and Mrs. William Chamberlain
401 North Salford Street
Philadelphia, Pennsylvania

Dear Mr. and Mrs. Chamberlain:

Dr. Cal Vanderwerf was in my office today and told me of the splendid visit he had with your fine family. He was greatly impressed by you good people and had nothing but praise for all members of your family.

In talking with Cal he asked the question whether or not Wilton's older sister had gone to college or desired to do so. He did not recall her name at the moment, but just raised the question of whether or not she had been to school or even desired to continue her schooling. He certainly thought she would make a fine student here at the University, so I thought I might pass this information along to you folks.

We have had some rains recently and the campus is awfully pretty at the present time. I certainly wish you could visit with us and see the beauty of our school. No doubt Wilton has told you how the school is situated and I am sure it greatly surprised you to learn that we are located above the city of Lawrence. Perhaps someday in the not too distant future you will be able to see for yourself what a fine campus we have.

Tell Wilton we certainly are wishing him a happy time at commencement. I know he will enjoy himself, for that is a high point in a young man's high school career.

I just wanted to write you this note to pass along the fine remarks by Dr. Cal concerning your family. He is one of the outstanding young professors in the United States, and such praise from him I know will mean a lot to you and your family.

With kindest personal regards, I am,

Sincerely,

Forrest C. Allen
Varsity Basketball Coach
Professor of Physical Education

FCA:es

---

depth of his being. He knew that Barbara Chamberlain could and would benefit from a K.U. experience as much, or more, than her brother. He was trying to make that happen.

The letter also reflects the respect held for Cal on the Kansas campus. Allen had a checkered history with regards to his own statements about race. But he respected Cal. Another indication is a copy of his autographed program for the dedication of Allen Fieldhouse, at K.U., also in 1955.

Allen Field House Dedication

University of Kansas

Vs

Kansas State College

March 1, 1955   Lawrence, Kansas

Official Souvenir Program
25c

The times, however, were unsettled. *Brown v. Board of Education*, Topeka, Kansas, had come down in May 1954 and this decision, which started the process of school desegregation in the south, had no little impact in Kansas. The schools in Lawrence were not segregated, but the community was. Blacks lived north of the Kaw River and there was a black school. The times are perhaps best set by a *TIME Magazine* article, titled "Wilt the Stilt" that appeared just after Chamberlain arrived in Lawrence.[119] I quote the article extensively below because it focuses on Allen, Kansas, and the setting into which Wilt was recruited. Wilt's sister Barbara told me that Wilt always liked the open spaces so she was not surprised by his choice of Lawrence, Kansas. But Wilt was just a young undergraduate when he enrolled at K.U. Though his impact on the game of basketball was to be legendary and he would change the sport forever, he was only nineteen years old. *Life Magazine's* article intimated this, and the *Saturday Evening Post* wrote "Can Basketball Survive Wilt Chamberlain?"

How could a nineteen-year old college freshman cope with all of that attention? Ironically neither Barbara nor her parents ever saw Wilt play at Allen Fieldhouse. Philadelphia was a long way from

---

[119] "Wilt the Stilt," *TIME Magazine*, December 12, 1955.

Lawrence, Kansas, and Lawrence, Kansas, a long way from Philadelphia. So when Wilt went to Kansas he was on his own. He was 7'2" tall and decidedly mature for his age; but he was still 19 years old and on his own making his own decisions.

He must have chosen the persons to whom he was to listen, his advisors, carefully. At K.U. that included Cal VanderWerf. One, I think, does not realize how important that must have been to him. And he always listened—according to Barbara—to his father. Barbara Chamberlain-Lewis shown at K.U. on a visit to the campus in February 2010, told me that "she saw Wilt play all during high school (which she still maintains was *the best*) since we were in the same grade. All of our family had the opportunity to see him play most of his high school games and when he was with the Philadelphia Warriors, 1959-62. I eloped to Seattle in 1960 so I missed the last two years the team was in Philadelphia and then moved to San Francisco and then to Oakland. Needless to say my mom and those of the family that lived in Los Angeles never missed a Laker game. My dad made his move to California in 1968 but was too sick to fly back to Philadelphia to see the Lakers win the Championship. But we were all really happy nonetheless that Dad was able to see Wilt win in Philadelphia."

Phog Allen, for his part, was doing his best to avoid a Kansas law that required state employees to retire at 70 so he could coach one more year and coach Wilt. Today, Phog could have coached as long as he wished given the relaxed retirement requirements now. Witness Joe Paterno. Phog had already received one age extension and was appealing for a second one. But in Phog's day, rules that actually date to German unifier Otto von Bismarck pertained, and the retirement age locked into most academic contracts, and university employ, was 65 years. But Phog wanted more. The Chamberlain visibility was hugely important to the making of his case.

But these were very different times in America. Look what was said about Phog at the time in one of the news magazines of the day:

"Forrest C. ("Phog") Allen, veteran basketball coach at the University of Kansas, turned 70 last month. As might be expected, he celebrated his birthday by watching a basketball game. It was quite a party. Phog saw his varsity soundly trounced, by the K.U. freshmen 81-7—and yet he was the happiest man in the jam-packed fieldhouse. Not that Phog likes to lose, but it was pure pleasure for him to watch the biggest freshman of them all, Wilton Chamberlain (7 ft. 2 in., 230 lbs.), dunk in 42 points all by himself. In 39 years of talking tall young men into coming to Kansas for their higher education, Phog Allen has never recruited a more promising student of basketball than 'Wilt the Stilt'."[120]

"The Philadelphia Negro is the main reason that Phog is still coaching. Kansas regents require that state college teachers retire at 70, but once Phog got his hands on the three-story Stilt, he wasted no time talking the regents into letting him stick to his job. 'I'm not going to miss the chance to coach this kid,' he said. 'He's the greatest basketball player alive today.'"

"While Wilt was still a student at Philadelphia's Overbrook High School, at least 140 different colleges shared Phog's high opinion of him. They offered Wilt

---

[120] Ibid.

180

the world—tuition, cars, free air travel home on weekends—but Phog outfoxed them all. After peddling Kansas' virtues to Wilt and his coach, he turned his charm on Wilt's mother ("Mrs. Chamberlain, now I see why Wilt is such a nice boy")."

"Phog then called in reinforcements, managed to enlist the help of 1) Negro concert singer Etta Motten, a Kansas alumna, who wrote to the Chamberlains, 2) Dowdal H. Davis, general manager of a Kansas City Negro weekly, who flew east to make his pitch, 3) Professor Calvin VanderWerf, of K.U.'s chemistry department, who passed through Philadelphia and called on Wilt's mother. Said Mrs. Chamberlain: "We've had many colleges speak to us about Wilton, but you're the first one who was a professor. I'm so happy to have someone talk about the academic side." By the end of May, Phog Allen and K.U. had won the Stilt sweepstakes. 'Wilton," he said, "I know you'll be happy here.'"

"When Wilt got to K.U. in the fall of 1955, Phog made sure that both he and Wilt stayed happy by working with the freshman phenomenon twice a week. One of the first things he did was to start Wilt reading Helen Keller's *The Story of My Life* 'to develop his sense of feel and touch.' Phog's current project: teaching Wilt finger manipulation, how to put English on the ball, how to spin it in from all angles when he is jammed in the bucket. The demand to see those big fingers in operation is so great that Phog has had to rearrange his schedule to put the freshmen on the program before each home varsity game. 'Everywhere I go,' says Phog, 'they ask me about Wilt the Stilt. I've seen them all: Joe Lapchick,* Clyde Lovelette, Hank Luisetti— all the top men, and this kid is the best I've ever seen. For 20 years I've used a twelve-foot basket in my gym; as far as I know, I'm the only coach who does it. Wilt can touch the rim of that basket on a jump. He can jump 24 inches off the floor. I've never seen a tall man in my life who could equal it. This kid actually slams the ball down into the basket. He uses two hands and just whams it right down in.'"

"Ever since basketball was first invaded by big men, Phog Allen has campaigned loudly to have that

181

twelve-foot basket of his made regulation; the regulation height is now ten feet. The big shooters, he has argued often, are killing the passing, the dribbling, the teamwork that makes basketball exciting. But now Phog has Wilt the Stilt. Says he with a quiet smile: 'Twelve-foot baskets? What are you talking about? I've developed amnesia.'"

Chamberlain, before he was 21, had been storied in *Life, Look,* and the *Saturday Evening Post* as well as *TIME*. One of the best discussions about the controversy in Wilt's time is a retrospective written by Aram Goudzousian in *Kansas History: A Journal of the Central Plains*.[121] Goudzousian describes the NCAA final against North Carolina in 1957 in which a white, 5'11" guard, Tommy Kearns, jumped center against Chamberlain who was variously reported as 7'1" or more. North Carolina won that game in what was the only defeat a Kansas team centered by Chamberlain suffered. Monte Johnson who did not play in the game, but was on the team and told me: "that was a boring game; North Carolina slowed the tempo in order to neutralize Wilt—and in the end—they managed to beat us. I didn't play, but neither did most of the K.U. team. Only seven players were used in the game."

Bill Hines, later Dean of the Law School at Iowa, reported a game in which he played as an undergraduate for Baker University, a small four-year school in Baldwin City, Kansas, against Chamberlain. For Monte Johnson that "big Chamberlain" was just fine; for Hines "big Chamberlain" was hugely imposing and impossible. "We didn't belong on the same floor."

Chamberlain's contacts with the businesses in Lawrence is a matter of discussion and debate. According to the VanderWerf children, and others, Margery Argersinger bristles when its claimed that Wilt integrated the theaters in Lawrence. "Not so," she says. Almost ninety, she tells the story of a group of L.L.P.D. members entering and being seated in a theater in Lawrence wherein a white couple, would sit with a black theater patron between them. "That was way before Wilt Chamberlain—perhaps even in the late 1940s."

Regardless, the story with the restaurants was entirely different. Many restaurants did not allow blacks to eat in them. According to the VanderWerfs, "Dad would talk about how hard he

[121] Aram Goudzousian, "Could Basketball Survive Chamberlain?" *Kansas History: A Journal of the Central Plains*, Autumn, 2005, pp 150-173.

and Mom and other families worked to turn things around, with limited success. And then Wilt Chamberlain came to town. Restaurant owners would practically stand out on the sidewalk and say, 'Wilt, come on in and have some pizza—it's on us.' Overnight things turned around because of Wilt."

Wilt stood on the shoulders of some real heroes. One was Wesley Elliott, the aforementioned black student, who took his Ph.D. in organic chemistry with Cal VanderWerf, whom Wesley considered "his role model." Wesley was born in Newton, Kansas, where his father worked for the Sante Fe Railroad. Like many aspiring chemists, Wesley had a chemistry set and set up a laboratory in the basement of his parent's home causing both an occasional smell and explosion. Encouraged by his mother, he entered the University of Kansas at the height of World War II (1943). There he pursued his interest in chemistry and graduated after four years with a degree. During his undergraduate, M.S. and Ph.D. studies, he was a close mentee of Cal VanderWerf.

Wesley Elliott was also an athlete and though he participated regularly in track meets with teams from universities where segregation was not the rule, he was not allowed to become a member of the K.U. track team because to do so would mean that the team could not enter meets with schools in which segregation was practiced, the most significant of which were those of the Kansas relays. Students and others protested the K.U./Big Six conference rules, and eventually the rules were changed, but this was not in time for Wesley and his colleagues. Though they opened the door for others, they themselves never did get the chance. Housing was also an issue but students living off campus formed cooperative rental groups, and two white students invited Wesley Elliott to join their housing cooperative.

Following his Master's degree, Wes took a teaching position at Southern University in Baton Rouge only to blow up his lab in an unfortunate accident. Though no one was hurt, he got fired, so he went back to Kansas to finish his Ph.D. degree. Wesley Elliott's graduate work started in Cal VanderWerf's laboratory but eventually shifted to Bill McEwen's laboratory where he worked on Reissert compounds. In fact, his work was partially supported by a grant from the U.S. Army awarded to Professor VanderWerf so that he finished his last year as a research assistant. His first teaching job was at Florida A.&M. University in Tallahassee. By this time, civil rights protests had begun in earnest in the South and he and his wife were involved in these. Eventually he was awarded an N.S.F. Fellowship to spend a year at Harvard, which he did. Following this he was appointed at Fisk where he spent the rest of his career.

Many others came to K.U. from Negro schools to seek advanced degrees. Subsequently, at least six students from Negro colleges in the south enrolled in the Ph.D. program in chemistry at K.U. Ivory Nelson, who served as a university president at three different schools for 26 years came north from Grambling in Louisiana, was awarded a Phi Beta Kappa key, and following a Ph.D. in analytical chemistry embarked on a hugely successful career in the academy.

Marion S. Barry, Jr. on the other hand, joined K.U. chemistry from Fisk University where he had been a student of Sam Massey's only to have difficulty with the program. Barry then joined the aforementioned James Farmer of C.O.R.E. and, in 1971 when the District of Columbia was given independent authority to form its own government, was elected to the school board, then served as mayor and finally served, and still remains, a member of the Washington District Council.

As I have studied Cal and Rachel VanderWerf's too-short lives, first in Lawrence and later at Hope College, I have been touched by how many lives they enabled, how many persons they interacted with, and how much good they managed. In race relations alone in Lawrence, their first home, they balanced raising six children, with the gracious help of Mrs. Wilburn who the entire Hope College community knew, with simple charm. Cal's presidency at Hope College changed the place in transformative ways beyond expectation, and he suffered of health and body because of it. But he left Hope better academically than he found it.

Lawrence was a small town when they moved there in 1942. Cal and Rachel thought differently and made a difference. The Wilt Chamberlain story is well documented and, though I could find no pictures of Cal with Wilt (or, for that matter, with Curtis McClinton), they both benefitted enormously from knowing him. They made a difference with their family, and they made a difference with faculty at their school and they made a difference with most everyone they touched. That each of us could leave what did.

Cal and Rachel VanderWerf were both committed to the ideals of that free country where one could live the American dream. They believed, with all their being, in the value of education and Cal, particularly, understood the academy in its deepest, and most

significant sense. Their commitment to those who had been oppressed, through no fault of their own, was simply who they, Cal and Rachel VanderWerf, were.

Cal took no prisoners when it came to educational standards, though, and much of what I knew of them came because they only expected the very best from everything they touched. They had no patience with periphery if it interfered with 'that very best' and, in some respects, Rachel was harder nosed about such things than Cal.

The VanderWerfs with Mrs. Wilburn

Chapter 11

The VanderWerfs Move to Holland

Cal and Rachel moved their family to Holland in July, 1963. By then, the VanderWerf family had grown to include Cal and Rachel, five daughters, one son, their house keeper and everyone's friend, Mrs. Wilburn, and one dog. Cal was to become the eighth President of Hope College, his alma mater with an inauguration being planned for November 9, 1963 by former President Irwin Lubbers, Vice President John Hollenbach, and Ekdal Buys Chair of the Board of Trustees. Like most Hope-ites, Cal was loyal to the school; perhaps less to the Church of which he'd not been a member since childhood. Though I doubt he had any idea of what he had gotten himself into, he would not only persevere in the job, but become the transformative president who was needed for Hope College as it grew from Depression-era and World War II-era to its survival and thriving. His tenure turned Hope academically to the great undergraduate institution of its future. However, the position cost him physically, emotionally and professionally. Van Zyl's chemistry department had gotten to close to the fires of the western Michigan church, and it was to be pushed back in a storm. But not before the transformative winds of change the school required to survive in the era of Sputnik had blown through most every department.

The VanderWerf Family: this iconic picture of the Vander Werf family appeared in all Hope publicity at the time of Cal's appointment as president.

Cal and his family left Lawrence in July 1963. As was often the case, when they got to Holland, Cal immediately left for professional travel and Rachel was left with getting the household settled. Irwin Lubbers who had served eighteen years, removed himself from the College during the spring of 1963 and John Hollenbach served as the Chief Administrative officer pending the arrival of a new president. Rachel VanderWerf said that she wanted Cal to take the job and pushed it. As he was to say in the letter to

Hugh DePree, Chair of the Board of Trustees, dated January 6, 1968, he gave up his career to lead Hope College. Cal had numerous other offers at the time that we know about now – at least two from DuPont. Ohio State had tried to hire him on several occasions as well. Some would say that he made a huge personal mistake in taking the position. However, his decision for Hope's sake was absolutely critical.

Before the VanderWerfs moved into the President's house there were discussions between Rachel and Henry Steffens, Hope's treasurer, about the location of the house. Remodeling needed to be done. Six children and a dog and two parents meant changes were necessary to protect the kids from the immediate neighbors. These were planned for before the VanderWerfs arrived. Rachel was concerned with privacy for the children living in a fish bowl on campus. The President's house was and is in the middle of the campus, and even then, it was not the best place for a group of young, active children to grow up. The Pine Grove, next to it, was a notorious place said to be used occasionally for notorious things. Nevertheless, arrive they did. The Hope campus was terribly dull in the summer, and even more so on weekends. When it was warm, any Hollander with sense would see the nearby lakes more interesting. But once the moving van was unloaded, and Sunday had come, the girls and Pieter were told by their Mother to get their rooms in as much order as they could and "then they'd go for ice cream." Dutifully, the children did as they were told until about 5 o'clock at which point they marched past the Pillar Church down College Ave. to Mills Ice Cream shop only to find that it closed! It was, after all, a Sunday! Gretchen said her Mother looked sheepishly around to make sure no one had noticed, marched the kids to the East side of the street, and took them back to their new house.

The administrative decisions Cal faced immediately upon arrival probably convinced him he should have listened to his own instincts, rather than Rachel's. The Dean of Women, Emma Reeverts, had retired and a search was on for another. One candidate was from the Board of Missions of the Reformed Church; another from Luther College in Iowa and said to be "colorless" by those who interviewed her. Still a third was from Carnegie Institute of Technology but not a serious candidate it turns out. Isla Van Eeenanam was selected, a home grown product, who had lost her husband. As it turned out, this was a superb, though non-traditional, choice.

Dancing on campus was a major issue for the then administration. Students could dance off-campus, and did. But the prior leadership had apparently made a decision that favored dancing by students in college organizations on campus under supervision. But the churches in the general Holland area had risen up against the measure and there was, according to a letter from John Hollenbach to Cal in July, a rather serious donnybrook in the offing. John was not sure "whether the churches were organizing the revolt or not" but it was there nonetheless. Cal inherited it immediately. The church/college issue was never far away though, and he had to get used to a situation he neither wanted nor cared much for.

What these issues speak to was what the College was when Cal moved there in 1963. In seven short years, and lots of sweat and plenty of tears, his presidency saved Hope College from total academic collapse in the 1970s and beyond. Students went from being worrying about social trivia, like dancing on campus, to the compulsory chapel, the war in Vietnam, Civil Rights, and racial equality, and the draft – who should go, and who should not go. From dancing to death by caused by one's nation, and its elders.

VanderWerf Inauguration

Higher education went through convulsive transitions immediately following World War II. Students flocked to colleges and universities because a generous G.I. Bill made it possible for them to do so. Society, too, was also transitioning from a rural society to an urban one. Though one could work in factories with a high school education, many persons wanted more than that so they wanted to go to college. Teachers were assembled from most anywhere during those immediate postwar year classes mainly because the size of classes in universities demanded it.

After this immediate boom was over, the academic community, and society as a whole, took stock. Alan Cartter wrote that there was a significant pessimism about the value of a college education in the 1950s.[122] Some thought the quality of college teaching had declined due to the gross expansions of faculty to accommodate the boom in enrollments. There was the McCarthy driven loyalty tests to root out the communists that faculty, particularly in the humanities and social sciences, found onerous. Also public universities in cities went through financial upheavals and universities in Buffalo, Cincinnati, Louisville, Omaha, Toledo

---

[122] Allan Cartter, "Retrospective View," *Ph.D.'s and the Academic Labor Market* (New York: MacGraw-Hill, 1976).

188

and Kansas City, to name just a few, sought absorption into state systems.

Cynicism about the value of a college degree abated somewhat in the late 1950s with the Sputnik challenge. Coupled with President Eisenhower's plan for more students, an increase in enrollments in the fields of science and engineering followed. By the time President Kennedy was inaugurated and the space race was begun, the nuclear arms race was in full swing with the nation challenging the Soviets as their boats approached the nearby island of Cuba in 1961. The National Science Foundation, founded 1950, was also starting to exercise one of its legislatively driven initiatives-to provide scholarships (and to improve the workforce) for the highly qualified to enter fields in science and technology. It was also increasing funding for graduate stipends and research. Those universities around the country sporting Ph. D. programs in the sciences were dipping their hands in this seemingly trough of infinite financial deepness with abandon.

Cartter's reports from the period were widely read, and deal with expansion of graduate school populations to address growing college enrollments by producing more Ph. D. graduates who could enter the fields of college teaching. The Cartter reports mostly address college teaching in the humanities, social sciences, education and the like because these fields were more effected by enrollment trends. The sciences were generally immune to general demographic trends because the Ph. D. degree had become the job entry card for the industrial laboratory, and getting to it – though it may have been an objective – was exceedingly difficult. As the State of Ohio's master plan for graduate education (1966) also remarks, industrial growth depended on graduate degrees in the sciences – not directly-but indirectly in that new businesses were proposed that might develop as a result of graduate studies in the sciences.

All of this collegiate growth had a significant impact on the individual states, and on the state universities in those states. The morale of the profession, teaching at university, had never been that good. California's university system expanded to include Santa Cruz, Santa Barbara, La Jolla, and Irvine among others. Late in the period, the State of New York, which had left higher education to the private sector, started the system known as the State University of New York (S.U.N.Y.). By that time, however, the expansion in graduate educations of the late 1950s and early 1960s had managed to fill the teaching ranks, and the job market in most professions, and even in chemistry, tightened. When one found a teaching job, events local and national had a way of damping one's enthusiasms for it. Students were mostly restless and that was seriously distracting. However, in the early 1960s, when I found my first teaching job at Hope College, I felt strongly that college teaching was an excellent way to earn a living. I respected persons in the field, specifically those that I knew teaching chemistry.

Hope's main idiosyncrasy was that the western Michigan Reformed church, which dominated Hope College, was rigid, reflecting the rural Dutch church traditions of a very small group of the uneducated at the time of the exodus (1840-1870). There were famines in Holland almost continually during those times, and the average life expectancy was less than thirty-five years. Dutch families in the old country would birth eight babies, and typically, only four would grow into adulthood. Almost none of the immigrants that moved in the 1840s and 1850s owned land in the Netherlands. Before Napoleon, many family names (specifically those with last names of two or three parts, Van der Werf, Te Winkle, Ter Meer, Van Dokkumburg, Van Lente, Van Eeenanam, and even some that were one word) were invented when young men were registered for the draft for Napoleon's armies.

For sure, there were no aristocrats among the immigrants. Most, if not all, were farmers, weavers, and other day laborers. The preachers needed to instruct them in the word of God and help them see the sins of their ways if they strayed socially. If one wishes to see the comparison between the elite and the peasants from a real Dutch perspective, read about Vincent Van Gogh's early years and, or visit Groot-Zundert where Van Gogh was born and, perhaps visit the village museum. There, Vincent's life story is compared with another, also born in Groot-Zundert on March 31, 1853. Vincent's father was a Dutch Reformed Church preacher; the venerated profession even of my childhood. And, for those that do not understand what Holland, Michigan became, one needs only read stories about Vincent's family home.

It was a Calvinist, church-dominated, domine-led social structure, and this reestablished itself in communities in western Michigan in the United States following the Dutch immigration of the mid-19th century. In many respects, the social mores thereby inculcated remained in Holland in 1963. There was open racial prejudice. When housekeeper Mrs. Wilburn moved with Cal, Rachel and the children in 1963, that undercurrent surfaced (or resurfaced). One of the more unsavory instances of prejudice occurred in Holland on Cal's appointment was the appearance of an anonymous letter posted on door fronts of businesses in downtown Holland. The letter was authored by someone, unidentified, at Holland Color and Chemical (then located on the north side of the River where Pfizer eventually was located). The author shares that a general of the Air Force had visited their plants on October 11, 1964 and told them that they would have to begin employing blacks if they were to continue getting federal funding. The letter points out that President VanderWerf at Hope College and Professor Earl Hall (sociology) were working to integrate Holland. These were idealists—"in fact, President VanderWerf came to Holland last summer and brought a 'colored maid' with his family."

Churches regularly held two Sunday services and many weekly events were common. Television was just coming on the scene, though, and Sunday evening programming – Ed Sullivan, George Burns and Gracie Allen, Groucho Marx, "What's My Line?" began to compete for the citizen's attention. Businesses in the communities were mostly closed on Sunday. Liquor "blue laws" were enforced, and though one could buy alcohol in Holland in 1964, none of the restaurants served drinks. There was a beer parlor on 8th Street, "Skiles," that became very popular later as the community changed. But in 1963, the college girls were locked into dormitory rooms by 10 pm on weeknights, including Sunday, and allowed to be out until midnight on Friday and Saturday night. There was a rigid check-in procedure, mostly administrated by students paid for their services and violators, or those who arrived late, could be rather severely punished.

The Dutch churches in the United States, and more particularly those in western Michigan, had an unhealthy pastoral underground too. In the first place, the pastors of most of the Dutch Reformed churches in Holland, and western Michigan, had grown up together. All the clergy had studied at Western Seminary in Holland, right across the street from the College or if they had not, they had at least been undergraduates at either Hope College or Central College, the other Dutch Reformed church college of that day. The American Dutch communities were small remaining resistant to changes from the outside, too. Collections of preachers from nearby communities would meet together regularly in formal events, and irregularly in social situations. Each pastor tried to out-do the other in surface piety. The congregations were made up of serious folks, and they were generous in their support. The Reformed Church was among the most benevolent of all Protestant denominations at that time.

191

This was like a cult, actually, and difficult for one, even of Dutch heritage but not from western Michigan, to break into. Cal was raised in Holland, at least in part, but he had the added deficiencies: He grew up poor, worked his way through College, was smarter than everybody else in his class, and he was genuinely concerned about his fellowman. He believed in equality, and not just for the races but also for the sexes (after all he had five daughters). One can see how his religious philosophies might be perceived as running counter to a rural, farmer's church given to counting external evidences of piety, as opposed to Christian actions.

This preamble sets the scene for what Cal stepped into in 1963. He faced a faculty set in their ways and comfortable because they were at Hope, in the vicinity of their home churches, and their extended families. Most, perhaps almost all, were not committed to scholarship in any competitive sense. Thus, the value of the education Hope College students received under their instruction was less than it might have been. Some were Doctors of Philosophy, a degree defined in the current sense in Germany at the Humboldt University in Berlin in the 18th century. Others, as scholars in more or less correspondence, non-resident Ph. D. programs, had completed courses of graduate study and written defensible Ph. D. dissertations. Almost none of the faculty Cal inherited were vigorous or active in their fields. The research drive and expectation to which a Ph. D. degree is targeted and intended had mostly been abandoned by them. They struggled to make ends meet teaching at a College with low salaries, poor or non-existent fringe benefits relative to other schools, and a mission that blurred the line between the values of a secular, academic instruction, with that of transferring, to the young, religious dogma.

Cal was a professor/scholar in the finest sense of the word. He understood the university and could translate the responsibility of the professoriate into simple words that everyone could understand. He was a well-known research scientist/educator. He had been President of the Division of Chemical Education in 1950 and his classroom teaching was universally recognized for its excellence. He was among K.U.'s first winners of the HOPE award: **H**onoring **O**utstanding **P**rofessor **E**ducators. Even though he was teaching organic chemistry to 100 or more students in a class, he would take a photo on the first day and at the second session call each student by name. Students selected H.O.P.E. winners, and Cal was extraordinarily popular with undergraduate students.

192

The University of Kansas is the flagship school of a state settled by tough pioneers in the 19th century. Citizens of Kansas take great pride in K.U., and the faculty take pride in their teaching, the students, and their chosen areas of specialization. It was then a school that represented most of what is good about American state universities, and it was from this background, the excellent state university that severely underpaid its faculty, that Cal moved to Hope.

What Cal found at Hope College in 1963 was a college on the brink. Persons were paid well below national average salaries for nine month's work. There was no contributory retirement plan. The plan that did exist consisted of a gentleman's agreement that if one were to "teach at Hope for 25 years, upon retirement, the College would pay one-third the last salary for the rest of that teacher's life." Tenure, the great perquisite of deemed successful collegiate professors elsewhere, was largely unheard of at Hope. There were no tenure decisions by department or collegiate committees nor approvals at the Board of Trustees level and new faculty did not have the onus of the tenure decision hanging over their heads. In principle, a faculty member of thirty years could be terminated on short notice. As academic expectations increased in the immediate future, that was particularly unsettling to some, with just Master's degrees, who had spent their lives at the institution. Hope's enrollment was also declining in the face of competition from state universities, and this was even though there was huge growth in the student pool. Higher education, too, was just starting to be of consequence for American young people, so the time was really ripe for all colleges to grow. Hope had no endowment of consequence at the College, and fund raising for the College was done on *ad hoc* basis because there was no Office of Development. Tuition accounted for the largest percentage of college operating funds though it was kept low presumably as an accommodation to young people from the Reformed Church. The lay Board of Trustees numbered around fifty individuals, chosen almost entirely from members of the various Classes and the Synod of the Reformed Church. Few were business people: most were clergy. Hope's management profile in 1963 reflected a different time in American academic history. The Church had strong institutional control but contributed little to the funding of College operations.

Operating budgets were small too. College departments, even in the sciences, had little instructional equipment of consequence, and no research equipment. Some, like chemistry, were

193

even more poorly accommodated for simple operational things like fume hoods, in a building constructed right after the Depression with local expertise. There was no controlled chemical store; student attendants could, and did, take chemicals from the store for their own uses – nefarious and otherwise. Those participating in the College theater, located on the fourth floor of the science building above the laboratories of the chemistry department and biology department, would turn off the fume hoods to the chemistry labs beneath when practicing their lines because the fans for the hoods were on the fourth floor of the science building and almost directly adjacent to the theater's stage. The mostly-donated science library collection was small, and none of the faculty members had research grants that paid summer salaries. A newly-constructed physics and mathematics building was too small on the day it opened, and had only minor laboratory space for contemporary physics experiments anyway. The College had borrowed the money for construction for the physics building from Old Kent Bank so, in addition to tuition covering most of the cost of instruction in those days, it also had to carry the debt load at 5% for the interest on the capital costs of this new facility. The admissions office was just that—it admitted students as they might happen to apply. There was little active recruiting of students save for the occasional choir or orchestra trip to the various parts of the then-Reformed Church. Even when active recruiting started in 1963-1964, recruiters faced an uphill fight—no history, few entries to regional high schools, and competition from state universities.

Hope was a college that "made do" with little, and very proud that it could and did. But as the 1960s were unfolding, making do was no longer enough and loyalty to the institution, town, and Church, though important, was not the only thing that counted. Professional credentials were to be increasingly important for the faculty members, even in liberal arts colleges.

Still, in the faculty, the ministerial class was predictably present. D. Ivan Dykstra held forth in a freshman seminar Philosophy 113. Dykstra had a B.Th. from Western Seminary and a Th. D. from Yale. The professors in Bible, later called Religion, offered the required courses called Old Testament, New Testament, a course called senior Bible, and even a course called non-Christian religions. To the extent the faculty in these departments published, it was only in *The Church Herald* – the Reformed Church house organ. Reformed Churches were encouraged by their ministers to subscribe in member plans for everyone, as a congregation, to the publication. And most did. So papers published in it were guaranteed an audience though the papers were not refereed in any sense of the word. *The Church Herald* was a journal in search of papers and Hope's religion faculty did their job.

Loyalty goes a long way, but it is not everything. Early in his tenure, Cal collapsed from overwork. He found himself trying to bring a college that survived the Depression and World War II to the point of competing in the post-*Sputnik* era and the job was almost beyond doing. On many occasions, he must have been convinced it could not be done. And he had no way of knowing that almost immediately after he was inaugurated as President of Hope College, the United States would encounter one of the most heart wrenching events in its history—the assassination of the young President John Fitzgerald Kennedy. But persevere Cal did and he became the transformative president that Hope College, at that time, dearly needed. Cal took Hope from a mentality of the Depression and World War II to a modern, academically-able, undergraduate liberal arts college.

Chapter 12

Cal VanderWerf's Return to Hope

VanderWerf Inauguration

Cal's World by 1963

Cal VanderWerf lived in Holland, Michigan as a boy, and studied at structured, religious Hope College. He left in 1937 to enroll in the Ph. D. program in chemistry at Ohio State, where, among the first organizations the young VanderWerf joined was Hillel, the campus organization for religious Jews.

Cal moved to Kansas as an instructor in 1941, where he continued his passion for others, in particular those with lessor opportunity. Curtis McClinton, the only black on the Kansas football team when I was a student in 1960-62, said "Dr. VanderWerf got him to the point of being able to get C's on tests in freshman chemistry; He tutored me. And when it became clear to me, that neither chemistry nor medicine, to which I targeted myself, was realistic, he helped me find another major." Curtis played five years with the Kansas City Chiefs and scored the first touchdown for an AFL team in the first Super Bowl in January 1965.

Cal's daughter, Gretchen, said her father wanted very badly for minority athletes to succeed, for fear of their being exploited. This exploitation was endemic in professional athletics at the time, as the picture below of a group led by Cleveland Browns star Jim Brown makes clear.

Curtis McClinton (center rear – behind Jim Brown) with other black athletes addressing the exploitation question.

Cal VanderWerf made the decision to go to Hope as its president because, according to his wife, "He held Hope in such affection because of all that it had done for him when he was unable, for personal reasons, to even afford a college education. He wanted to pay that back." As Vic Heasley would say, affectionately about Hope when he, Vic, was quite old, "There was a certain Hope ethos." Cal had that ethos and he let that ethos make a decision for him at mid-career.

The Vander Werf inauguration on November 8, 1963 was a big event of the College. John Hollenback was in charge of the arrangements. The reports in the *Holland Evening Sentinel* and the Hope College *Anchor* suggest it was a very nice event (Cal's inaugural address is retyped and reprinted in the appendix). In it, he identified a number of clairvoyant insights about higher education that, when put to practice, brought the College quickly up to speed as an academic entity. It was his minor remark on nerve gasses in this address that suggested to me that he knew a lot more about this than he ever let on. It also brought into context work he had done in his laboratories at Kansas on the resolution of tetra substituted phosphonium salts.

Hope College in 1963 – VanderWerf Meets New Reality

The Hope College of 1963 was severely limited according to the report of the North Central Accrediting committee, dated April 26, 1964. Most of the faculty were oblivious to the demands of the academy as were most of the administrators. In the report, the Director of Admissions focused almost entirely on dealing with policy and applications (presumably from students wishing to enroll).

197

The admissions staff included one assistant who had partial duties as a recruiter, along with serving as the church constituency liaison. There was no sense of recruiting, or seeking, student applications. The business manager in 1964 found the facilities adequate and in "first class condition." A building and grounds staff of hardworking old men did an amazing amount on a low budget, while the "food service worked cooperatively with a reasonable amount of satisfaction among the students." The treasurer found that the Church had supported the College, very well by contributing in excess of $125,000 for operations and this is "so predictable that if capitalized in the form of endowments it would, in effect, resemble amounts in excess of $2,500,000 at present (interest) rates." Vigorous appeals to alumni "caused an increase in their giving to the college to nearly $100,000." Businessmen were approached with increasing intensity, contributing $45,000 in the year. John Hollenbach, the vice president for academic affairs recognized the serious shortcomings and limitations with Hope and its preparation for the future but did not have any remedies.

Worrisome was that the shortfall in endowment "led to a drop-off where faculty research was concerned": the college supported research work at this time from its own funds. The ratio between students and teachers was reaching record levels, and students were suffering from a lack of individual attention. The library was service minded though the building was new in 1961. But the catalog of the collection was not complete and the book collection was far below the standard set by the American Library Association for undergraduate college libraries. The chemistry library in the science building had some good journals, but not enough of them.

One of the highlights for Hope College then, as now, was its successful Vienna Summer School, an in-residence program in Vienna for interested juniors and seniors. The program was among the first international studies programs in the country, and an excellent bell-weather for things to come. It was also Hope's first, creative step, recognizing the rebuilding of Europe after the devastation of World War II and the Holocaust. The incentive and driving force for this was himself a victim of the Holocaust, and alumnus of the Nuremberg trials, Paul J. Fried. Esther Snow, a widowed professor of German language, was a strong supporter of the Vienna program, and counselor for the program when it started.

VanderWerf Inauguration

Mrs. Esther Snow is in the front row. She, along with others of the faculty of long standing, found this new young guy a very different president than President Lubbers.

In his report on Curriculum and Staff, Dean of the Faculty William Vander Lugt described the realities of the Institution at the time – the faculty were non-engaged and the chemistry department was responsible for the reputation of the school in the academic world. "It would be false, however, to conclude that all departments have followed its example. The framework of the institution contributed to the [success of the chemistry department] so credit should both be given to the institution and the department."

The remainder of the Dean's report dealt with administrative policies with regard to curriculum though, in the end, he points out that the department chairs as a group, "were only partially attentive to the good counseling of their students."

Vander Lugt and John Hollenbach were the only two to recognize some short comings with Hope at that time. The rest, including all the chairs, were almost entirely self-satisfied, and complacently so.

The majority of the Central report was left to short reports by the individual department heads. These reports range from sensible, as in chemistry expects that "each faculty will engage in research with the students, and to seek funding from the Petroleum Research Fund, Research Corporation, or National Science Foundation"; to nonsensical, as in the Biology chair's "overall, the number of majors is increasing, faculty morale is very good, our pre-medical and pre-graduate school programs are improving, and overall this department is in jolly good shape." This made no mention of student or faculty research. The report from the Department of Religion and Bible notes that most of its students were required to take its courses by the overall curriculum requirements of the college. There was no mention of original studies, research, or creative activities which, in the end, probably did more harm to the church that most anything else. The Philosophy chair wrote more words, mostly flowery, than most. But the overall impression left was that the department existed to prepare the liberally educated person by

199

providing instruction to that student in his/her freshman year. It served as a service department for student majors in other departments. Its three-person staff did not help it much because it was seriously understaffed for the teaching responsibility, mostly self-imposed. Some programs were espousing nonsense – as in the department of education's thoughts about a five-year undergraduate program for teacher preparation.

The Response and Committee Evaluation

North Central Association sent two liberal arts scholars to do the evaluation of the report and of Hope College: Robert T. Blackburn, Dean of the Faculty at Shirmer College – University of Chicago, and Chandler W. Rowe, Dean of the College at Lawrence College. Both schools were traditional liberal arts colleges; places where scholarship was expected, but there was a vigorous focus on lifelong learning as well. Shirmer College was founded near Chicago as a replica of the University of Chicago and early in its history adopted Robert Maynard Hutchins' Great Books program as a curricular guide. Blackburn wrote distinguished books about university faculty late in his career, one of which was *Faculty at Work – Motivation, Expectation and Satisfaction*. Lawrence College (now University) in Appleton, Wisconsin, had a distinguished reputation, a strong music conservatory, and a working relationship with the Institute of Paper Chemistry also located in Appleton. Rowe was an anthropologist, and had written extensively about Indian burial grounds in the Milwaukee area.

These reviewers were as knowledgeable as they were tough. Regardless of the College's non-critical thoughts of itself, and the committee's doffing of its cap to its history, "with a college whose path is patently clear: a high quality institution is moving steadily forward."

This section highlighted the belief that the college had produced a "high quality liberal arts education in a Christian atmosphere has been and will continue to be Hope's guiding light." Parenthetically, this motivation makes no mention of pre-professional education which was, as much as anything, the goal of the chemistry program. Though it offered only an A. B. degree, the numbers of hours of chemistry taken by its graduates paralleled the number of hours taken by most Bachelor of Science graduates.

The administration had issues: the lines separating different offices and functions was not well-drawn, and communication was always an issue. In addition, there was a duplication of duties

200

between the Dean, the Vice President of Academic Affairs, and the Assistant to the President.

The assessment of the instructional program was basic: "The College's instructional program is traditional," "departments vary considerably in size/quality," and the "faculty teach normal loads for the typical undergraduate college." Attention was paid in the report to the low salary of the faculty, and the College was in danger of becoming non-competitive if nothing was done to raise the salary. As always, the lack of an adequate retirement plan was an issue.

While remarking that the class of 1964 was honored with seven Wilson Fellows, two Danforth Fellows, and one George C. Marshall scholar, the assessors noted that the student body at Hope was remarkably homogenous: "More than 60% of the students are members of the Reformed Church, fairly conservative, young people who know why they are there and what they will do with their education when it is finished." While the school was to be commended for its good student body, there were "shadows on the horizon" for a negative downturn unless action was taken.

The assessors then listed the eight internal and external forces that threaten Hope's dreams, the first in terms of finances:

1. The cost [of higher education] is increasing markedly.
2. Faculty salaries are increasingly non-competitive [at Hope]; this particularly impacts new hiring.
3. Beginning in the fall, with the debt service for the physics/mathematics building coming on, every $100 in additional income will go to service that debt.
4. None of the tuition dollars from a (presumed) tuition increase will end up increasing the value of the College.
5. No increase from the Church can be expected. The denomination is not growing, nor is it a particularly wealthy Church; besides supporting Hope College it must also support Central College and Northwestern Jr. College, both in Iowa. Realistically Hope cannot expect much increase from its principle donor.
6. Hope's alumni are not wealthy. And there has been no significant courting of that base.
7. New faculty will cost more than retiring faculty in several ways.
8. The next generation of Hope students will be very different. Society's becoming more mobile; old loyalties are changing; the Reformed Church has a limited demographic; alumni

will join other churches; the net effect, "though we're not sociologists" is that the alumni's loyalties will become decidedly divided.

The problems in demographics facing the College:

1. Hope's largest graduating class was leaving in June 1964 and the replacement classes were not big, and had at the time, not made their deposits to retain their spot. This was seen as a negative portent, especially on a school operating at its highest attendance, a figure it could not retain or support.

2. An increase in tuition would result in less retention and there was an increase in tuition planned (to pay the debt on the physics/mathematics building).

3. The majority (80%) of the operating funds came from tuition. A drop in the freshman class could result in a situation that would approach a crisis. A physical plant, faculty, and staff had been budgeted for the current size.

4. Few Hope faculty had terminal degrees in their fields. Many had Master's degrees, and those with doctorates had generalized education doctorates.

5. Hope College's admissions staff was not recruiting outside of Holland very well, and when it did, only went after places where the Reformed Church was strong. The Church did not have a great deal of wealthy families, and increasingly those students would be better served by attending the growing state universities. The niche that Hope College could offer would be replaced by hard economic decisions.

6. A fifth year requirement for public school teaching?

7. Hope needed to devote resources into its finances to identify ways to save money (a paradox, to be sure), and examine other instructional options.

8. "Tomorrow's Hope is, in part, present today. A distinguished alumnus, Calvin VanderWerf, has acutely addressed Hope's situation; read the dire predictions; knows Hope is a quality institution; but knows most importantly Hope cannot maintain that if it stands still fighting a rear guard action."

Continuing, the Report stated: "(VanderWerf) knows that if Hope is to continue to deserve its high reputation, it must be a distinctive and unique college. President VanderWerf has carefully and forcefully presented his analysis to the Board of Trustees and this unsettled many of them." The Board was noted for its conservatism, and its opposition to radical ideas to reform the school to meet present needs.

Hope College would need to reflect Cal's vision, both to stay open and to have a future. In turn, hiring new faculty would require finding those who shared Cal's vision for Hope College, and instilling in students and faculty alike the idea of a shared academic community fit for the 1960s. No longer could the school rely on the Reformed Church for their students: it was at once a dangerous venture, when students were looking elsewhere, at more affordable schools, and a limiting idea, as Hope could benefit from the infusion of different-minded students. The central question: Could Hope College retain its Christian heritage and charge while becoming a competitive contemporary college?

To be competitive, Hope College needed to appoint someone who could be charged with obtaining grants and someone who could continue development of educational programs. These appointees could reach out to companies, whole industries, and foundations (not limited in scope by religious affiliation) for support.

The examiner ended by stating that they did not have all of the answers, and merely suggestions for ways Hope could begin to improve. They closed with a quote from Antoine Lavoisier: "Thus while I thought myself employed only in forming a Nomenclature, and while I proposed to myself nothing more than the improvement of a chemical language, my work transformed itself by degrees without my being able to prevent it into a treatise upon the elements of chemistry."

"Lavoisier laid the foundations of chemistry. The results of President VanderWerf's labors could well be a new contribution to higher education."

The Report

The North Central Report emerged as Cal's first year as president was ending. In the fall that year, seventeen new faculty arrived, including three in chemistry. That they subscribed to the expectations outlined by the North Central examining committee is demonstrated by their records. By the end of the decade, three faculty members in chemistry were winners of the most prestigious awards in their field, the Camille and Henry Dreyfus Award for teaching and scholarship, and the prestigious Alfred P. Sloan Fellowship. Within the decade, Hope College had made enormous strides to become a better and more competitive school, mainly because of the leadership and direction that Cal VanderWerf provided.

The trail

To anyone alive in the 1960s and old enough to know what was going on, the tensions of the decade were difficult to wade through and even more difficult to properly confront today. It took a human being of extraordinary faith and dedication on an ideal to survive in a leadership position during these years. John Kennedy had been assassinated; Lyndon Johnson was elected President and had his domestic agenda hijacked by international troubles; the war in Vietnam progressed, unadvisedly, killing and maiming American youth and the Vietnamese; Civil Rights protests were breaking out across the nation, leading to the assassination of Martin Luther King, Jr.; students were seeing the draft reinstated and hated it while older patriots were telling them "do your duty" in a war they detested. A generation gap that had always existed became unnavigable.

Hope had its unique problems, most of its own making. As the North Central Report said, Hope had been mostly passed by and was undertaking measures to improve and catch up with colleges around them. Compulsory chapel was soon to go but before it did, students would be expelled. Female students increasingly were demanding more of a voice on campus, along with increased responsibilities and freedom for their choices.

Money remained an issue. The Student Center was a dream, but not yet funded so that its construction could not be started. The Physics/Mathematics building was sufficiently underfunded so that every new dime in student tuition was needed to pay its capital construction costs. New faculty were focused on making progress in their respective fields, and much less interested in running the College than had been their predecessors. The "comfortable Hope" of Biology chairman Phil Crook's assessment for the report was about to disappear, and Crook with it, as the school and his department became more focused on advancing the field and making the College stand-out academically. Faculty who were not on board would be replaced.

Over the next decade, Hope became the Hope College of which Cal VanderWerf dreamed.

Chapter 13

Cal VanderWerf's Dreams for Hope

Cal VanderWerf did not expect to be a College president; and it was only because Hope called that he became one. When he got in Holland, he was overwhelmed by the work he saw as necessary to build the place to a college of which he be proud. As the North Central report said about Antoine Lavoisier who came to his position to make better gunpowder, what he found when got there was the need for a [chemical] revolution.

VanderWerf's inaugural address lays claim to what was to be the overriding goal of his administration: his intention was to build the best liberal arts college in America. He saw the sciences as a corner stone for this best and turned his attention to rebuilding chemistry, and building physics in the first year. But every part of the school was in need of a reconstruction. That does not mean that departments did not do good teaching, and did not have good teachers. But, World War II changed that. The instrument/N.S.F. revolution that followed almost immediately made teaching in the sciences a completely different exercise than it had been before the war. The Hope faculty extant was built before the war.

A big gorilla in the room was the Dutch Reformed Church. By 1963, though it was still laying large claim to it colleges, its presence was losing sway with the younger generations. There was not a great exodus. But young people that left rural churches for college study, and did not come back to those churches or those rural areas. Nor did they continue in Reformed churches. In most urban areas there were none. Their rigidness was a turn off; and television watching could easily substitute on Sunday for a second, evening church service.

But at Hope, and in Holland, the church continued its domination as though nothing had changed. This was an enormous challenge for the President of Hope College. At the time he took over the presidency, Cal VanderWerf found a Board of Trustees of more than 50 ministers most of whom were Reformed Church preachers. His first action was to get the Board of Trustees to vote itself out of office; by 1966, he had reduced the number of Trustees to less than 20. And for the first time in its history, that Board was contributing more to the alumni fund, than it cost to bring them to Holland, Michigan for their semi-annual meetings. He stocked the Board with business people.

Cal never hesitated in complimenting others whose aggressive creativity he admired. Irwin Lubbers came in for many congratulations for his insight and strong representations on behalf of the College and was in Cal's envy for talking the other eleven presidents in the soon-to-be-formed Great Lakes College Association into a membership for Hope College. At the time, Hope was not at the level of Oberlin, Denison, Kalamazoo, or Earlham Colleges in academic breadth, endowment, or the credentials of its faculty. Yet, Hope became a charter member of the Great Lakes College Association because, Cal always said, Lubbers told the other presidents a pretty good story. Admission to the G.L.C.A., though, it may not be familiar to many, placed Hope College academically where it wanted to compete—even if it could not compete at the time. That objective was Cal's objective, and it was to be among the best liberal arts colleges in the Midwest.

Cal dreamed of a Hope College whose academic success and excellence would justify a position among the best colleges in America for the money. His vision for the school pre-dated the *U.S. New and World Reports* on Colleges by at least 40 years, but their objectives to be #1 were his goals for Hope. He wanted Hope College to recruit the best students for the best college in America. We, in chemistry for sure, worked hard to make that happen. This was not the case with other parts of the institution. For reasons I never really understood, the coaches and athletic department did not think it their job to recruit athletes for their teams. As a result, in those competitive days when state universities were emerging as academic forces at the undergraduate level and Hope's entire operating budget including faculty salaries was based on tuition, one entire part of the College—a part that could offer good competitive opportunities for student athletes—said the "status quo" was good enough. Hope College would never be a University of Kansas—where coaches, alumni and faculty would give their right arms to recruit a Wilt Chamberlain. But Hope could be a lot more than it was with the best students in the classroom and the best athletes on the fields and gym floors. Cal was a strong supporter of athletics, and troubled because he could not convince the coaches and sports administrators to work toward this end.

The single development that took Hope to a premier position among colleges in the United States was its attention to research publication. Its youngest faculty treated their position as being no different than was the position of their university colleagues. It was publish or perish, and Hope's young faculty published their work at a regular, in some cases, prodigious rate. Today, Hope's focus on research with undergraduate students is its biggest academic plus. Then, it was first Van Zyl and the people in chemistry, but by the end of the 1960s, a pervasive attitude across the sciences, the social sciences, and the arts.

Cal made some mistakes here – he did not really need to talk about this. His younger faculty were publishing because it was the expected norm in their fields. But he did talk about it as, in his view, a way to encourage the non-publishing faculty to publish. But that did not work. Most of them had nothing to say. So when Cal and his dean pushed on the faculty to write their best thinking, some claimed "We are teachers, we are not researchers." These faculty members somehow resided on higher ground in this academic institution because they did not offer their creative thoughts on a piece of paper so that others could understand them, appreciate them, and maybe criticize them. Cal always said "It takes tremendous courage to publish," and almost by saying this he was calling some of the more entrenched of those that did no research, and published nothing, "gutless." Actually what some younger colleagues thought, and some of us were alumni of the institution, was that those teaching faculty were lazy, and that thought did not pass them by either. One sad thing about this was that unmotivated or not, they still taught students and their teaching impacted next generations.

Cal VanderWerf's presidency was not particularly damaged by the Peters affair save it reinforced his failed attempt to hire his dean. Bill Mathis had arrived on campus the self-proclaimed apostle of the liberal arts college administrator. Liberal arts college devotees of those days had it all over us mere scientists because they could speak with many tongues, or so they said. They prepared students for "life-long learning." It seems that they were taken by the Wilsonian attitudes that the Princetons of the world educated men, while the M.I.T.s trained technologists.

Later, I came to agree at least at some level. But at that point in time, I was more concerned that the College prepare students for professional careers, and the professions, than about the periphery. So that Dean, like many other deans later in my life, was a preacher for his cause, though he did not necessarily know that others – from very different perspectives – knew that cause as well as did he.

There were stronger negative vibes between those in the humanities and in the sciences at Hope than at any other institution I have taught, probably because scientists knew pre-professional education was part of their responsibilities. Humanists did not know a molecule from a mole, or benzene from bromine, but they could easily point out how poorly educated the scientists were because, in contrast to themselves—generalists or Renaissance men—mere scientists were too ensconced in our fields. Maybe the liberal arts devotees looked at their scientist colleagues and thought there goes the prototypical M.I.T. nerd. Maybe this was a residue for the Van Zyl years. But it was surely prevalent at Hope, and it was unpleasant. Mathis had crystallized the "liberal arts over all" opposition.

Young scientists, like Dick Brockmeier and me, thought Mathis had good ideas and strongly supported his efforts to make Hope into that best of show institutions in that field. But Mathis, the administrator, made serious political errors some of which reflected on the President. His most important mistake was releasing all the faculty salaries to a faculty committee, because that was "the way it was done at his former school". This meant that a professor in history with twenty years of experience, or even more terribly a professor in Bible/religion, could find out he was making less than a new assistant professor of physics. Regardless of the fact that there were few, if any, Anglo-Saxon Protestant physicists looking for jobs—at least physicists with research credentials —and Hope had pretty strict hiring policies, even then, one just did not do such things. It was also an affront when a young physics professor with a Ph. D. from Cal Tech or M.I.T. was hired at a higher salary than the old physicist with a Master's degree who taught physics from Schaum's outline.

Salary disparities such as these did not sit well with the older faculty and further eroded VanderWerf's position politically. Cal was seeking administrative help in his attempts to bring the College to academic competitiveness in the evolving collegiate marketplace of the 1960s. He thought he had found that help in an experienced Dean, Bill Mathis. But he made a mistake by hiring Bill Mathis, and that mistake translated into more administrative pressure on the president. Mathis departed after less than a year, and Mathis' departure, with all pious pomp and circumstance, further convinced those in the humanities that President VanderWerf was clearly wrong for the College.

Cal told me once, "I try not to make too many mistakes, but as Bill (Mathis) proves, when I make one, I make a big one."

Church Relations

Cal's ideas about the Church are hard to assess mainly because he took over a campus at the beginning of what became the most turbulent time on college campuses in American history.

Though Hope's student body was mostly obedient and conservative, there were plenty rebels waiting for a good opportunity. During the Vietnam War, protests which began in earnest throughout the country in the late 1960s, Hope had no serious student unrest, no riots, no bomb threats; the College was open even during the darkest days of the protest movement much to the credit of the then administration and those administrations that immediately followed. Most of what the College president had to do in the 1960s was keep the lid on and Cal's presidency did this admirably.

I am sure the Church corporate, particularly the ministerial force residing within a fifty-mile radius of Holland, was uneasy with Cal's presidency. He was the first seriously-credentialed academician to head the institution and some found fault with that. He was a scientist. Preachers of that time were proud of how little they knew about science. He was the first new president after World War II, following a previous president that had had an extensive tenure. His family also brought a black woman to the town, and that made many in the community nervous. Cal was also a scientist. And though a native of the Reformed Church, he was a religious outsider, too, in that he was a member of a Congregational Church in Lawrence, Kansas. Never mind that there were no Reformed Churches within miles of Lawrence. He could at least have been Presbyterian!

Among the first things I heard from a fellow faculty member was: "Look at that! His father was the minister at First Reformed Church and he/his family joined Hope Church!" The Holland, Michigan of the time was so inbred that even a simple matter like which church the college president belonged to was discussed in a flurry.

The Reformed Church in America, chauvinistic and conservative, had then, and still has, a difficult time penetrating both suburban and urban areas. Denominationalism in America was on the wane in the 1960s but the Church still persisted. While the Netherlands was at time mostly homogeneous, the US was a melting pot and girls of protestant Dutch ancestry were falling in love with Methodists, Episcopalians, and, perish the thought...an occasional Roman Catholic. Young people going to college expected music, art, social action, concern for the oppressed, and help in the moral education of their children from their church and unless they lived in western Michigan they mostly were not getting it from the Reformed Church. Also, in former generations, the church was the social outlet in most towns; theaters, entertainment villas, and even television were becoming alternatives. Sunday evening services were giving way to Jack Benny and Ed Sullivan. Mobility, too, meant persons could shop around for their religious experiences. Americans, as had their Dutch counterparts, were starting to do that. As a result, the R.C.A. did not grow rapidly outside western Michigan and few other isolated areas of the country. The Reformed Church was essentially a church of most Hope person's childhoods, but not – unless they lived in western Michigan – of college educated adults.

Hope's faculty, too, though mostly protestant, became less and less of Reformed Church origins. With growth in the value of academic credentials over religious emphasis as criteria for hiring new faculty, the unreal bond that had held the College to the church through two wars and a depression dissolved from within. So, staffing academic departments in the 1960s, except in isolated fields, became a non-Reformed Church operation.

There was no question on which side the President had to be. He understood the university in its broadest, most beautiful sense. And he put the classical faculty values of teaching, research and service into practice through the persons he recruited and hired to take his alma mater, Hope College, into the future. The Hope College of the 21st century inculcates academic values, without question, as part of its mission. This century's Hope College presents its students with the plethora of opportunities required to succeed in their chosen fields. Were it not doing so, it would not be able to compete for student's tuition dollars today.

Compulsory chapel attendance was discontinued during Cal VanderWerf's tenure as president, too. Was the President against compulsory chapel? I am not sure. But leaders of the student body pushed hard to have the requirement lifted during a time when campuses across the U.S. were burning in protest engendered by a misguided war in Vietnam. Compulsory chapel had reached the end of the line at a time when campuses were restless, and it would have been dropped by whoever was president at the time. That just happened to be Cal VanderWerf.

Actions During the VanderWerf Years

A college has four things to offer—educational content, faculty expertise, facilities, and campus lifestyle/environment. No matter how much Hope thought it was doing for its students, a Hope education had grown non-competitive by 1963. Its faculty and its facilities were outdated. Cal's presidency was driven by the long view. First and foremost, Cal VanderWerf built the faculty.

210

Faculty

When he left Hope in 1970, the faculty had grown by at least 50% and the leadership that brought undergraduate research prominence to Hope College in the 1980s and beyond, was mostly assembled. Though time blurs my memory somewhat, I name a few individuals below knowing there are many others:

Sheldon Wettack, later Dean of Science and President of Wabash, Executive VP and Dean of Faculty at Harvey Mudd-chemistry.

Dwight Smith, later Chancellor University of Denver – chemistry.

David Marker, later President Cornell College Iowa – physics.

Mike Doyle, later President of Research Corporation, now Chemistry Chair at University of Maryland-chemistry.

David Myers, widely recognized author – psychology.

Cotter Tharin, 1st Chair – geology.

Ralph Ocherse, later Dean, Arts and Sciences IUPUI – biology.

Del Michael, nationally recognized painter – art.

Jim Tallis, organist – music.

Roger Davis, organist – music.

Bob Wegter, playright - theater.

(I hesitate to mention myself, a McMaster Distinguished Professor and Executive Director/Founder for the Center for Photochemical Sciences at Bowling Green State University.)

When the President introduced the seventeen new faculty individually at a fall meeting of all faculty in 1964, he talked about the publications they had authored and the research grants they had received. It was the norm in the new chemistry department to have regular departmental meetings and luncheons to discuss student progress and program opportunities. I probably went to chapel occasionally, but do not remember it if I did. When we arrived in the morning our objective was to prepare for our classes, advise our students, and, as time allowed, go into the lab to run a reaction, analyze a product mixture, or study, in some other way, a new set of compounds we (and the granting agencies) considered of consequence. We were driven by academic values, not history.

In fact, if one carefully analyzes the objectives of Hope's founders, the collegiate academic experience was foremost in their minds. In the 1850s, college experiences universally in America were closely related to churches and denominations. But this did not stay

211

that way even later in the 19th century. I am certain Hope's founders would also have expected it to stay current with American higher educational tides.

New Faculty

In many respects the younger of the new faculty in 1964, were incredibly naïve. When we met first as a faculty *en masse* at Kalamazoo College, in that hated of all hated activities by young family men—the faculty retreat—we must have been told about faculty fringe benefits. Eventually, the College offered two health insurance plans: one which provided general coverage with a small deductible, and another that was a major medical policy with a deductible that was five times or more as large.

We were also told at the retreat by, I presume, Henry Steffins, about the College retirement plan. It was the previously-mentioned "teach here twenty-five years and when you retire, we'll pay you one-third of your highest salary for the rest of your life." My dad, who was by this time a high school music teacher in western New York had a fit. "That's no retirement plan."

By year four of his Presidency, Cal had convinced the remodeled Hope Board of Trustees that it needed to improve the faculty fringe benefit package. TIAA-CREF was added to faculty compensation in the fall of 1967.

New faculty changed the campus environment from one which was church/community centered to one that was discipline centered. In his first two years at Hope, Stan Harrington and Del Michael started an art department, Roger Davis and Jim Tallis were added to music where they taught organ and, perhaps even more important, played at morning chapel. Bob Wegter replaced Jim Malcolm (on leave) in the department of theater. The new faculty in all departments, including the sciences, forced the College to spend money on programs and equipment. I ordered journals through the librarian that the College had needed for years but did not have. If one looks to this day, you will find that many journal subscriptions date to 1964—the year of my first appointment. These acquisitions went through no committee, and the librarian complained a bit about all of the money we were spending on his library but knew "the president was a chemist so it was probably alright." Maybe it was fine and maybe it was not. But our students and our faculty now had the use of subscriptions to about fifteen new, current journals.

Facilities

Instrumentation was also entering the research university arena. Small colleges, like Hope, found they needed to expose their students to the latest instruments for their students to be competitive in graduate study and their faculty able to compete for research grants.

I remember Irwin Brink telling the assembled chemistry faculty at one of our regular meetings in the spring of 1965 that he had just met with Dean Vander Lugt to explain to him why the department had completely outrun its budget in January. For the last five months, Irwin had been spending someone else's money at an ever-increasing pace. The Dean opined that this was a rather bad idea, though he eventually found the cash to cover Irwin's malfeasance, and Irwin encouraging us to do our best to reinvent chemistry instruction in the instrument era at Hope College. On what was Irwin spending? He was spending funds on new equipment mostly for the three faculty members who had joined the chemistry department in 1964.

From Irwin on that day, I learned the first principle of being a good chairman: It is easier to beg forgiveness than ask permission. Also, run over your budget until somebody stops you, and if nobody stops you, all the better so keep right on spending. I cannot recall how many faculty labs I set up at Bowling Green with last year's equipment money thanks to the lesson Irwin taught me about running a department! "Did I do that? How sorry I am!" But I never said I would not do it the following year again if I could get away with it!

When I graduated from Hope College, there was not a functioning instrument on the whole campus. Before he left, Doc Van Zyl had received funding from the National Science Foundation for instructional equipment. When we got there in the fall of 1964, we found a series of boxes from Beckman Instruments and other companies that sat mostly in the entrance way to the first floor of the science building. We opened the boxes after a while, only to find mostly items that somebody else did not want dumped on Hope by bad-guy salespeople. That which we could not make work to our specifications, we returned to the vendor at, of course, a substantive discount. Soon, Hope College, through the various mechanical hands of my colleagues, found itself with an ultraviolet visible spectrometer, an infrared spectrometer, and a gas chromatograph, and best of all, they all worked.

The other science departments grew, too. Two alumni, Dick Brockmeier and Jim Van Putten, joined David Marker in vitalizing (notice I did not say revitalizing) a physics department that had only one faculty member with a Ph.D. degree in 1964, another who taught introductory physics from Schaum's outline, and still another who mostly made radios for missionaries as his research activity. The biology department transformed itself, too, but more slowly. It had certain entrenched mid-career faculty with an "I teach" philosophy of undergraduate education, and some of these had to leave before biology could get on its way. That only happened really with the hiring of Jim Gentile. A geology department was started and Cotter Tharin hired as its first faculty member chairman. Computation also was being discussed. I had taken courses in computer programming as a graduate student, but found them far removed from my research at the time. I do remember, though, Dick Brockmeier's seminar at some time during his first year at Hope when he talked about computers talking to computers talking to computers. What he envisioned, from his graduate study at Cal Tech, became the internet.

Competitiveness

In testimony to the attention the VanderWerf Hope College paid to competitive competency, the seven VanderWerf years brought to the departments of chemistry and physics about ten new faculty. Some stayed at Hope for their entire careers; others for decades. Still others moved on more quickly. As I have already mentioned, three became college presidents, one college dean, one the president of private, non-profit, two became major university administrators, and one a Distinguished University Professor. During that faculty's years at Hope, Sheldon Wettack won a Camille and Henry Dreyfus Teaching Research Fellowship from his work; and I remain the only person in the history of the awards whose work was solely in a four-year college without a Ph.D. program to have been awarded an Alfred P. Sloan Fellowship.

The faculty transformation continued through the 1960s. I instigated the hiring of Mike Doyle, arguably the most successful faculty member in chemistry at Hope in its history, even though he was an active, practicing Roman Catholic. Mike arrived in 1967. Dwight Smith came to Hope from Wesleyan University and he recruited Cotter Tharin. Biology, too, eventually transformed with the result that Hope, in the sciences at least, was set for the rest of the century based on the momentum established in the VanderWerf years.

Money

Another transformative came about through money and its expectation. By the fall of 1970, Cal's last academic year at Hope, every faculty member among the new hires had research support. There were two, I believe, URP programs at the school (Undergraduate Research Programs proceeded Research Experience for Undergraduates; more about that later). We made regular visits to the National Science Foundation for major equipment funding. When that did not work for some reason, or if other sources arose, we went there. There were at least two regular N.S.F. grants in the departments as well.

Development

Money for general operations was also given campus attention. There was no development office when Cal became president. Every autumn, Rein Visscher and Henry Steffins would call a group of faculty together and have them fan out over the Holland/Zeeland area in search of contributions to the College. As an alumnus, I was one chosen for this duty.

My colleagues in chemistry, at least some of them, had a fit. "Pandering for Hope" was their term for the exercise. In retrospect, that is exactly what it was. Each faculty member was partnered with a community leader, and these teams assigned a few "calls"— businesses to visit in search of some loose change.

I was assigned to work with a stockbroker, Nelis Bade, and visit at least two businesses that I remember, Russ's Hamburger Shop and the Macatawa Bay Yacht Club. On a dark November day, Bade picked me up at the science building and we drove to Russ' Drive In in his new Lincoln. After a bit of chit-chat, we made the important "ask." As I recall, Russ gave the College $50! I am not sure but I probably had spent more than that on cheeseburgers in Russ' drive-in myself over my several years as a student. Then Bade and I visited the Yacht Club. As it turned out, Bade had had some issues with the management there, so this was an uncomfortable visit for him. I do not remember if there was a gift to the College or not.

I did this once and then, I believe, this Depression-era opportunity faded into the past. Soon, Cal hired Larry ter Molen, captain of the 1959 football team to head his new development office. As I remember, there had been someone assigned to this duty in the past, but he was either fired or left before I started teaching at Hope.

Larry was in the process of assembling a staff when I noticed that a former fraternity friend from Hope days, Lee Wenke, was running for Congress as a Democrat in a Republican district north of Lansing. Lee was all of 25 and had been an extraordinary, fun, free spirit while at Hope. His main *malefaction* as an undergraduate was to steal the pulpit from the Chapel a day before baccalaureate and graduation in June 1960. Actually, he and his conspirator did not steal it; they hid it in the Clarence Kleis' room in which the graduation robes were stored under the west end of the Chapel. But hide the pulpit they did, and for at least twenty-four hours, the campus was substantially more upset than it would be ten years later during the Vietnam War.

When the malefactors were caught, they confessed and showed the powers where the pulpit had been stored. Each was expelled from school for graduation though I guess even Hope College could not prevent the awarding of degrees—so Lee got his degree three weeks later in a solemn ceremony in the pine grove in the dead of night. But I thought his creativity superb and that he would make an excellent development man for Hope's new effort. I recommended Lee to Larry and a few months later, Larry and Lee were the Hope development office.

Larry, Lee, and others began aggressively seeking funding not only in western Michigan but also with Hope alumni around the Nation. One of Cal's crazy ideas was that he would sell shares in the College to celebrities. He even had a few dozen golf balls imprinted for Bob Hope's Hope College in this effort. But Larry ter Molen and Lee Wenke began Hope's development office and effort. To them, the College owes significant thanks for the effort. Both incidentally spent their careers in development retiring at the appropriate times from the Art Institute of Chicago and University of Montana, respectively. One major initiative Lee managed was that initiative which brought George F. Baker scholarship money to the school; this was a major breakthrough for the College and Baker Scholars a still a part of the campus scene.

Research Funds

All faculty in the sciences at least looked hard for research and program funding. National Science Foundation funds started to flow into the Department through programs like Undergraduate Research Participation (U.R.P.) and Student Research Fellowships. P.R.F., N.S.F., and other regular programs funded summer salaries for faculty. By August 1965, faculty were in their offices in the science building. Cal wrote a Summer Institute proposal to the N.S.F. that brought forty or more advanced placement chemistry teachers to

216

the school in the summer. Gene Jekel took it over, and ran it but for eight weeks the labs were buzzing with high school teachers. When it came time to send their best students to college, they thought of Hope. Some of our most able students over the subsequent years came from that institute.

Though most new funding went to undergraduate student stipends and for equipment, faculty summer salaries also became common in the sciences, and almost every grant had an additional "summer salary" for the Principal Investigator. This created financial disparities among faculty in that scientists were paid more than humanists, and there were a continuing bone of contention in the Kletz. The Kafe Kletz was a small coffee shop in the basement of Van Raalte Hall to which was annexed an area past the furnace room where faculty could go for conversation, and not incidentally, smoke cigarettes, cigars and pipes. The aforementioned Dick Brockmeier was a financial manager of nit-picking proportions. When he returned to Holland to teach at Hope, he managed to buy a former banker's house at a relatively nice location on Lake Macatawa for $41,000! How often did we hear in the Kletz from Lambert Ponstein, among others, about "that Brockmeier who went out and bought/could afford that *banker's house*." Most of the rest of us worked all summer on one or more supported research projects being paid one-third our annual salaries for the effort.

In 1966-67, at least three major program grants across the sciences were received by the various departments, and by the division, and this meant the addition of even more new faculty. The largest grant the College had received to that time was a college grant to the sciences for curriculum revision and faculty hiring funded by the Alfred P. Sloan Foundation. Dick Brockmeier and I wrote the Sloan grant with Cal's help in August 1966. Dwight Smith wrote a Research Corporation College grant; Sheldon Wettack wrote a major N.S.F. grant. All three of these were funded within a period of a few months in 1967. Suddenly, a College for whom a grant of a few thousand dollars was a lot of money less than a decade earlier, found, from the competitiveness of its faculty and their ideas, more than $1 million in new money being committed to it by the nation's most prestigious granting agencies. The Sloan grant itself was $375,000— more than the aggregate total of all College grants awarded before.

In retrospect, Cal foresaw this growth of support for programs in small colleges but none of us did. He used to say "the universities (in chemistry) need you (to produce qualified undergraduate students). You don't need the universities." He had a

217

lot more confidence in our abilities, personally and collectively, than we did in ourselves. All we were doing was working as hard as we could to do the best for our Institution, our students, and our careers. And he had also been in a university so he knew just how a smaller, non-Ph.D. granting institution could compete—*and win* in that competition. He armed us with confidence in the always-increasing competition of the funded research rat race.

With research money came the expectation of publication. In the year I left Hope, a seven-person chemistry department published forty-one papers in peer reviewed journals.

Hope/Holland Town/Gown

Young, new faculty had different political ideas, too. Town/gown matters in Holland first surfaced consequentially during the Presidential election of 1964. The Kennedy presidency brought a heightened intellectual presence to Washington, television brought Washington into many person's living rooms, and for the first time politics was finding the medium more important than the message. Kennedy was smooth and articulate with a ready smile so even if one disagreed with his policies, one had to like the way he presented them. He also gave young Americans an icon to revere. He used the television camera skillfully and almost casually. One felt, in some ways, that he was sitting on one's living room couch as he made his points to those in his TV audiences.

His opponents and political competitors were less adroit. One of the more outspoken was Reformed Church minister Norman Vincent Peale who led the Protestant clergy's effort to prevent the Kennedy election, because J.F.K. was a Roman Catholic. Peale did not want the Pope running the country!

After the Kennedy assassination, an old Washington returned, and they brought a political presence of a former era. Lyndon Johnson was crude, inarticulate, and unattractive. Americans, through the television news, found that they had gone from a president that made the literate feel good about their country, to a Texas farmer who could barely put two words together without making at least one grammatical error. That great step forward and a government beginning anew evaporated with an assassin's bullet.

The election of 1964 pitted what was to become the great Republican right wing under Ronald Reagan against a big-eared Texas cowboy. Barry Goldwater was the Republican nominee in this election foreshadowing what was to become. Whatever one says about that first Arizona presidential candidate, Goldwater made the message of the right, constructed by clever writers, abundantly clear—too clear.

"Extremism in the cause of liberty is no vice," he screamed from the convention podium in San Francisco in August 1964 where he had also invented "tactical nuclear weapons." The right's mantra was that the military could "win in Vietnam" if given the power to go forth with all their might, and the next president of the United States, who they hoped would be Goldwater, needed to empower them to do just that.

The intellectual community was not buying Goldwater's message about a war in Southeast Asia and the election was almost entirely about that. Though all the intellectuals had as alternative was Lyndon Baines Johnson. The halo that lay over Kennedy carried a hope that Johnson would avoid further escalation toward a full scale land war in Southeast Asia. That hope was wrongly placed as is now clear, but the Goldwater war mongering rhetoric virtually assured that those who wanted peace could only vote for Johnson.

In Holland, a conservative Republican stronghold, this tension escalated, with a full-page ad in the *Holland Evening Sentinel* that said, essentially, "We, the undersigned, support Johnson/Humphrey for president." There were about seventy signers, including nearly fifty Hope faculty and administrators. Some were known Democrats, like Ivan Dykstra and Al Vander Bush. But most were the new guys-Hope faculty who had been on campus for just a few years. Cal was guilty by association.

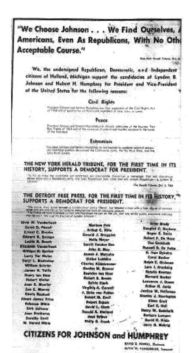

These were the guys (and girls) that new President VanderWerf was bringing in. Bill Hillegonds, the pastor at Hope Church was a signatory of the ad for Johnson/Humphrey. Though Bill would always deny it, signing that letter in support of Lyndon Johnson probably caused him to leave Hope Church to become Hope's first chaplain a few years later. Bill, of course, was that preacher with whom Cal's family found affiliation when they joined Hope Church in 1963.

Robert Donia, whose opinions in the *Anchors* contained great insights, had the Johnson election analyzed correctly: "It would be a mistake to say that Lyndon Johnson had attracted a great majority of the American people to himself on Tuesday. It would be more correct to say that Barry Goldwater drove them (away and) to him. True, the net result of November 3, 1964's election will be tremendous legislative programs passed by Congress, increasing emphasis on peace in foreign policy (how wrong he was on this one) and, in short, probably the most liberal period in American political history since the New Deal. But this doesn't change the fact that the election was a resounding rejection of Goldwater Republicanism."[123]

The world had been shattered by Lee Harvey Oswald's bullet, and the peace we had so short a time to enjoy went up in smoke as Lyndon Johnson's presidency roared into the next gear. Johnson did nothing to quell Vietnam and the U.S. firestorm it created. Instead, as the late Robert McNamara was so clear to confirm, he virtually alone escalated it.

Vietnam, a place most of us had not heard before Americans starting dying in a war there, was on everyone's lips. Television brought the war to our living rooms; spin doctors interpreted it for all to hear; and well-meaning citizens regardless of their politics were helpless to do anything about a war that was frustrating them more all of the time. One had the impression that the government was in control of everything, but not able to control the one thing that mattered—the deaths of young Americans in a land war in Southeast Asia.

College students, for the first time in American history, rebelled. The America of their parents was irrelevant they said. They were old enough to die they pointed out, but not old enough to drink beer or even vote. The Constitutional amendment that gave the right to vote to 18 year olds was a direct result of the national angst/guilt over Vietnam.

Though Hope's campus was never threatened, its students were understandably upset and markedly different during those years. I have no doubt that the student's reactions in relation to the war affected the way I viewed undergraduate teaching for the rest of my career.

---

[123] Robert Donia, Opinion. *Hope College Anchor*, November 6, 1964.

To give the devil his due, Lyndon Johnson's presidency did provide positive changes in America. The Civil Rights act of 1965 *finally*, after almost 100 years, guaranteed the rights of blacks to vote. The first public health legislations of any consequence, Medicare and Medicaid, became law.

Still superimposed on the Vietnam protests were those for the Civil Rights minorities in the southern states. These were the years of Orville Faubus, George Wallace, James Meredith, and the others who played roles in integrating public facilities in the old Confederate South.

Hope College was relatively unaffected by Civil Rights protests, directly. But the Detroit riots of 1967 scared everyone in the state of Michigan; Muskegon to the north and Grand Rapids to the east had industrial bases and significant minority populations. One could barely turn on a newscast without some mention of these important matters. Try as Hope might, and there were some initiatives in this direction, recruiting inner city minority students to study in Holland was a tough sell. So Hope remained, and remains, a campus of mostly white students.

All of this turmoil culminated in a week in early April 1968. Lyndon Johnson announced from his Texas ranch that he would not run for a second term as president on March 31, and four days later Martin Luther King was assassinated in Memphis! Suddenly, it seemed, all hell was breaking loose. Where had that domestic tranquility gone?

I was in San Francisco attending an American Chemical Society meeting that week. Eugene McCarthy, the other Senator from Minnesota (Hubert Humphrey was the Senator of first notice from Minnesota) was running for President, challenging Lyndon Johnson at about every stop. McCarthy was speaking at a Democrat fundraiser at the San Francisco Hilton, so I went up by elevator from the floor where my meetings had been, and stood with Sherry Rowland, Nobel laureate in 2000, and my professor from K.U., to listen to McCarthy. The Democrat congressman from the Bay area was one of his active supporters. In this speech, I first heard J. Edgar Hoover criticized in public, and the standard fare of criticisms of General Louis Hershey, the administrator of the draft, was also part of his speech.

On the way back down on the elevator, someone had a copy of the San Francisco Chronicle announcing the assassination of Martin Luther King. Robert F. Kennedy was still in the race, but he was to be assassinated in Los Angeles in June. So America was being driven to the brink by single shot gunsters, and our ability as a nation to handle it was being severely tested. Eventually, Hubert Humphrey was chosen at the Democratic convention in Chicago in August, but that seemed eons away at this time. The years do not even seem connected somehow now, but they were and the future was to hold even more lightening.

At just about this time, the Cultural Affairs Committee of Hope College, the chairmanship of which had been thrust on me in summer 1967 by Morette Ryder (who then left for a year's sabbatical in Eugene, Oregon), was gearing up to host a several day conference on the "Crisis in the Cities." The event had been the brain-child of one or more former ministers (not Reformed Church ministers by the way) occupying positions in Hope's history department in reaction to the riots in Detroit the summer before. I often said the reason my hair is white was that week in April 1968. The highlight speaker of the event was a comedian turned civil rights commentator/activist—Dick Gregory. The conference went forward with few hitches, and I survived, but barely.

Even though chair of the committee hosting the event, I avoided most of the conference because I was much too nervous about its outcome. I dreamed of the Hope College of my unrequited loyalty, burning down much as had the stores in Detroit the summer before. Congressman John Conyers from Detroit was one of the speakers at the conference on the cities, and I did go to his talk. I remember pushing him on why the Congress could not get us out of Vietnam. His answer—"I'm just a Congressman. What do you expect me to do?" Students graduating from Hope were being sent in harms' way by a draft process one could not control, and our own Congresspersons were unable to figure out what to do about it? Clearly, we had more than just a crisis in our cities—we had crisis in government that did not get resolved until Richard Nixon resigned in disgrace six years later, and Gerald Ford restored some order.

The Hope event "Crisis in Our Cities," in retrospect, was a great success. Persons with little or no prior awareness of the College became aware of it in a good way. There was a feeling that maybe we were doing something—even if an ever so small something—to alleviate what was a continually evolving reinventing of America. This event was positive Christian activism and I can say now that I am pleased to have been part of it.

After nearly fifty years, it seems almost a miracle that the United States did not come apart during these turbulent times. Still to come in 1968 was the assassination of Senator Robert Kennedy in Los Angeles, and the riots of the Democratic convention in Chicago that August. The national patience was wearing thinner by the month and the only thing that seemed inevitable was that the U.S. needed a political sea change. Though the election was close and, for sure, his election was not that much of a change in a grand scheme of things dominated by the Vietnam War, Civil Rights, and a national mindset that government was one's enemy, Richard Nixon became the new man in Washington to despise, and he served that purpose admirably.

Of course it did not take long before Nixon proved to be the same old government only worse. His Vice President Spiro Agnew was soon found a criminal and in only a few more years Nixon would become the first president in history forced to resign. But for six long years, our evening news broadcast was bathed in the blood of young soldiers, and filled with the meaningless words of presidential leaders no one could respect.

Board of Trustees: Lay Governance – The VanderWerf Years

When I returned to Holland in the summer of 1969, the campus was a different place. The Mathis affair had taken a toll and Cal VanderWerf's political supporters were younger with limited affiliation with the power structure on the Board of Trustees, or the Church. Try as we might to support him, and I surely tried to the best of my ability, it became increasingly clear that he had lost support at the Board level. He would describe Board meetings to us on occasion, and point out persons who made comments about this change or that next event who should have been locked in step with him and his objectives. But they were not. This was troubling.

Hope's campus tranquility was, as I have intimated, unraveling too and I am not positive why. Cal's presidency was under attack mostly from a few faculty who were middle aged and the power structure ascendant when Lubbers retired. A few critics like Phil Crook in biology were scientists, but mostly the criticism came from the humanists. No doubt, most of the unrest was caused by the nearly 50% of the College faculty hired since 1964. These new faculty (and I was certainly one), were more interested in their disciplines than had their predecessors been. It was not good enough simply for them to have a job, any job. Most were professionals in their fields and mobile. They wanted the best possible job that they could get and did not necessarily have to remain at Hope if it did not have high aspirations for its future. Hope College came under pressure from its younger faculty to change in the direction of more scholarship and professional visibility. The *Chronicle of Higher Education* became more generally read than the *Church Herald*.

Hope was undergoing academic maturation. The Church was not the dominant force it had been and the Bible, religion, theology and philosophy departments were parts of a much bigger whole. Faculty in these departments came under pressure to publish in peer-reviewed journals, as opposed to church organs, just as their counterparts in departments across campus were under such pressure.

Hope's Board of Trustees when I was first a faculty member consisted of around fifty members. If one reviews the oldest records of the institution, it seems that when Hope was founded, each of the General Synod of the Reformed Church and each of the Classies had one or more members on something called the College Council. This extends at least to the time of President Kollen. Most of these representatives, though not all, were preachers and professional church personnel.

By the 1960s, it cost the College many more dollars to bring the fifty or more church representatives to Holland, house them, and feed them for each of their two a year meetings than they contributed to the enterprise. At some stage early in his presidency, Cal undertook to reduce the number of Board members and in one of the proudest political events of his presidency, he got the group of fifty-five to vote that the number be reduced to twelve. "Never before," he maintained, "did a group so large vote itself out of office so quickly."

I am not sure how much this reduction in the Board saved the College in dollars, but according to former Vice President Robert De Young, reducing the size of the Board to a more manageable number was Cal VanderWerf's crowning achievement. Whether his most important achievement or not, it was certainly an important step in smoothing College governance. From that time on, the College could function more and more like the business that it was. And, in the general scheme of things, that is what saved the Institution.

I have since found out that the President of the then-Board, Hugh DePree, was even encouraging younger staff persons at Hope—persons making significant contributions—to move on. Cal had a vision, the young faculty had a vision, and that vision, for better or worse, has become the Hope College of the 21st century. Unfortunately, some of the most progressive people on his Board of Trustees, persons who should have shared that vision, were not onboard. It was a situation that would have to ultimately fail.

Chapter 14

My Hope College

America in 1960 – the year I graduated from Hope - was finally at peace. World War II had been over for fifteen years and the Korean War was behind us; Dwight D. Eisenhower was finishing his second term; the military draft remained but most graduate students in the sciences received a 2-S deferral. Though we were living a calm before many storms, we did not know it. Life was good.

Kansas in August is *hot* and as newlyweds just about to start our first jobs, Sue and I could not afford a Sunday newspaper let alone an apartment with air-conditioning. When we went to Cal and Rachel's house to introduce ourselves to Cal in the fall, 1960 (we had met Rachel in June on that recruiting trip), I recall how cool their house seemed. Cal was bare-footed, and the VanderWerf house filled with kids.

Universities, specifically their science and engineering departments, were beginning to reap war's benefits. Funding for research, equipment and to train advanced students following the Soviet's *Sputnik*, flooded the graduate schools. Small schools were trying to lay claim to some of the money, too. Conant and Bush made it particularly clear that one of N.S.F.'s missions was to prepare future scientists. Undergraduate schools led much by Cal VanderWerf's awareness of such matters, were seeking N.S.F.'s support. To earn that, schools like Hope had to make a case that they had a niche in science education. For Hope, part of that became, naturally, research with undergraduate students. Cal was connected. The Petroleum Research Fund (P.R.F.), administered by the American Chemical Society, was also providing grants for research for small school faculty called their Type-B grants; Cal was on their advisory board. Through his initiative, a new program for young investigators, the Type-G program, was started in 1964. Research Corporation, which had funded much of Doc Van Zyl's work in the 1940s and 1950s was in high favor, too, and Cal was on their Board of Directors. The National Institute of Health was funding medical studies though it was not nearly the force it became a decade later.

At the university level, federal funding was making a real difference. Many more students were in college as the products of war time marriages reached college age. Graduate enrollments in the sciences were increasing too, because science was a sacred cow to high school graduates of those years. The K.U. chemistry department was thriving with all of the new money coming to research colleges/departments in universities. When I enrolled in the fall of 1960, ours was the largest first year graduate student class in the history of the department of chemistry. It was grown particularly large by the expectation of much more money from the government. The department of pharmaceutical chemistry, a near neighbor, had several large N.I.H. supported professional training grants. Pharmaceutical companies were becoming research and development enterprises. Ed Smissman, its head, had taken his Ph.D. with Hope alumnus Gene Van Tamelen at Wisconsin, so his department was mostly an organic chemistry/natural products department by another name. Their programs were particularly thriving.

Instruments made research faster and easier. Many had resulted from government weapons research during the war. Recording infrared and U.V./visible spectrometers were commercial—one could buy them. The first nuclear magnetic resonance (N.M.R.) spectrometer (1957). Varian Associates, a California company located near Stanford University that had made klystron tubes (radio wave generators/detectors) for radar during the war, started making N.M.R. machines for sale in the industrial and academic markets. In the hands of Cal Tech chemists, Jack Roberts in particular, the Purcell/Block observation of the nuclear magnetic resonance effect (1947) became J. D. Roberts' N.M.R. experiments on simple organic molecules soon meant Roberts showed those hydrogen atoms that one wrote on blackboards in the 1950s as part of chemical structures, could actually be envisioned through a spectroscopic observation. It took another twenty years, but by the 1970s, Paul Lauterbur at S.U.N.Y. Brookhaven ignored the *cogniscenti's* assurances that it would never work, and invented nuclear magnetic resonance imaging using a Varian A-60 N.M.R. just like the one pictured. The first M.R.I. installations (M.R.I. because no medical laboratories would wish to use a technique with *nuclear* in the name as in nuclear magnetic...) found their way into medical school research laboratories by about 1980 and those Hope students to whom I taught N.M.R. spectroscopy in the 1960s ("Why do we have to learn that?") would return in the 1980s telling me N.M.R. was the only thing I taught them they really used as physicians! The invention of M.R.I. used for non-invasive diagnostic imaging, was probably the single most important invention in medical science of the last quarter of the 20th century.

Cal taught his courses at Kansas, advised his three remaining graduate students, and chaired the recruiting committee for graduate

226

students. But one could clearly see when one talked to him about chemistry that he was not interested. His mind was miles and miles away from chemistry. Cal's last Ph.D. student was Vic Heasley, recently retired from Point Loma College, and Steve Taylor's undergraduate (Hope emeritus) preceptor.

In spite of his absenteeism (or maybe because of it), his colleagues elected him chair of the chemistry department in 1961. It took Cal a long time to accept this offer in part, his faculty colleagues believed, because he was negotiating with the Deans for more faculty lines. Though he did not tell me this directly, I suspect that he was considering opportunities elsewhere because he frequently entertained offers from other universities as well as industrial research directorships. Nonetheless, accept the offer of his colleagues to be chair he did – a position he assumed in the summer of 1961. Cal's first move as chair was to recruit one of the best teachers at Lawrence Junior High, where Sue taught, to be his administrative assistant. Gretchen (9th grade) and Klasina (7th grade) were students there, and Bernita Hartnett had been Grechen's eighth grade teacher. Soon, she was Cal's administrative assistant.

Sue and I thoroughly enjoyed living in Kansas, the chemistry department and most everything about our K.U. experience. My impression formed over the years is that the University of Kansas of my generation was about as good a combination of quality teaching and quality research as could be found in a state university. The future Nobel laureate F. Sherwood Rowland, with whom I almost took my Ph.D., and with whom I had many stimulating conversations over the next forty years, taught the honors freshman chemistry course and I was lucky enough to serve as its teaching assistant.

My graduate career went by quickly. Examinations went mostly smoothly and personal research took off, after a slow start. So sometime in mid-1962, Earl and I concluded I should study for preliminary exams and write my dissertation a year ahead of my class. Earl knew I was targeting an academic career and felt a good postdoctoral appointment would enable that objective. He also assured me that I had done well enough so that I could expect an appointment at Harvard, Cal Tech, or some other school of like reputation, and that this would be a good next step for me.

In retrospect, this was extraordinarily generous of Earl—I do not think he wanted to get rid of me quickly necessarily because research was going pretty well and research success meant more publications for the mentor. But, as a result of his generosity and probably my pushing, I accelerated by taking the preliminary

227

examinations a year before my class, and passing them in the fall of 1962 with the to-be-graduated Ph.D. class of 1963. When last in Lawrence a few years ago, I was told by the then chair of the department of chemistry that my record for finishing the Ph.D. degree in chemistry in less than three years still stands.

Sometime during the summer of 1962, when I was laboring at home over still another German organic chemistry text, a friend stopped by to tell me that Dr. VanderWerf had been appointed President at Hope. I loved Hope and thought this a terrific move by the College so I made a special appointment with Cal the next day to both congratulate him and to thank him, as an alumnus, for accepting the presidency. Though K.U.'s loss, it was definitely Hope's gain and though I was really quite surprised by this turn of events, I was pleased that my alma mater had reached toward one of its most able alumni to be its next president.

I was completely removed from the Hope experience by then, but in the winter 1962-63 Cal called me to his office and after a few quick personal questions told me "Doc is retiring; and the chemistry department could use some good energetic blood like yours: Would you be interested in returning to Hope as the new head of the chemistry department?"

I about fell off my chair because I was all of 24 years old and Cal had a lot more confidence in me than I had in myself. Teach at Hope? I talked with Sue about it, and she was not that enthusiastic, so we put the idea on hold while I completed my work on my dissertation. Kansas also had a minor requirement at the time, and I had foolishly chosen to minor in mathematics. It went mostly smoothly but that fall I was counseled to enroll in a class for which I had none of the prerequisites. Rather like George Hammond's Harvard experience with quantum mechanics (he got a D from E. Bright Wilson) I found myself way in over my head. I survived with the only D of my university/college career and considered myself lucky to get that. I could tell it took papal dispensation to resolve the "minor" issue though, because Cal came to my labs one morning in January and told me "not to worry"—he was the head of the department. I was too naive to have worried about it, but probably should have worried more. As it turned out, an offer for a post-doctoral appointment came shortly from Harvard and I felt I could not turn this down. So I accepted a position Fellow of the Harvard Corporation. I had no idea what that meant save Harvard. So after I left Kansas, I went to Harvard as a postdoctoral fellow where I worked with Paul Bartlett. Bartlett was a former student of the

228

former Harvard President and Science Advisor to the President during World War II, James Bryant Conant. Paul described Conant, in his National Academy Memorial to Conant, as odds and away the smartest man he had ever met. In many respects this postdoctoral experience at Harvard formed the basis for the rest of my career. The people I met in Bartlett's group became the scientific cornerstone of my professional future and we remain colleagues and friends nearly fifty years later. Hope College benefitted, too, because I did eventually return to Holland to teach and two of the organic chemists Hope hired, following my return, came from recommendations from colleagues with whom I worked in those busy days at Harvard in the early 1960s.

Sometime during the winter when we were in Cambridge, I got a call from Cal suggesting that, if I could now be interested in returning to Hope, Sue and I could meet he and Rachel in New York to discuss the idea. I had no idea what I wanted to do for a living, but was sure that I wanted to teach, and for a teaching career to develop I had some amount of research to get done "with my own hands." For sure, I needed a job and returning to Hope for a few years while I did my own personal research seemed attractive. The deal was consummated on that New York trip in a fine Swiss restaurant and, though I had no formal interview, at Hope, I was to become appointed assistant professor of chemistry to begin in the fall. The lack of an interview was not unusual for the times—several of my colleagues were placed at major universities without having interviewed there first. My only demand, if I remember correctly, was that I not have to teach freshman chemistry and I insisted that this be part of the deal.

A few weeks after our New York meeting, I attended my first meeting of the American Chemical Society in Philadelphia. During that meeting this guy I had never heard of kept leaving messages on my hotel phone, saying "We've got to meet." That guy was Jerry Mohrig. He had also been appointed an assistant professor in organic chemistry at Hope too. We both started our teaching careers at Hope in the fall of 1964.

My offer from Hope included a salary of $8000 and Cal added "1/3 of that for the first summer." I was so naïve that I had no clue what he meant by "1/3 for the first summer" and assumed it meant my salary would be $8,000 + 1/3 when I got to Hope in the fall. When I received my first pay check, it was for 1/9 of $8,000 so I went to Henry Steffins, Hope College treasurer, and asked him where the 1/9 of the other 1/3 was? Henry said "That must be between you and Prexy, because I don't know a thing about it," whereupon I learned the first of many early lessons about the academy. One has to negotiate a deal, and then one has to renegotiate the same deal over and over again until finally something happens to make it real. I did get the extra 1/3 but had to go to Columbus, Ohio and teach organic chemistry to pre-dental students at Ohio State for the first summer to get it.

Cal's promise to me – not a Hope College promise - did not make any difference – $8000 was still more than $6600 so I was satisfied. But hence endeth the first lesson – words are just words and unless one has it in writing. When summer came, I got a call from Cal suggesting that I contact Dr. Newman at Ohio State because he had a summer teaching position that I might like to consider. I was being pushed to take my 1/3 summer salary by teaching organic chemistry for pre-dental students at Ohio State, to save Hope money. Cal's motivation, however, was obvious. For twenty-five years or more, Van Zyl had supported students to work in the labs in the summer, but they were almost entirely unsupervised. He was never around and for good reason. He was paid for nine months' work, and was "off" in the summer. He had to work at a second job to pay his bills. Yet, his boys continued in the labs. The list of papers published from Hope labs with Van Zyl as a co-author started in 1949 and continued through 1965. But, with my accepting the job at Hope, call saw a way to change this – he would pay my summer salary, and then coach me to get my own funding in the second year. I did not need much coaching as it turned out, and his idea worked – there was no question that Mohrig, Dave Klein and I, the new faculty hired for the fall of 1964 were set up to change the Hope College science summers forever.

The original photochemistry Gordon Conference in the summer of 1964 was an exceptional experience for me. When George Porter, the 1967 Nobel laureate in chemistry, spoke about benzophenone photo-reductions, and indicated that p-dimethylaminobenzophenone underwent no photoreaction under certain conditions in his labs, but might under other conditions, Saul Cohen, from Brandeis rose and interrupted to say "We've done that," and finished the sentence with 10 minutes about the specific chemistry he and his students had done. Orville Chapman (Iowa State -- and a rising star) talked about organic nitro compound photochemistry; Sean McGlynn (LSU Baton Rouge) talked about whatever he talked about —and he made it sound so smart that the front row of the conference was on pins and needles. This conference, as much as any single event, set my first few years' scientific program. Years later the dapper Porter would be the dean of British chemists, and assertive Cohen Dean of Arts and Sciences at Brandeis, where he was married to Henry Kissenger's first wife.

The fall of 1964 presented interesting times for me. Cal VanderWerf hyperbolically introduced Jerry Mohrig and me to the faculty assembled at a retreat held at Kalamazoo College in the fall of 1964 with "Doc Van Zyl was so great, that it took two young guys to replace him." Jerry and I worked together well. We shared teaching organic chemistry that was being taught to sophomores for the first time in the fall of 1964—he, the lecture, and I, the lab. Jerry and I were in the first group to receive P.R.F. Type-G grants for young scientists. There were only twenty or so of these awards, and Cal was quite right in pointing out that for Hope to have two award winners said some good things about the future College and its faculty selection processes. Jerry's career and mine went side by side for another two years. We were both awarded Research Corporation funding at the same time. The grants were not big but they were big enough. Jerry eventually was recruited by his wife's alma mater, Carleton College in Northfield, MN, and he left in the summer of 1967. It was a great loss for Hope and a loss of a good friend for me. It took us a year or more to find his replacement.

Common sense or Dean Bill Vander Lugt prevailed, and Irwin Brink was named head of the department in Doc's stead in the fall of 1964. Irwin did a great job putting up with several young guys each trying to find his way in the world that was becoming known as academic research chemistry. He managed to keep us from making too many mistakes, though I am sure he wondered at times how he did it. But he saved me from anything that would have been huge mistakes early in my career.

Important curricular changes were about in the Hope chemistry department that fall of 1964 – first, Gerrit Van Zyl had retired, so the department had taken advantage of this to remove the normal second year of analytical chemistry from the curriculum, and place organic chemistry courses in the second year in its place. That meant the normal organic chemistry class size was to be doubled for a year and, with two new organic chemistry faculty, Jerry Mohrig and me, the department's plan was to have each of us teach one section. Jerry and I talked about that and quickly decided that we would divide things differently – he would teach the lecture in the fall semester, and I would manage the labs, while we would switch in the second semester; I would lecture, he would do the labs. Because I was familiar with Hope's idiosyncrasies in chemistry, and it sure had them, I found managing two laboratory sections meeting two three-hour sessions each week to be a possible load. I am not sure I would see it that way now; but I did then and we proceeded with that teaching schedule the fall semester. As had been the case when I was a student several years earlier, we made extensive use of, and gave an opportunity to, students of the classes ahead to serve as laboratory assistants. I chose the textbook for the course as I recall, though there was not much choice – *Morrison and Boyd's Organic Chemistry* was in its first edition and most everyone was switching, or had already switched, to using it. We chose *Organic Laboratory Experiments*, by Brewster, McEwen and VanderWerf, for pretty obvious reasons. I have already related how on the first day of one of the sessions, three boys were aligned at the west end of the old organic laboratory in the then Science Building, Fred Van Lente, Paul Schaap and David Anderson. On first meeting Schaap, he told me that he really liked the last part of general chemistry because they had had a "research project" and that was the "best part of the course." My reaction, though I did not tell him, was "yea, right?" and counted him immediately as a brown nose. I knew of Fred's father, Ken, who was my Uncle's colleague at Southern Illinois. A few years earlier I had met his sister, Cynthia, and his then brother in law, Harold Ward. Harold was a professor at Brown University.

At Hope, at the time, every senior chemistry major had to enroll in a research project directed by a professor. There were six faculty members taking students, Professor Harvey Kleinheksel took no students, so the eighteen or so senior majors were divided among six people. Paul Bartlett, during my year at Harvard, had assigned an undergraduate to me, Margaret Nothier, so I had Bartlett's advice on how to direct undergraduates in research "to give undergraduates a project about as long as it would take you to do in a day" ringing in my ears. (Peg Nothier, by the way, did not get much done with me though it was more my fault than hers, and the main thing I remember about her was her Radcliffe book bag. But she later became an extremely successful IT entrepreneur.) Three seniors were assigned to me—Jim Fargher, David Stehower, and John Stam. I assigned each a simple project, realizing that they would be with me for just a few hours for a very short time and that none would work more than two semesters. I do not remember at all what I assigned to Fargher or Stehower, but I assigned Stam a project involving the ethyl ester of cinnamic acid. I do not remember much about that either, save John was on crutches having been injured on the football field to the point of requiring knee surgery either that year, or a year earlier. John was one of Hope's amazing people – he was majoring in chemistry; the year before he had played football, while working several hours a week in a Holland shoe store, because he was married, with two little boys. John still graduated in four years, went to graduate school, earned a Ph. D. degree, and spent his entire career working for Pfizer in Groton, CN. Such was also the inherent nature of most Hope undergraduates at the time – they were career driven, and hard working.

I spent a lot of time that fall advising the majors on graduate schools; Fargher was a struggling student, but hard worker. He went to Southern Illinois on my advice, and earned a Ph. D. in 19 71 working with Herb Hall in organic chemistry. He spent his entire career at Owens Illinois in Toledo, and I had the pleasure of reconnecting with him when I moved to Bowling Green. Stam went to Iowa State, also my advice. Tom Kinstle, his advisor at Iowa State, was pretty important in my future life but the first tim Iowa State directed organic students to their advisors and John was assigned to work with Tom. Years later, I met John again when I visited the labs of Pfizer, Inc. in Groton, CN. John spent his career there, retiring some years back and, though I have no idea what he worked on with Pfizer, he clearly used his Hope undergraduate experience in chemistry to his best advantage. He worked for many years for a company that made plastics for injection molding. I know that because I also consulted for that company. At the time, they were looking for opportunities to use photochemical processes to encase wires for use in electronics in automobiles. These wires are all the same, but color coded; so their idea was to make the wires with color change coatings so that one wire could be used to make wires of several colors just by virtue of the colors to which the coating might be driven when exposed to light of specific wavelengths. It was an interesting idea, impossible to carry out. So the company and I talked, but nothing came of the project, save I found out where Dave Stehower, who took his Ph. D. at Michigan, spent most of his career.

One of the things I observed early on about Hope undergraduates in chemistry was that they were most receptive to conversations, advice maybe, about their next career steps – for example what would be in their best interests for their graduate careers. I knew a bit about all of the seniors' undergraduate records, and to the extent they wanted my advice, gave it to them freely though I targeted them to institutions that I thought would be good for them and which fit their skills. Among seniors in the class of 1965, I know that two completed degrees at the University of Colorado, another went to Cal Tech, one to Michigan, one to Iowa State, another to Utah, one went to Kansas but flunked out, another to Illinois and Jim Fargher went to Southern Illinois in Carbondale. And there were several more. With one exception, they finished their degrees and had – as far as I know – excellent careers.

The fall of 1964 passed without major incidents in the laboratories – always a good sign for a starting assistant professor! It did seem to both Jerry Mohrig and myself that the expansion of lab hours happened because organic chemistry experiments do not run by the clock, caused young, aggressive students to work far more hours than the allotted lab time. Somewhat later we tried to make amends for this by changing the work hours for the organic labs. But at the beginning things went smoothly. Morrison and Boyd, according to Mohrig, was a successful and teachable text.

The second semester went well, too, as far as I know, but it started almost on the first day with a major emotional display in my office by one of the young ladies in the class. This young girl came to talk to me and she told me, in no uncertain terms, how much she hated organic chemistry. As a young assistant professor, I did not quite know what to do because – at some level, I thought it was my fault though I had yet to give my first lecture in the Hope course she was enrolled to take. So I just sat there and suffered internally from the bullets of being at fault because I was an organic chemist, while a good student dropped the course during the first week because she hated it (I blame Mohrig!). As it turned out the girl's father was a chemistry professor, so she took chemistry because she thought he wanted her to do so. She really wanted to major in English.

To the best of my knowledge, when this girl settled down, she knew that her problems with organic chemistry were of her background causing, and not because of the instructor. As for me, I did not know about the former, but I sure as heck knew what ever her problems with organic chemistry, I did not have time to cause them! This was one of my first steps in academic maturity. When a young lady comes crying in your office, there are probably more things causing it than just your last awful lecture in chemistry – so grin and bear it. She will figure that out, too. (This was extraordinarily useful advice to me, by me, when I later became a department chair!) And so went what was to become a continuous flow of tears from almost my first year of teaching until my last was something I learned to live with. Young people were under so much pressure and stress, and organic chemistry – as well organic chemistry research – really brought it out of them.

234

My research career started simply enough. After I convinced Irwin Brink to buy my labs a simple gas chromatograph, and I – with great help from a young freshman, Jim Hardy, installed it - I was in business. In the words uttered by scholars more eminent than I, one's research accomplishments become less important – like mountains in a rear view mirror – as one gets farther from them. And so it was with my publications. As a matter of fact, and this will come later, of all the work my students and I did over more than 47 years in the laboratories, only about five concepts remain of consequence today and they were mostly the results from the years when I was closest to the lab. After I had Ph. D. student responsibilities project direction deferred more and more to them, and their creativity was a lot less than mine. It is an extraordinary Ph. D. advisor who remains creative throughout his/her career. The rest of the data, and procedures in my many papers, merely added to the flotsam and jetsam of science, though one can frequently be surprised when one or more of these results show up as important in a new project in a diverse field. The important idea is to express what one wishes to say clearly, and to be careful that errors in results, preparations and data are as infrequent as is humanly possible. In the case of the academic chemist, the students trained during the process are the most important product and I always kept that at the forefront of my work.

My first publication was a short four-page communication in *Tetrahedron Letters*. At this stage and time, it seems so trivial as to not even mention what it taught, but trivialities never bothered me and this one does not either. For it was really the first paper – ever – published in chemistry by a faculty member at Hope College working on his own. Van Zyl, with several students had published work starting as early as 1949 and had, according to my count at least 15 papers in the *Journal of the American Chemical Society, Journal of Organic Chemistry*, and *Canadian Journal of Chemistry*. But it was the student's work. My work, was done by me, a young faculty member, and took on some significance over the long run because it was. As I said, Cal had targeted my colleagues and I to change the culture of Hope's chemistry department from one where chemistry was taught during the school year, to a professional department that lived with chemistry all year around. That happened during those first years; the initiative invaded other departments in the sciences one by one, and today, it pervades most of the culture of the institution.

My first paper, "Behavior of (α-Hydroxy)cyclopropylcarbinyl Free Radicals" was published in a then-out-of-the-mainstream-journal-of-organic-chemistry, *Tetrahedron Letters*.[124] Hope did not subscribe to the journal before I managed a subscription with the then librarian, John May. So Hope's library's collection of Tetrahedron Letters began in 1964. Now, of course, with everything online, it is no longer an issue what a library does, or does not, subscribe to. But it was then and among my first objectives during that first year was to improve the chemistry library. *Tetrahedron Letters* was one of several new publications started after World War II by the brilliant, famous, crazy Robert Maxwell. Maxwell, who died by falling a boat into the Mediterranean years later. He had become incredibly wealthy using his insight that scientists wanted their papers to appear quickly to get the publications of young scientists in print quickly. His life and career went in the dumpers later, but that's another story. On inspection now, my paper is self-standing and rather interesting but derivative over things I did as a Ph. D. student. What it did mainly for me was to put my name, alone, on a paper in organic chemistry before the fall, 1965 A.C.S. meeting – so some of my colleagues in organic chemistry could see that even though Hope was a four year liberal arts college, I did not go there to die professionally. That was a very important point to make for the College too, and luckily the school learned from it. Professors in liberal arts colleges have to have a "publish or perish" mentality. In chemistry at least, I believe we set that standard at Hope in the 1960s.

Funding for research programs in small colleges was mostly limited to funding from Research Corporation and the Petroleum Research Fund (P.R.F.). The former directed modest efforts to younger faculty, and had a strong bias – though this was not an exclusive *sine quo non* for Research Corporation to fund four year colleges - at least not then. P.R.F. had two programs that became important to the early stages of my career; the P.R.F. Type-G program which funded young faculty members – Mohrig and I both received funding from this in the fall of 1964, and Cal bragged that Hope was the only school in the country where two young faculty members had successfully been funded by P.R.F., Type-G. The Award was $2000 and was for one year. It could not be renewed, and there was no salary included for the PI – principal investigator. Research Corporation also funded my program early too, but only after I received the first rejection of my career (there were to be many,

---

[124] DC Neckers, "Behavior of (α-Hydroxy)cyclopropylcarbinyl Free Radicals," Tetrahedron Letters, **1965**, *23*, 1889-1896.

many more!). The reason for the rejection, which Brian Andreen, eventually to become a good friend, explained, was that Paul Bartlett did not understand one of my proposed experiments. This was a first lesson for me and important one. One has to write clearly and accurately if one wants another to understand what one has written. As it turned out, I rewrote a few sentences, and resubmitted the same proposal which was funded on a second attempt. So I started out with funding from two separate sources.

The P.R.F. Type-B program, which was started in 1956 and provided modest funding for faculty in four year colleges. Cal was on the P.R.F. board at the time, and I am sure had more than a little bit to do with the fund's recognition of four-year college faculty. Van Zyl was among the first to be funded; and in the same year, or a year earlier, he won the then Scientific Apparatus Maker's award for chemical education administered by the American Chemical Society. Cal wrote that nomination for him, too.

I used my P.R.F. Type-G funding to buy some chemicals from Aldrich Chemical Company. I remember questioning the purity of the compound I received because, "my gc trace" showed their materials to be impure. I was a Hope alumnus, so my entire mind set in starting my career there was to "do so as inexpensively as possible." Like summer salary, start-up salary was an unknown to me. The only things I carried with me as my start-ups for laboratory were two microdistillation rigs purchased from the Harvard glassblower, and a bottle of D_2O. Harvard bought this from Israel in quantity for its chemical stores. Why I thought I needed it is beyond me, but it seemed like a good thing to have so I bought it. I say "bought" because when I went to Bob Vanelli, the Harvard chemistry department's financial administrator, and asked him what I needed to do "because I'd taken a job and wanted to buy these 'things' for beginning in that position," Vanelli had said "bring me a purchase order."

Unfortunately, I had no idea what a "purchase order" was, so I sent President VanderWerf a note saying "Harvard says give them a 'purchase order.'" What came back was a letter from the indomitable Henry Steffens saying, "the Prexy says Harvard wants a purchase order; just buy what you need and tell those Harvard guys we'll pay for it."

When I took this to Vanelli he laughed, and he talked to me for at least ten minutes about how he wished for a job in a place so simple. They gave me the distillation rigs, and the D_2O. I carried them with me everywhere because they were so precious. I used the distillation rigs but save for a couple of experiments in my early career, never did use the heavy water.

Hope had purchased, or, rather, Van Zyl had gotten N.S.F. instrument funding and purchased, from that same salesman at Wilkins Anderson, a Beckman D2 absorption spectrometer, as well as a Beckman infrared spectrometer. When we started at Hope, the entrance facility to the Science Building looked like a Russian apartment complex in Soviet days, filled with boxes containing things to be used sometime, somewhere, but no one knew exactly where or how. We collectively—Klein, Mohrig, and I—unpacked the instruments. Klein was the most savvy instruments guy, and knew – as did we all at some level – that the Cary U.V. visible spectrometer was the class in the field. So we convinced, or Dave Klein did, Irwin to return the Beckman D2 spectrometer and this he did during that first year. We had to take a "restocking hit," as in the cost for the company to take it back, but soon a new Cary showed up and uses were developed for it. I have already mentioned my experience with the gas chromatographs. Gas chromatography was in its infancy at the time, so we were using facilities that would be considered dark ages rudimentary now. But there were first generation instruments with all of the implications thereof. Nevertheless, if the demands were not too imposing, they worked and worked well. If compounds were reasonably volatile, they lent themselves to its analytical powers. So another part of what I chose to work on at the beginning of research career was to ask the question just how amenable is this project to simple, rudimentary gas chromatographic analysis.

Qualitative and quantitative information was relatively easy to get. The standard approach for identification by gas chromatography was to obtain an authentic sample, and confirm identity by comparative retention times (as in the time after injection required for the sample to be transported through the column). In the case of my first paper, this was simple because the principle products expected were two similar organic compounds, phenyl cyclopropyl ketone, and butyrophenone. I remember I made the butyrophenone because for the synthesis, I had the starting material, butyryl chloride. I carried out the preparation on a Saturday morning when few others were in the science building. Butyric acid smells like rancid butter and once one gets it on one's hands, it takes several days to get rid of the smell. The synthesis of butyrophenone is trivial Friedel Crafts acylation – excess anhydrous aluminum chloride, carbon disulfide as solvent, 1:1 benzene with the benzene added first, and then the acid chloride. After the vigorous evolution of hydrogen chloride ceases one boils the mixture for about an hour, pours the whole gemisch into an ice bath, separates the organic layer, distills the carbon disulfide away, and is left mostly with pure butyrophenone.

This is an easy synthesis, but it was hard to think of one that makes a bigger impact because every part of it is bad news; it stinks, the components are flammable, and carbon disulfide is also cancer causing. The hoods in Hope's old science building were self-made – as in Van Zyl, when the building was being built, went to a plumbing contractor, or HVAC contractor (though it was long before the acronym was used – 1940) in Holland. The directions were "something like this is a fume cupboard, can you make such a thing in the various labs in the building?" As I remember now, only one of the hoods in the old science building worked at all efficiently, and that sure was not the hood in Doc Van Zyl's office at the top of the second floor stairs.

What is clear to me now that the projects on which I chose to work in my first months as a faculty member at Hope College were excellent. They were conceptually easy to explain even to the youngest students, the research work was simple to do and the results easy to quantify. Which brings me to the next point about gas chromatography. In order to find out "how much?" as opposed to what was being produced, or used up in a chemical reaction, one had to integrate the area under the curve of the product and then relate that integrated area to the integrated area of a standard reagent either added before the reaction in question had taken place, or immediately afterward and before analysis. The standard had to be non-reactive under the conditions of the experiment if added before reaction and similar in chemical characteristics so it's response to the thermocouples of the gc did not differ that much from the reactants or products. It also helped that it had a boiling point similar to that of the reactants and products. In order to use it for quantitative analysis, one had to calibrate its thermal response hopefully developing a linear relationship between its concentration and the areas under curves that were obtained in response. So, as the standard was used, it could be quantitated directly. Alternatively, it could be quantitated against the product's response as in 1 mmol of product gives a response of 100 while 1 mmol of standard gives a response of 90. Therefore, if in a reaction, one has a response of 90 from the standard, that must mean that whatever the response of the product, it must relate, 10/9 to the standard's response. One obtained the quantitative information in any of several ways. The most common at this time, and the one I used, was to cut the peaks represented by the responses out and weigh them. This assumed that the graph paper of the recorder had about the same density/in^2 everywhere – which it surely did not so that introduced errors.

If one looks at each of my papers published exclusively from work done at Hope College, the gas chromatograph used was that Wilkins Aerograph, the recorder and recorder paper, from a Sargent recorder, and the experiments quantitated by cutting out the peaks and weighing them. Every experiment was done a minimum of four times in what became, eventually, mindless work. But it demonstrated the required effort required to do publishable laboratory work. The experiments being done were easy to understand.

I would say this mantra on advising young assistant professors: that work be done is easy, and could be understood by college undergraduates. Simple ideas dominated what I did in my more than six years at Hope. Fortunately, I did not make in huge mistakes, though I could have. The work was done carefully so it has stood the test of time.

Campus Politics in Hope in the 1960s

Multiple forces were imposing on all colleges and universities by the time I decided that leaving Hope to work with Ph. D. students would be in my best interests. It became my goal. Eventually I managed this with some spectacular successes though that took a while. Hope's political environment had become corrosive, particularly toward the end of my tenure. See the letter to me about promotion to full professor Appendix 11. As a chemist, I was under continual pressure to be more involved in campus, as opposed to just chemistry, activities. I was added to two faculty committees, International Education and Cultural Affairs, by their respective chairs as much because my campus colleagues knew me as a student, and knew my respective interests in European history and music. I liked these appointments mostly, but both chairs left for sabbaticals in 1967-1968 and left me in one case with the chairmanship, and in the other case with a more extensive role than I had expected or wanted.

I took over as chair of the Cultural Affairs committee in the fall of 1967 and suddenly found myself in the middle of political controversies national in scope that neither I, nor anyone else, understood. In fairness, Cal had told me that I did not need to take on this responsibility. But I felt the pressures from my colleagues and did so as much because of that as for any other reason. Much of what went on that year was interesting and enriching. The music faculty brought organists Marie Claire Alain and Jean Langlais to the campus; they also brought John Cage about whom I remember two things. He put his honorarium check in an envelope and mailed it back to the College – by mistake. And within fifteen minutes of the start of his concert, the Chapel was nearly empty. But the biggest event of the year was a symposium in the spring entitled "Crisis in our Cities." The summer before, in 1967, Detroit had erupted in flames in what was among the worst urban riots in American history – the so-called 12th Street riots. I had been left with a scheduled event involving the, I thought, comedian Dick Gregory by the previous chair. I was put in the difficult position of knowing what to do in what might well become a volatile situation. So, at Dr. VanderWerf's suggestion, I called a few colleagues with strong interests in racial equality to a meeting, and said "what can we do to reflect positively on the College given that Dick Gregory is soon to arrive." And this was clairvoyant because by the time he arrived, March, 1968, it was a few days before Lyndon Johnson had chosen not to run for reelection, and Martin Luther King had been assassinated.

But that was much later. At some time in our first year together at Hope, Dave Klein, who was several years more experienced and also older than Jerry Mohrig and myself, introduced us at a faculty meeting to the idea of the N.S.F. undergraduate research participation program. He had even gone so far as to write out a brief pre-proposal to share with all of us. Faculty meetings in the chemistry department at the time were in the Kleinheksel/Jekel office suite, across from Van Zyl's office. They amounted to six or seven of us talking – periodically though not regularly – about the great issues of the day. One of the things I really enjoyed (I know that sounds like a ridiculous word to use in the context of "faculty meeting") about these sessions was we spent a lot of time discussing the individual progress of specific students. We also often talked about recruiting students, and what we could do to assist the Hope admission office in those activities. This was great fun in that it brought us all into the never ending search for students for our classes and our programs.

But in this case, as in this particular meeting, the issue was the undergraduate research program. Dave said "I've thought about who the Director should be, and I don't want to do it." I was pretty sure I did not want to do it either so that left Jerry Mohrig – or more appropriately Jerry Mohrig was really the best and most logical person to take on the duty – and this is what he did. Jerry wrote up a

241

more detailed proposal and we, collectively, spent several sessions working and reworking Jerry's ideas. I do not recall there were any major issues of consequence, but there was a lot of nitpicking back and forth over small things. Eventually, Jerry put the proposal together, and sent it off in – I think – the spring of 1965. All we had to do now was prepare to get the program started because, for sure, we would get the money.

But we were in for a huge collective letdown when a letter arrived from National Science Foundation saying our proposal had been rejected. Rejections are a fact of life in the sciences, and were to become a regular event in my career. But for my colleagues, and for me, this U.R.P. rejection this was as tough blow. We had worked so hard on the program only to have our peers say it was not up to their expectations – at least it was not to be funded.

Eventually we met as a group with Cal, and he advised to call the program director, find out what was wrong, and prepare to submit for the next deadline. Jerry did this. He found out that there was essentially nothing wrong; N.S.F. just did not have enough money. "If we were to resubmit," and this is what Jerry did the next year, there was a good chance the program would be funded. And in that year, I think 1966, the program was funded. As I think I heard on last visit to Hope, the chemistry department had had U.R.P., later R.E.U. funding, almost – but not exactly – every year since 1966. At some level, that must be a national record.

I funded several students in my research program during those early years from Research Corporation funding, from N.S.F. funding and from P.R.F. Type-B funding. I had grants from P.R.F. for a number of consecutive years, Type-B, Type-C, Type-A, and the original Type-G. Fifteen years later, P.R.F. brought in a new, young administrator and I found him knowing more about my research than I did (he thought.) So, after two consecutive rejections, I did not apply to P.R.F. again. By that time, I had N.S.F. funding, O.N.R. funding, and an occasional N.I.H. grant too with some industrial support. So, the small amount from P.R.F. was not worth the effort. But in those early years, P.R.F. was what kept my research program alive.

I have thought a lot about why research was so important to me, and to my colleagues in chemistry in those early days in a presumably teaching, undergraduate, institution. Though that question has many answers, and others might answer it differently than I would, its main value in the liberal arts college setting, as it was in university settings, was that it kept a faculty member's intellect acute and alive. It presented an opportunity to work one to one with students where they could do "chemistry as chemists do it." Personally, from the first day of my career to the last, and beyond, I was working on something to which I did not know the answer. Were these matters important and/or of commercial value? Most frequently they were not. Did they function to prepare students for advance study? They did. Were they the best things to be working on? Sometimes they were; other times they were not. Were the projects safe? Were the students taught good safety practices? And in most cases the answer to that is "no." We did not know what good safety practices were at the 1950s and 1960s Hope College. Fortunately for everybody, there were no serious accidents nor long term effects. Though I had several close calls as will be related in a moment.

However, the most important part of those years for me was that I learned how much I liked to teach, and how to work with student research associates. Those two facts controlled the rest of my university career.

I have often related an incident that happened my second year at Hope as being transformative. Every Hope senior was assigned a professor with whom to do senior research. The second senior class of majors (class of 1966) with whom I came in contact was small relatively so I remember only one student assigned to me from that class—Tom Elwood. Tom does not know it because I have not seen him from the day he graduated to the day I am writing this, but he played a huge role in how I was to work with students. As part of my working with undergraduates in senior research, I always remembered Paul Bartlett's advice – give them something you could do yourself in a day, and let them be frustrated by it. Help them, but do not do things for them. So it was with Tom Elwood. I do not remember now what project I assigned to him, but it was a simple idea that could be completed in an afternoon I was sure.

Two days later, Tom came back with a result that was entirely different than I had told him to expect to find or to find. Tom was older than most majors and had two young children at home because he had been in the military. He was not about to back down from a young assistant professor who was almost as young as he was. If he knew, he knew.

Thank goodness for that. I told Tom "to do the experiment again" because I did know everything. So he did this – with the same result. Finally, after a third try, resulting in the same report, this dumb assistant professor figured it out. The student was the observer; I was the mentor. If the student said "this is the result" that, within reason and as long as care was in evidence, was what was the result. From that point on, I stepped back a notch from my student's lab work. From that point forward, we became partners in pushing back the edges of the small bit of nature we were trying to observe. The mentor-mentee relationship became a partnership of equal partners – at least at some level. I had learned to manage research – not directly enforce it.

Later, as my career progressed and my group of Ph. D. students and post-doctorals reached a high of twenty-four in 2006 or 2007, my colleagues felt I might have been too far removed from my students' work. But I had a system of multiple collaborations set up – one in which more advanced students worked closely with beginning students – and it worked beautifully for the entire time I directed research. As my students have continually attested, there was a mutual respect engendered that worked to everyone's advantage. For that methodology, I have always thanked Tom Elwood, Hope class of 1966. My directing research and how I did that resembled the way English cathedral choir directors manage the boy sopranos and altos.

Student investment: Publishable research

Unless one were to decide that one's career was to be fixed in the four-year college, the primary goal had to be publishable work. In my earliest Hope years, I had no clue what I wanted to do when I grew up, so publish or perish claimed me more than just a little bit. I was sure, after two or three years at Hope, that this was not for me long term. I did not feel comfortable in Holland, Michigan for religious reasons – and that was a primary driving force; but more I did not see myself teaching just undergraduates for the rest of my career. It was really too limiting. I wanted to work with Ph. D. students. And by the time I made the move to university, I was mature enough to do so.

During those early years at Hope, I had several inquiries about my interest in teaching at bigger schools. Following my summer of teaching organic chemistry for pre-dental students at Ohio State, I visited there as a candidate for a permanent position; but they hired John Swenton. John was a young man I knew a bit during his undergraduate years at Kansas and he was brilliant though his research career never reached its full potential in my view. As I remember, he had taken a Ph. D. with Howard Zimmerman at Wisconsin and done post-doctoral work at Harvard with P.D. Bartlett. Rumors back about the Bartlett group of those days suggested it much less genial than it was when I was there but these

244

are only rumors. I also interviewed at the University of Cincinnati, and was offered that job. I went to Cincinnati when it was spring in southern Ohio; still winter in Michigan. And, of course at that time, Cincinnati's basketball team was the best in the country. So this job was very tempting. In the end, I felt I would not have been fair to Cal or to Hope to leave so quickly. So I turned the job down. Eventually Cincinnati hired Marshall Wilson who, in about 2006 decided he had had enough with that place, and pleaded with us for a position at Bowling Green. We afforded him a low paying research professor's position and as of this writing, he's serving as interim director of the Center for Photochemical Sciences which I started. Funny how it all turns out!

The only other job I seriously considered was one at the University of Massachusetts in Amherst. Bill McEwen, my former professor at Kansas had taken over as head of that department in the mid-60s, and he asked me if I would be a candidate. I thought about that too, and in the end, decided not to consider it. UMass at the time was struggling to find out who it was in chemistry. Its polymer initiatives had not yet begun or were just beginning. I had extraordinary loyalty to Hope College and to Cal VanderWerf. My sixth sense said I had to stay a few more years out of an honor to my family and to their extraordinary loyalty for the institution. So this I did.

Everyone is haunted by roads not taken. And I surely was. I kicked myself many times over for not at least more seriously considering the Cincinnati opportunity. But in the end, even that turned out much in my favor. I will be writing the history of the Center for Photochemical Sciences in a later chapter but with that, and with the extraordinary insight of those developments, developments in photopolymerization and in three-dimensional printing my research career took off in ways I never anticipated.

Cal had been president of Hope for a year by the time I started and he had determined that no matter the costs, the faculty had to grow. He said, in announcing a class of new faculty in the fall of 1964, that "every department that came to me this year in need of new faculty was allowed to make a new hire." He followed that pattern for at least two more years. In essence, he turned the faculty at Hope from one generation to another in the period from 1963/64-1966/67, and greatly expanded it. The school was financially strapped as his often pointed memos to the Chair of the Board of Trustees would point out, but he decided, with the Board, that Hope had to grow with new blood to survive. And he challenged his departments to get good people so it could.

By the time his presidency was finished a core of new faculty had been assembled. This included Sheldon Wettack and Mike Doyle in chemistry, Jim Van Putten, David Marker and Dick Brockmeier in physics and Ralph Ockerse in biology. These persons, and some of their colleagues, were determining a future institution where research with undergraduates would be its strength. Cotter Tharin had started the department of geology, and when Ockerse left to become Dean at I.U.P.U.I. in the early 1970s, Jim Gentile came to take his place.

Chapter 15

Creative Cal VanderWerf

Cal VanderWerf was arguably among the most creative persons I have ever met. He did not see problems; he saw opportunities. Others would say "It can't be done." He would say, "We can do it. It may just take a little longer." Cal's forte was in finding unique ways to have others help him accomplish the finest of things for Hope College.

Pressure for the DeWitt Cultural Center at Hope came about because Cal called the President of the Student Government to his home one night and said "You students are too complacent. You know we need a student union. So start a petition drive targeted at the President (him) and Board of Trustees demanding a new student union be built."

Rachel and Cal VanderWerf

It took a few years, but eventually Cal talked friendly Zeeland chicken ranchers into giving the then Hope College enough money to initiate the building of a student union, now known as the DeWitt Student Center.

Recruiting students in the East was a major issue for the College in the mid-1960s too. Neither New York nor New Jersey had state university systems at that time, and many high school graduates from those states went west for their collegiate experience. But Hope was losing out, and Cal felt Hope was unrepresented in the east and not well known. So, he recruited Larry ter Molen, a former football team captain and graduate of the class of 1959, to become Hope's one-man recruiting effort in New Jersey. Larry told me recently that he set up an office in the east and loved the job but because he did so well, Cal moved him back to Holland where he became director of development. Recruiting passed to those who did not mind spending airplane time. Since Larry knew the territory, that became him as, for a time, he did two jobs!

Phi Beta Kappa was another VanderWerf initiative. Cal always regretted that his alma mater, Hope College, did not have a

Phi Beta Kappa chapter because he had nearly a 4.0 grade point as an undergraduate, and would certainly have been a member. He was also likely jealous of Rachel's Phi Beta Kappa key so he pushed and shoved the members of Hope's faculty who were members of Phi Beta Kappa, with Rachel among the leaders, to petition the national honor fraternity to consider starting a new chapter at Hope. This did not happen during his presidency but it did a few years later and Cal was the only member of Hope's alumni inducted by Hope's new Phi Beta Kappa chapter when it did.

My own experience in providing funding for the first major instrument on campus—a nuclear magnetic resonance spectrometer—followed a similar pattern. Though I have told this story in Mike Doyle's *Academic Excellence,* it bears repeating because it shows the ends to which we went to make *our* Hope College competitive.

At some time in late 1966 or early 1967, Cal called the chemistry department to his office to discuss "new instruments for chemistry instruction." I must have made a good case that we really needed a nuclear magnetic resonance spectrometer because after discussing other things for a time, Cal came back with "I really want you to work hard to get an N.M.R. spectrometer for your programs." Cal knew the drill, because the first N.M.R. that Kansas chemistry had showed up during his department chairmanship.

I was assigned the task to find the funds for an N.M.R. machine. In those days, an N.M.R. was a huge investment not just for a college but also for universities and for a granting agency. At the time, one 60mHz commercial N.M.R. spectrometer cost more than six Cadillacs. Local businessmen, including a couple members of the Board of Trustees had worked their whole life for just one Cadillac. How could it be possible that *that College* to which they gave what was left of their hard earned dollars as contributions needed to spend that much just for a bunch of chemists?

But I proceeded to write a proposal to the National Science Foundation that did not get funded. In those days, N.S.F. did not have to tell a principal investigator why his proposal was not funded; it was good enough just to send a letter to the President saying "This proposal cannot be funded this time."

I must admit to having second thoughts myself. Though I believed strongly in Hope College, and its mission in the sciences, I knew that undergraduates could never make as much use of a nuclear magnetic resonance spectrometer as graduate students could. Were I a member of the National Science Foundation, I would use the money that I had for N.M.R. spectrometers on instruments for graduate institutions, not undergraduate institutions. What did I know about running an N.M.R. lab? My experience at Kansas was almost non-existent, and one of the only times I used one of Harvard chemistry's two N.M.R. machines, I broke my N.M.R. tube in the magnet insert because I was so nervous.

At some time when this search for funds was going on, I ran into Henry Steffins in Van Raalte Hall and he told me about this new federal program meant to improve instructional equipment for teaching. The great influx of students into universities was causing huge instructional dislocations, and educationists felt that better equipment could be of benefit for instruction of the masses. By better equipment, these educationists meant more electronic means such as internal television monitors that would allow a single lecturer to reach many more than a single classroom full of students.

Steffins gave me a copy of the proposal guidelines. I found them to be manageable and proceeded to write a proposal for an N.M.R. machine to this mass education competition. Remembering that this program was meant to provide instructional equipment mostly for large sections of undergraduates (mostly freshmen) in universities like Michigan State, I wrote the proposal for an N.M.R. spectrometer rather making Hope College look like it was teaching N.M.R. spectroscopy to thousands of freshmen students! I do not think I counted any that sold burgers during Tulip Time on 8th street as being part of the great masses that would be impacted by this new N.M.R. spectrometer, but I came close.

The College had to commit 50% as a match and this meant that the commitment had to come from the President. After my friend Henry read the budget and found a few addition mistakes ("Boy, you'd be a lousy accountant Neckers"), he passed the proposal for an N.M.R. spectrometer on to Cal. Cal signed it committing the institution to about $18,500 in matching funds for a new N.M.R. spectrometer. In other words, the fear of those Board members could come true—*those chemists* at *that College* were using their charitable donations to buy a spectrometer that was worth three Cadillacs. (At least *that President* did not commit funds the equivalent of six Cadillacs!)

As I remember, the turn-around time from program announcement to the proposal being due was quite short, and this was a state program so we had to get the proposal when it was finished to Lansing. With all of the signatures required, we just made the deadline.

On a hot summer day, Larry ter Molen and I packed the proposal in its multiple copies in his car and headed for the State Department of Education in Lansing. I do not remember much about what we found on the other end, but I do remember this as the first time in my life I was in Lansing, Michigan. Larry was a Grand Rapids native, and I do not think he'd been in Lansing very often either, because Lansing was the other side of the world to the west Michigan Dutch. But we got there, delivered the proposal, and went back to Holland to await the outcome. It took a very short time as I recall, but one day later in the fall my friend Henry called and said "The Title 6 proposal is going to be funded. You've got the money!"

There was real joy that night but that joy lasted about twelve hours. The next morning, I went over to Cal's office to tell him the good news and ask him about the matching money, when he told me, "Okay Doug, now you have to get the matching money."

Suddenly I was on the hook for the next $18,000 or so. The next lesson about administrative life in the academy: a good administrator looks at capably people and says if you really want that, you're going to have to really work for it. And then he supports those that can do that.

Fortunately, I had a good teacher in Cal VanderWerf.

The manager of the local Holland Color and Chemical Co. plant, by this time called Chemtron, had had a summer experience at Chautauqua Institution in western New York and his kids played with my brothers Bruce and Craig at Lighthouse Point on Chautauqua Lake. Though I had met him (his name was Ace Candee) only casually in Mayville, NY, that casualness was good enough for me to make an appointment with him for a conversation about an N.M.R. spectrometer. I went to his office to talk about that only to meet, for the first time, Frank Moser, Ph.D. I had read papers in the literature by Bachmann and Moser so I knew Frank's work and immediately felt among scientific friends. Frank was a most loyal Hope chemistry alumnus. What a break this was for the College.

I needed the matching money for an N.M.R. spectrometer, so before I got out of Candee's office I had agreed to teach a course on spectroscopy to local chemists most of whom were working for Chemtron. But I still had no commitment for funding for the N.M.R. machine from Chemtron or anyone else. I called Cal and told him of my meeting with Candee, Moser, et al., expecting he would say, "Let's go see them." Instead he said "Bring them to my office." Now I thought that pretentious—here we were asking them for money, and they had to come to his office to talk about it?

But I called them, they obliged, and a few days later Candee and Frank Moser came to Cal's office. After small talk, Cal said to Ace (who he had never met) "Ace, we all know what N.M.R. means to research in chemistry, why doesn't Chemtron commit the matching money to our purchase, and you can use it for a prescribed number of hours every month." I about fell off my chair. He was asking this company for a contribution of $18,500 to my N.M.R. spectrometer!

Eventually, they gave the College $10,000 for the spectrometer. Cal confessed afterwards that he did not know how high they would go, but had he asked for $5,000 he would have gotten $5,000—by shooting high, the company came back with $10,000. Another lesson learned in academic entrepreneurship: Ask for more than you expect, and you will get a lot more than you would had you asked for just what you needed.

Raising the rest of the matching money was not really a problem as I recall so then we had to order the instrument from Varian Associates. In this purchase, I learned that even big companies do not make a large instrument without a down payment so the College had to come up with half the purchase price on submitting the purchase order. The spectrometer took six months to build so Varian had our money for quite a while before we saw the machine.

One snowy day in February I was teaching organic chemistry lecture on the second floor of the Science Building when Irwin Brink came to the door and called me out of the class. "Global Van Lines just called and the N.M.R. will be delivered this afternoon. I've talked to the driver and he says we have ninety minutes to get it off his van or he'll take it back to California."

Welcome to the next stage of life in big science! Not only does one have to get the money, and place the order, but one has to be prepared to lift the item off the truck when it comes as well! Even if the damned thing weighs half a ton. At about 2:30 p.m. a Global Van Lines truck drove down 10th St. and parked in front of the Science Building. Irwin, who always managed a cool head said, "We'll call Henry Boersma" and in ten minutes or so Henry showed up with the campus maintenance employee, Jim the janitor. In the meantime, the Van Lines driver was prancing around like a steed in search of mare in heat telling all who would listen that we had to "get that thing off his truck or he was taking it back to California." Henry and Jim, the janitor, cool heads that they were, walked over to

Gilmore Hall (still under construction), found a front loading Ford tractor, and convinced the construction foreman to "let us borrow it for a short time."

A few minutes later the construction company operator drove the tractor with its front loader onto 10th Street, unloaded the magnet for the spectrometer (it weighed almost 1500 pounds which was the reason for all the consternation/worry) and put the magnet on the rear loader of Jim's Jeep (At that time as I recall the College had two vehicles—a Jeep to plow snow, and a truck to haul trash). The last thing I remember was Jim's Jeep driving down 10th Street with my N.M.R. magnet (worth at least 5 Cadillacs) hanging off its backend.

"Not to worry Neckers," he told me with confidence, "that box won't fall off my Jeep."

A few days later (it took that long to find movers that could handle a 1500-pound magnet), we got the spectrometer in place on the first floor of the Science Building (still another bathroom bit the dust) and, after a month or two Dick Homan—a wonderful Varian engineer and long-time friend from that forward—finally managed to get it installed.

For the rest of the time until I went to Groningen, the Netherlands in 1968, September, my office was also in that former bathroom. When Mike Doyle arrived, I gave the honored position next to the N.M.R. spectrometer to him but all he did was complain—and he does this to this day—about my giving him an office in the N.M.R. room! He should be so lucky I thought. Had the N.M.R. not been there his office could have been in the first floor woman's bathroom!

Chapter 16

The 1960s Begin to Unfold

The college community had been, mainly, enthusiastic about the VanderWerf's arrival. Cal and Rachel brought six children to the campus. The VanderWerfs were very different than had been the Lubbers – the presidential family immediately preceding them. The oldest daughter, Gretchen, was a senior in high school – a transition, as in moving during her senior year in high school, that for her that had to have been traumatic. Gretchen enrolled at Hope the next year, took her degree in 1969, won a Wilson Fellowship, went to Michigan and eventually the University of Colorado to become an oil and gas lawyer. The others, four girls and a boy, Rachel worried, were being made to live in a fishbowl that was the President's house. As reflected elsewhere, changes were made in the house to accommodate that.

But even before he was inaugurated, Cal collapsed from the job at hand. He was inclined to get up early finding his best hours were those before noon. But when he got to Hope he was trying to wrap up his scientific career, and overwhelmed by what he faced at the school.

Cal understood higher education from its roots and had excellent insights into what Hope could become. Rachel was not an unseen player. With their housekeeper, Mrs. Wilburn, to help with the house and the children, she was able to interact with the community. She and Cal were close friends and excellent for each other. And, most importantly, she was Cal's intellectual equal.

When Dr. VanderWerf's appointment as Hope's president was announced, I went to his office on the second floor of Malott Hall in Lawrence, and offered my congratulations in the clumsy way that most any graduate student might manage. I said something like I was "Sorry for Kansas but delighted for my alma mater." In the back of my mind, though, I wondered if this would work, and if it did, how it might work? Cal was involved in everything at Kansas. He was recognized as a master teacher both of freshmen in the general chemistry course, and in organic chemistry; he was vigorously involved in matters of racial justice both for blacks and for those being recovered from World War II in the community and in the greater university at large; and he was a strong proponent of campus athletics of the kind that supported the mission of an academic

253

institution. How, possibly, could he move to an undergraduate college in Michigan that still retained denominational control that was close to irrational? Hope's athletic programs were good for a small school. Its music programs were almost totally church oriented and the programs suffered from too many that had religious inclinations on college graduation. My sense was he had too much pizazz for the Hope of my fathers. He was too academically inclined for a college that, at best, did well with a homogeneous group of hard working students.

I can imagine how Cal felt on his inauguration day. It was well planned and a great celebration. Cal's father died when he was very young and he barely knew him so he knew that his father would have been really proud of him that day. Inauguration day had many special contributions: President VanderWerf, in his speech for the day, was uncharacteristically pessimistic, at least at the outset, though mixed with his brand of humor: "A college president is a person who attempts the impossible while awaiting the inevitable. A college president is paid to talk; a faculty paid to think, and a dean to keep the faculty from talking, and the president from thinking.

He continued with his mission: "Society entrusts to our fellowship and tutelage its most precious and priceless possession; its youth." VanderWerf recognized founder Van Raalte, who uttered his prophetic words about the act of faith that was starting a College: "This is my anchor of Hope for this people."

Van Raalte had said in prophetic words for the situation VanderWerf was now entering: "I am determined" Van Raalte said, "that my people will not become the fag end of civilization." The College had thrived, VanderWerf noted, with accomplishments that were too great for its size. Graduates were making unique contributions in the arts, the sciences, in business, in the professions, and in religion far out of proportion to the size of Hope College.

When he turned to research, he noted how far man had come and how far students had left to go. "Ladies and gentlemen, our freshmen today know more mathematics than Descartes, more physics than Newton, more chemistry than Madam Curie. In the last fifty years, there have been more scientific advances, pure and applied, than in all the previous ages of man's existence."

He was direct in discussing the future of higher education: "The numbers of students are growing at an ever increasing rate, and by the year 2000 we can expect 25 million students, but mostly in tax supported institutions."

VanderWerf laid out the unique challenges ahead: "While we, as the pioneers on the frontiers of the new Age of the Intellect, are faced with the question of the Christian liberal arts college. Many observers of the American scene believe the combination of factors I've described spell doom for it, and think that by 2000 it can be written off completely. Hope College, I declare, will be there to offer a continued liberal arts education with distinction and excellence, to the able and the ordinary, whom come to us. We promise the ambitious with a desire to learn that come us will find the door of Hope open."

The entire address was published in the *Anchor* for students and the community to read. Few knew how clear Cal was laying out his vision for Hope's future and future contributions, but VanderWerf had the misfortune that his charge to Hope College was published on the morning after John Kennedy was shot in Dallas. The age of American innocence had passed.

Hope went on normally or at least the *Anchor* said nothing about the Kennedy assassination after the November 27 series of articles in the Hope *Anchor*, and the prayer vigil at the Chapel on that Friday.[125]

[125] Hope College *Anchor* (Volume 76, Issue 12), November 27, 1963. Copyright © 1963, Hope College.

Cal's second year ran into the presidential election of 1964. As the presidential campaign progressed, Barry Goldwater continued to stick his war-hawk foot in his mouth. *The New York Times* reported in August 1964 that Goldwater suggested that the White House had authorized the use of nuclear weapons in Southeast Asia. The White House immediately denied the report, while they scurried to find out who said something that might be so interpreted. But an *ex post facto* view fifty years later has to believe there was some truth in it. Military advisors remained the same whomever the President, and there were serious military hawks about so they could, indeed, have gotten White House buy in to such a scheme.

Cal's immediate postwar research caused significant questions in his mind about the direction of the nation at that time. He was, after all, a conscientious objector who had done chemical research for the war effort. With the election over, Hope College went on. Students, because there was no student center, had to find their fun where they could find it. Holland, for its part, remained conservative and stilted. The Holland Theater showed recent movies but had to be careful not to assault the prudish sensitivities of the community.

The Diem government in Vietnam, which the U.S. was supporting, was tenuous at best. The Viet Cong were everywhere, and America was heading for that land war. President Johnson, for whom most of us voted because of his peace stance, turned out to be a peaceful president only when compared to Barry Goldwater. Within weeks of his election, he was sharing with Secretary of Defense Robert McNamara that he intended to "win" the war in Vietnam regardless of the cost.

One can understand where that strategy might have come from. Twenty-five years earlier, America and the world had stood together to defeat militarism in the Pacific and the Nazis in Germany. Winston Churchill died in January 1965 and Paul J. Fried, a Hope professor of history knowledgeable on the history of Nuremberg, wrote in the *Anchor* (January 22, 1965) that Churchill proved:

- That in an age of mass society there remains room for the uncommon man of imagination, courage and individualism;
- That in a world that places a high premium on security and conformity it remains possible to seek adventure;
- That in a society which accents youth and rejects the old, a man can have his finest hour after others have retired

- That in a climate of totalitarianism of the right and left there is a desperate need for those that believe in individual liberty
- That in the face of growing nationalism it is possible to serve the nation best by promoting the unity of nations
- That despite growing emphasis on greater and greater specialization what we need most are men of broad interest, versatility and vision.

"In the best sense of the word Winston Churchill's life is that of a Renaissance man who combines an interest in art and music, war and philosophy, history and politics, social reform, loyalty to the King, profound scholarship and good food – capable of lofty idealism and down to earth reality – he is easily contented by the best of everything."[126]

What Fried missed was science and technology. Churchill knew he was indebted to his technical colleagues: chemical weapons research was stronger in England than in the U.S.; the British supported the Manhattan project as best they could; radar came from England; Turing was developing the skills that led to digital information explosions and the computer; and even penicillin had its genesis in London's St. Mary's Hospital.

Fried also missed that Churchill must have had a healthy respect for the methods of worship of the Anglican Church of England.

When Neville Chamberlain returned from Munich, September 1938, with the document that meant "peace in our time' Churchill's speech of criticism was delivered as an Anglican psalm: Every speech of consequence he made thereafter was also written as psalmody:

"And do not suppose that this is the end.
This is only the beginning of the reckoning
This only the first sip; the first foretaste of a bitter cup
Which will be proffered to us year by year unless –
By supreme recovery of moral health and martial vigour
We arise again and take our stand for
Freedom as in the olden time."

"We have our own dream and our own task.
We are with Europe, but not of it.

---

[126] Hope College *Anchor* (Volume 77, Issue 15), January 22, 1965.

We are linked but not combined.
We are interested and associated but not absorbed."

An important Hope campus issue was hotly debated. It regarded the liberal arts. What were they? Could the study of liberal arts remain relevant? Could allocating money towards those programs be justified?

The questions were age old, truly American, and always in conversation as the degree of undergraduate specialization became more prevalent. Hope had the added complication of denominationalism. It is amazing that all students that came through the College at the time left as anything but Dutch Reformed preachers. That denominationalism remained that intense. But I never considered a Hope degree as a pre-seminary degree. James W. Neckers, my uncle, had serious concerns of conscience about it though. Almost on his father, Albert's death bed, he confessed he had disappointed him. Albert Jr. wanted him to go to the Seminary. What Jim did not know was that Albert, Jr. had also been a disappointment to his elders. Letters I found from this uncle, J. W. Warnshuis, in 1891 told how much Albert, Jr. had let his sponsors down because he left Hope, not as a preacher, but to become a western New York businessman.

When I started teaching at Hope, the almost irrational denominationalism had been augmented by a similar intensity about the liberal arts. We, as in the sciences, were defined as some sort of aberrations because we had pre-professional, and even professional, education as our objectives. We were accepted as long as we kept our distance. But the campus academic leadership at the time was sure Hope's main goals for the future were in becoming a quality liberal arts institution.

This meant it would remain a four year, less professionally oriented, and educationally non-specific, institution. In retrospect, that might have been its best choice at the time. Though the Holland of today would be much more economically vibrant had it developed different, more professionally oriented objectives in the 1960s it did not have the resources, or the will, to exercise Phillip Phelps dream in the mid-20th century any more than it had in the 19th century.

On a personal note; I have a liberal arts education and cherish it because it gave me a life along with science. But, to the extent, chemistry instruction fits in the humanists' definition of being a liberal art, I fail to see it. Even understanding chemistry at a cocktail party conversation level is beyond most, even the very well-educated, humanists. Historical novelists like Lauren Belfer (*A Fierce Radiance*) do pretty well. Perhaps historical trails, as I am trying to do with this, represent the scientist's way to liberally educate the population?

Cal's inauguration speech, published in its entirety in the appendix, is a trail through his own life, from boy, preacher's son, Holland native, through a scientific career that took him to chemical weapons doorsteps, to a realization in the value of the liberal arts initiative in higher education.

Chapter 17

National Politics and Hope

The election of 1964 was nasty – not as bad as would be the election of 1968, but it was contentious nonetheless. The contentiousness in western Michigan was brittle. The area was Republican, mostly law abiding, and respectful. But more venom was unleashed by Republican partisans before and after the election of 1964 than at any time older citizens could remember previously. Students held mixed opinions: I was a peripheral bystander, unimpressed with a big eared Texan who butchered the English language but even more scared of an Arizona senator who thought nuclear weapons were battlefield ready to best the Viet Cong in Southeast Asia. Robert Donia, writing in the student newspaper, *The Anchor* (October 23, 1964) captured the essence. The economy was good; a major Democratic selling point. The Republicans thought they had God on their side. Goldwater's handlers were sure the press was not treating him fairly. Campus straw votes made it clear that the straw in the wind depended on who was in the room before it was distributed. Goldwater won one vote by at least 4 to 1, but this was at a Republican campus meeting consisting of less than 75 voters to be. At that time students had to be 21 years old to vote. A few days later, a mock election held by student council named Johnson the winner by 12 to 11 over Goldwater. In a room across campus a political science professor was calling Goldwater "a reactionary fascist."

Two history faculty commented on the campaign in the *Anchor*, that "Conspiracy in Pursuit of History is No Virtue" picking up on Goldwater's acceptance speech. They claimed they found his acceptance speech highly corrosive, and aggressive, which it clearly was. The speech was also aimed at splitting the electorate after Nixon's loss in 1960 and that it did. It was not clear whether the Hope College status committee (the committee in charge of faculty evaluations and salaries) gave the two historians credit for a publication in the *Anchor* or not; that likely depended on the politics of the committee members. But it was clear that the political nation was polarized politics was headed downhill and presidential elections would never be the same. During the same time, a professor of history from Wayne State, Alfred P. Kelly, spoke to a few hundred students on the campus and categorically stated that Goldwater could not be elected: the nation had come to an agreement that the New Deal of the 1930s was consensually accepted and that the nation *politic* was government together. Neither party had the upper hand so each contributed to the whole. Though senators like Goldwater thought differently, he was frequently on the outside of decisions within the government.

Grand Rapids' Gerald Ford was rising as a potential leader in the House. That body was apparently insecure with its Republican leadership, even at that time, so Ford brought one of the Republicans later butchered by the Nixon political machine for his stand on Vietnam, Charles Goodell, to Holland and Hope for a visit. Goodell, along with Representative Robert Griffin, later Senator from Michigan, managed the campaign to get Ford elected minority leader. Goodell's job, according to a story in the *Washington Post* was to bring a positive image back to the Republican Party.

Bob Donia heard Goodell (whose five sons included future N.F.L. commissioner Roger) speak about Vietnam. Donia said "Goodell said we have two extreme alternatives in Vietnam: either all-out war, or we get out. If we don't accept some alternative in the middle, those are the only possibilities." According to Donia, however, he gave no alternative ideas.

Donia had accepted the position that America was in Vietnam on the invitation of their government to stop subversive, communist activists; and it was in our interests to stop for surge of communism because if we did not everything would "fall in Asia to the communists bringing the frontier to Hawaii."

Time would tell this to be not true; but by that time Charlie Goodell's political career had come and gone, he settled north of New York City, and was making a good living as a lawyer. Goodell, a congressman from western New York, was appointed by Governor Nelson Rockefeller to the Senate seat held by Robert Kennedy when Kennedy was assassinated in June 1968. He was one of the first Republicans to become a critic of the Vietnam War, and earned attacks from vice president Spiro Agnew, as a "radiclib": a radical liberal.

When Goodell stood for his first election in the Senate in 1970, Agnew attacked him during the primaries for being a "nattering nabob of negativism" because of his stance on Vietnam. Consequently, his opponent in the primary, Richard Schlesinger, won and was later elected. Agnew would eventually resign the vice presidency, and Grand Rapids native Gerald Ford began his ascent to the presidency.[127]

---

[127] Hope College *Anchor* (Volume 77, Issue 16), February 12, 1965.

In 1964-1965, the Republican Party was in a state of disarray. They had lost two elections after President Eisenhower was retired; a close one in 1960 with Kennedy prevailing over Nixon, and then a landslide in 1964.

Meanwhile, some were oblivious to all this politics. I published my first paper "The Behavior of (Alpha-hydroxy)-cyclopropylcarbinyl Free Radicals" independently in the Spring 1965 issue of *Tetrahedron Letters* (as noted previously), a journal that Hope had subscribed to just a year earlier.

Tetrahedron Letters was started by the infamous Robert Maxwell as part of Pergamon Press. *Tetrahedron Letters* published camera ready copy which meant it could appear in print much more quickly than could a journal set in type. Pictures were taken of each individual page of a publication, and published in sequence. It also saved the arduous, and costly, printing processes that had grown up at the time. However, it also meant that the author was responsible for all presentation of the work. Errors in the typing, or the drawing of the structures stayed in the final paper. I found none in my particular first paper, but one can see that it was typed with a typewriter equipped with a cloth, carbon ribbon. The type was poorly resolved; structures were drawn by hand or typed. It was always difficult to find secretaries/typists who could copy the technical writing of chemistry well.

The picture came from an early 1965 *Anchor* and I include it in this biography because of the role model one of the students in it, Graham Lampert. Graham took organic chemistry from Jerry Mohrig and me. But the first time through, he managed a solid F. He failed the course. After that he came to me and wanted to try again, because he wanted to go to either graduate school or medical school. He took the course again and did an excellent job the second time through, giving a lesson about how much students can change and how much they can achieve when motivated.[128]

MONEY

Hope had deep-seated financial problems. It was teetering on a precipice because there was no endowment of consequence, and its principal benefactor, the Reformed Church, had little or no money so most of the operating budget came from student tuition. Any campus improvements, whether for capital upgrades, to take care of

---

[128] Hope College *Anchor* (Volume 77, Issue 24), April 24, 1965.

maintenance issues, or to upgrade the faculty, had to be paid for by the students. This included the funds to support the football team. At the time, players loaded in the back of the school's only truck, and were carted to Riverview Park for practice. When they were finished, the truck loaded up and brought them all back. Since the Science building and my office looked south toward Carnegie Gymnasium, I used to see them come back from practice muddied and beaten.

I was an alumnus and thought this was the way it was supposed to be in colleges. But colleague Norm Norton in biology got upset by this, so he went on a campaign to get the football team a bus. He succeeded, and soon, the football team was carted in a bus. They won no more games on being so well-treated, than they had when treated like cows headed for a slaughter, but Norton at least felt better.

In early February 1965, the students started to analyze where the money for their education was coming from—who was paying the bills? What they managed out of the Board of Trustees and the long-standing secretary of the Board was that students at Hope were paying for 80% of their education, while those at Michigan State or Western Michigan were paying only 12% of the costs of their tuitions. I was never certain how much influence, particularly on the topics involving money being discussed in the *Anchor* in Spring 1965, that President VanderWerf had or how much of a role he played in stimulating the actions of students at the time. He certainly was behind the student activism that led to Student Center activism, and I suspect he was behind stories that started to appear about tuition. It was likely VanderWerf who got the *Anchor* to interview the long standing secretary of the Board of Trustees, William Wichers to talk frankly about money. Cal was a political person, but he avoided controversy and confrontation. He did so by getting others to be the front person. As VanderWerf knew it would, these stories brought the student paper editors into the fray. On a practical level, his point was, and would be, that if the Church wanted this College so badly, the Church should be willing to pay for its ownership. An *Anchor* editorial, dated February 26, 1965, concluded 1) the college cannot continue to operate and maintain a high degree of excellence without more money from somewhere; 2) if tuition is raised students of moderate means may be squeezed out and enrollment will go down; and 3) the Church should contribute more funds toward the operation of its finest school.

The editorial then posed a question many were considering—given the parochial nature of the college did this prevent it from seeking funds from sources that were clearly targeted at non-sectarian institutions, should the college remain affiliated with the Reformed Church?[129]

It is unclear how Cal would have answered that question in private. Like many who grew up in the Reformed Church, he had grown away from it. There was an identity pattern on campus though that remained in the 1960s: more than 50% of the students were from Reformed Church, or Reformed Church-like origins.

But activist changes were occurring. Students were beginning to meet to evaluate campus departments and the teachers in them. Criticism was leveled when the new college computer was a bust—it was anointed to be used for the registration of returned second semester students and flopped terribly. By the end of the 1960s at Hope, it was being used by Clark Borgeson, with Dick Brockmeier's help, to calibrate the student's responses to questions about their teachers. It was Hope's first attempt at a student evaluation of teaching.

On February 26, 1965, another article in the *Anchor* by Bill Wichers discussed where the money of the College went. Bill first said that watching the money is the Board of Trustees job, but that was just a theory. The Board of almost 68 members from the Church could only do so much. At this time, the Board could not/did not include faculty or staff members of the College, or anyone who was not a member of the Reformed Church. The Board was not doing its job, but it was not necessarily its fault.[130]

In the next article on March 5, 1965, Wichers talked about new expenses, new building initiatives, higher faculty salaries, corrections of deferred maintenance to facilities, and new additions to other on campus buildings. Wichers pointed "the Church seems to be paying little for the right to a lot of authority – it distributes its funds equally between three church colleges; not on a per capita basis but on an equal basis. So Hope, with many more students, gets less per student than do the other Reformed Church schools. The domination of the Board by members of the Reformed Church means that persons with experience from other denominations cannot be tapped; nor can one attract those from outside the ranks who might have sources of funds."[131]

Hope College, by all accounts, was fiscally conservative. An

---

[129] Hope College *Anchor* (Volume 77, Issue 18), February 26, 1965.

[130] Hope College *Anchor* (Volume 77, Issue 18), February 26, 1965.

[131] Hope College *Anchor* (Volume 77, Issue 19), March 5, 1965.

estimate of the chemistry department budget at the time, including salaries, was that it was no more than $100,000 and Cal's salary never exceeded $25,000. Understandably, Wichers was asked "Where does the College money go? And who decides the budgets?" Blaming the papers of incorporation, he explained:

- The Board has the responsibility but no power.
- The administration does all of the spending.
- The Church has all the control but gives no money.
- The students get a good education, but pay little for it.

Of course, the students focused on faculty salaries as being the root of tuition increases, showing a picture in the *Anchor* of Henry Voogd, a professor from the Religion Department and philosopher, D. Ivan Dykstra, cashing on their wealth from their 10% raise (They probably had salaries less than $12,000 for 9 months!).

These issues were absolutely critical at this time, because the Hope College of the future was to be much different than the Hope of the past and the students, and more of the campus community needed to be involved.

In the March 5, 1965 issue, another Hope College legacy emerged: the Christian evangelist. Hope College claimed two – Norman Vincent Peale, the senior minister at Marble Collegiate (Reformed) Church in Manhattan, and Robert Schuler, before the Crystal Cathedral ministering in Garden Grove, CA. Peale visited Hope's campus on April 22, 1959 with his wife, plus Stanley and Mrs. Kresge. They spoke at morning Chapel that day but I found it more or less an institution in its pandering best. President Lubbers said "This is surely one of the outstanding events for Hope and Holland for the year. To have either of distinguished guests visit the college would be honor enough. But to have both men bring us their insight and inspiration is an exceptional tribute to the community." Peale had led the opposition to the election of John F. Kennedy because he was a Catholic.[132]

When Robert Schuler came to Hope College in spring 1965, he had just begun reaching the unwashed California masses and his assistant pastor was on the Board of Trustees. In some respects, Schuler represented to the College what many wanted the Church to be by making the Church visible and recognizable. Naturally, he was greeted as a hero by the school.

---

[132] Hope College *Anchor* (Volume 77, Issue 19), March 5, 1965.

But another California was also around the corner. That California came to Hope soon, and again, in the next years. In the March 12, 1965 *Anchor*, the resignations of President Clark Kerr and Chancellor Martin Meyerson, from their positions in the University of California system, were announced. The issue that forced them to resign was community reactions to the Free Speech movement on the Berkeley campus. Students were arrested for using obscenities in speeches about the draft, the Vietnam War, and other things. Essentially this was the beginning of protest movements on campuses that would become encompassing in the next few years. Kerr and Meyerson saw the reaction of the university as government invasion of the academy so they quit.[133]

Meyerson's replacement at Berkeley was Holland native and Calvin graduate Roger Heyns. Heyns was a psychologist and faculty member at Michigan before rising to the rank of Vice Chancellor in Ann Arbor. He was selected from outside Berkeley and served as Chancellor at Berkeley for 6 years – 1965-71. Heyns was considered a calming influence – a matter specifically commented on by former President Kerr shortly after Heyns arrived. After retirement, Heyns was President of the William and Flora Hewlett Foundation.

The *Anchor* portrayed a campus with issues that were the same as were those at the larger, more uncivil universities. But Hope's students handled them in a peaceful way. Civil rights protests that were also beginning even in Holland and at Hope were driven toward the same ends. In the week of May 14, 1965, more than 700 students signed and 300 students marched to the Holland mayor's office with a petition to the let the people of Mississippi vote, to effect voting rights legislation, and to develop a system of voter registration that had integrity. Civil rights legislation was passing through Congress, James Meredith had already been enrolled at the

University of Mississippi through Justice Department efforts, and specifically through the work of the late John Doar.[134]

John Doar showing papers to James Meredith on the University of Mississippi campus, 1962.

---

[133] Hope College *Anchor* (Volume 77, Issue 20), March 12, 1965.

[134] Hope College *Anchor* (Volume 77, Issue 27), May 14, 1965.

At Hope, the Dutch prince, prince Bernhard, stole the Hollander's heart. He was in Holland, Michigan, for other reasons, to dedicate De Zwaan, the *molen* on Windmill Island but Cal VanderWerf never missed a publicity opportunity. Bill Wichers was charged with buying the windmill, traveling to the Netherlands to purchase one.[135]

The visit of Prins Bernhard to the Hope campus excited everyone - students, faculty, staff, town - everyone. The Prins was most gracious and Holland put on its finest for him. President Vander Werf was his host on campus. I remember his speech at the Chapel but, in all honesty, not much of what he said. He was an icon of identification of the Netherlands for the time its citizens spent under German occupation. And everyone knew he himself was German. The Prins Bernhard scholarships meant a great deal to Hope for years.

In fact, that fall, Johannes Huber from Utrecht arrived funded by a Prince Bernhard scholarship. The *Anchor*'s story says a Dutch professor of chemistry, Hans Wynberg, who knew President VanderWerf, encouraged Johannes to apply for the scholarship. Later that year, Hans would wander through Hope's science building late on a Friday afternoon checking up on Huber because "he knew his parents and he wanted to make sure all was well." Soon, Paul Schaap was invited to Groningen for a year's study.

Another issue was on the horizon. That was the student activism that would surface most fully after Lyndon Johnson's acceleration of the war caused the reinstitution of the draft – or at least it made students more susceptible to being drafted. Students, in just a couple of years, would become very different, much to their detriment and to the detriment of America's civilized society. The *Anchor* (October 1, 1965) reported on a public policy forum in the chapel that featured the president of Students for a Democratic Society (SDS) speaking against our continued involvement in Southeast Asia, and two graduate students from Michigan State defending the policy. Carl Oglesby, representing SDS, asked questions that would become the standard of those opposed to the war: our nation is losing its social conscience; the U.S. in Vietnam is serving as a counter-revolutionary power helping to maintain the status quo, and we should be allowing the Vietnamese to decide their own political fate.[136]

---

[135] Hope College *Anchor* (Volume 77, Issue 27), May 14, 1965.

[136] Hope College *Anchor* (Volume 78, Issue 2), October 1, 1965.

The counter points argued the domino theory: "If the U.S. does not stand against communism in Vietnam, the entire region could fall. We could have communists knocking on Hawaii's door." As time would tell, this argument did not hold true. Vietnam fell in 1974 and though it took years for it to be rebuilt, it would seem that there was little truth to the "all will fall" position.

Student dissent itself was becoming an issue. An *Anchor* interview with three wounded soldiers at Great Lakes Naval Station in Chicago presented the involvement story. The *Anchor*'s editorial said "Dissent is a basic ingredient of democracy or of any government that claims to represent the people... It is disturbing to see a government concerned with annihilating criticism and quiet adversaries. Using military induction for this recalls the abuses of power of the Nazis in Germany." Later, they wrote "A nation that silences those who dissent is lost. It deprives itself of the power to correct its errors."

In letters to the editor, four students defended the President's war without declaration in Vietnam by saying it was just following a policy established by the Korean War; and, mind-bogglingly, because Japan occupied Manchuria without declaring war; and Hitler "did all those bad things without declaring war."

Harvey Kleinheksel's death in early 1966. Death happened as Harvey flew with his wife to Dallas to spend Christmas with their daughter. Gene Van Tamelen commented on Harvey, the teacher. "All of us that studied at Hope from 1927 through 1966 knew him; none criticized his teaching that I am aware of. His style was to ask questions of the students. He did not lecture, really, but he just presented the material in a pleasant way." President VanderWerf who also said in Harvey's classes said that "in his quiet, simple, unassuming way he had the rare gift of being able to bring out the best in every student."

Cal had that gift, too.

The departure of Hope's director of development prompted the *Anchor* editor to start a campaign knocking President VanderWerf. Years later, after a serious fall from grace, that same person found himself much more humble in presenting his next offering to the public.

Hope was undergoing transitions. some appointments were good; some new appointments were bad. So that added to the confusion. Students picked this up as the *Anchor* cartoon below.

Editorial cartoon, Hope College *Anchor* (27:2), October 1, 1965, Copyright © 1965, Hope College.

Hope's *Anchor* dealt with another topic of increasing interest in February 1966: artificial birth control. Fred Oettle, a chemistry major that later took his Ph. D. at the University of Kansas wrote a hot letter regarding "the pill," blaming theologians for their head-in-the-sand, non-scientific philosophies because they did not accept the pills. Fred spent some of his career at DuPont; later he was an administrator in chemical education at the National Science Foundation.[137]

Subsequently during the spring, a new college catalogue appeared. This was a job for Wilma Bauman, a mother of five who returned to work in the President's office where her first project was to edit a new catalogue. Hope had changed dramatically since the old catalogue and a new one drastically needed. Wilma was an excellent typist, but she could also write, so the new catalogue was a decided improvement over the old one.

Hope students put a peculiar spin on what it was. Two students wrote a strong letter to the editor about the new catalogue's lack of statements about compulsory chapel and the campus drinking policy. This was followed by a much more detailed and cogent letter from the head of the Art Department pointing out that the new catalogue was much better than the old, because it actually mentioned

---

[137] Hope College *Anchor* (Volume 78, Issue 17), February 18, 1966.

269

that Hope had an art department. Besides, campus social policies belong in a student handbook, not the catalogue of courses.

This student response, however, stated that Hope's problem from its first days was still there. It did not know whether it was an academic institution, or a school for and of the (rural, poorly educated, 19th century) Church. Henry ten Hoor, who we all loved because complaining was his wont, also wrote in the *Anchor* of the day ruing the lack of Dutchness to Hope's centennial celebrations (1866-1966). Henry's complaints came at a time when at least three Dutch art historians were on campus sharing information about Dutch art history. Henry's complaint, in part, was that the college of that day offered no Dutch language courses while Columbia, Michigan, and Harvard did. The College taught little Dutch history, evidencing either trouble recognizing its heritage or the lack or persons who could teach such. This he blamed on Holland's heritage; the people who immigrated were poor, uneducated, and in search of better economic opportunity. The sooner they could forget the old country, the better.

Those early settlers did not forget the Church they had hoped for, and this translated at Hope to a dichotomy of purpose. During the VanderWerf day, it was a growing academic institution developing its own reputation. Almost immediately thereafter, it reverted to the closed, self-appealing, church driven entity that turned off as many as it attracted. As Marshall Anstandig, a Jew, stated in a letter to me while I was preparing this biography the Hope of the 1970s had returned to its stand against diversity that seemed the antithesis of the academic institution that it could be.

John Piet, a professor at Western Seminary, writing in the *Anchor* of April 29, 1966 addressed the question "What is a Christian College?" Piet pointed out that the two ideas, Christian and college, are contradictory. "Christian faith is subjective at core," and "a college is a congregation of people committed to the pursuit of truth." Piet was not alone. Father Theodore Hesburgh, then appointed President of the University of Notre Dame, said almost the same thing to the Roman Catholic hierarchy about Notre Dame.

Piet's words were unheard. "Clearly, a Christian college cannot serve both masters" – the subjective identification of religious opinion, and the objective search for truth. Hope mostly tried to be fair with the balance but at times, particularly after VanderWerf, it forgot the objective, becoming more subjective. Students surely suffered.

William Hillegonds, the senior pastor at Hope Church

located about four blocks from the campus, was continually recruited by Cal to be the College chaplain. This started almost from the time Cal appeared on campus. Eventually, Hillegonds moved to the campus to work with the students, and teach a religion course or two. He stayed there for twelve years – well beyond the VanderWerf tenure. Then, the urge to "marry and bury" overcame him and he returned to the pulpit, as senior pastor of the First Presbyterian Church in Ann Arbor. Hillegonds had been burned at Hope Church by his support for the candidacy of Lyndon Johnson in the presidential election of 1964. His consistory, the managing body of Dutch Reformed congregations, took umbrage at his political position. Bill did not really belong in a mainstream Holland church anyway: He was too rebellious. So he moved to the campus where the students seriously enjoyed him.

By the spring of 1966, Hillegonds started a student church. For the first time in its history, the never-on-Sunday Dimnent Chapel had lights and an organ playing on Sunday morning. This caused, according to a story written by the *Anchor* editor, Tom Hildebrandt, consternation in the Reformed Churches in Holland that students mostly attended – Hope Church on 11th Street; and 3rd Reformed Church on 13th Street. Jack Walchenbach, who was serving as the Pastor at Hope Church in the interim following the Hillegonds departure was the most exercised. The congregation at Hope Church had dropped so much from the absence of students that the Church was considering reducing from two Sunday services to one. Of course, what Walchenbach did not mention, but surely knew, was that Bill Hillegonds had a great following at Hope Church that moved with him to the campus church when Bill moved. Russell VandeBunte at Third Reformed Church had a different spin: "His congregation saw the students as in an ivory tower six days a week and they think this should not be extended to the 7th day. They need to be experiencing the living world." … and a trip, three blocks away, to a Sunday church would do that.

As I have read these *Anchors* from the distance of years, I am continually impressed by the experience of discovery, as in who came to campus then and why that started with the VanderWerf energies. Hope's young faculty was out and about professionally and that meant they had connections in a world outside Holland, and often outside the United States. This translated to an extraordinary presence of young scholars on campus. We have already seen that with the impact in chemistry that a connection the organic chemists – VanderWerf, Earl Huyser, and others (including yours truly had),

271

which brought a Dutch professor to the campus for a short visit. This took Paul Schaap and later Jim Hardy to the Netherlands.

In following this thought, the April 29, 1966 *Anchor* reported that Professor James Tallis in the department of music had started a motet choir. Jim was a keyboard artist; he had studied harpsichord in Amsterdam with Gustav Leonhardt; there he had met a number of young musicians of that day. It was not that long after the devastation of World War II, and though the European economies were emerging, Amsterdam's guilder was almost 4 to the U.S. dollar: a night in a fine hotel with breakfast cost 50 guilders; the trams were almost free; a trip by train to Groningen, about as far away as one could get in Holland cost 15 guilders; a round trip flight to the U.S. was relatively expensive but, in reality, cost less than $200.[138]

Jim's incentive to form a small group was a faculty Christmas party in his first year as a faculty member, 1964. He realized that his first year music theory class of sixteen voices had exactly the composition and balance of vocal parts. So they started singing together as the Motet Choir without accompaniment; by April 1966, they sang for the campus fine arts festival. At that time, one of Tallis' Amsterdam friends, then studying with Leonard Bernstein, was Edo de Waart. De Waart had heard the concert of the motel choir in Holland, and made extraordinary positive comments about it.

De Waart would return to Amsterdam to become assistant conductor of the Concertgebouw orchestra, then take over as conductor of the Rotterdam Philharmonic. Later in his career he served as principal conductor of the St. Paul Chamber Orchestra in Minnesota. Now, among other things, he's musical director of the Milwaukee Symphony. De Waart was at Hope when he was very young. Almost at the same time, Maurice Durufle came to campus to conduct his soon to be famous Requiem mass.

Tallis had a tragic end. He had been a graduate student at Union Seminary in New York and decided a few years later that he had to finish his doctoral degree. So for a year he commuted every week to New York, completed his study, and then wrote his thesis. His wife Joan, and two young girls, stayed in Holland. But this represents the dedication of the young faculty. When finished, Jim was almost immediately recruited by Southern Methodist University in Dallas. Salaries at Hope were awful at the time so money surely was part of the Texas incentive. Sadly, Jim developed a brain tumor shortly thereafter and died tragically and soon.

---

[138] Hope College *Anchor* (Volume 78, Issue 25), April 29, 1966.

President Johnson's full term brought a large number of social changes to the United States. His victory was a landslide for the Democrats, so he could insert numerous social programs in the American scene. Less known among these was the granting of federal aid for the cost of facilities' improvements at colleges and universities in the United States. The State of Michigan took the initiative, too; the State Senate passed bills that would provide tuition aid for Michigan residents studying in private colleges and universities in May 1966.

Never far away, though, was the war in Vietnam, and the draft. On May 13, 1966, a story about Hope alumnus, A. J. Muste one of the better known "peace advocates" chronicled a bit about his life: he had just returned from a peace mission to Saigon, where he protested U.S. involvement in that war. At the bottom of the same page, there was a story about the administration of a test to determine an individual's qualifications for the draft and who might qualify for an academic deferment. At Tulip Time that year, a group of ten student protestors of the draft crashed the Tulip Time parade without the appropriate permits. Their immediate target was the draft test, but secondarily, they were opposed to the war itself. One of the protestors, in answering the question, "Why did you do this?" indicated what many of us felt. The citizens were accepting the war as though it was alright. In this student's opinion, it was not, and the people were too accepting. Aside from angry catcalls from the parade watchers, the students survived, were not charged by the police, and probably made their point. The campus dynamic transitioned into a community's concern for today's youth.[139]

---

[139] Hope College *Anchor* (Volume 78, Issue 27), May 13, 1966.

Chapter 18

September 1966

Administration

President VanderWerf inherited an entrenched, nearly dysfunctional administrative team. The administrative structure was not clearly defined. There were "two persons doing some things; and none doing others." There was no development office, but two persons – being paid good salaries – were to worry about the money that came in. The admissions office managed applications when they came to the College, but did nothing to increase the pool of applicants. The 1964 graduating class was large for Hope. Even though Rick Smalley left after his sophomore year, several others became successful academic scientists and corporation leaders.

The problems with the Hope administration were enhanced by an entrenched, easily-fed, gossip mill. Failures in the then administration building, Van Raalte Hall, were announced and magnified by a small group of faculty members that met regularly in a basement smoking room near the Koffee Kletz. If they were bad enough, or this group thought that they were, the got the students involved, via the student newspaper. This aggravated the problems. All administrative changes would have worked out save for the fact that the College had no money. So the replacements had to be people that would work cheaply. Most of the time, hiring for a low end dollar gets one less able people, and that's what happened in too many instances for the President's good.

Changes and more changes

One sore need was for the school to have someone on the road telling its story, raising money for its programs. In the fall of 1966, President VanderWerf appointed a director of development, William Hender. He also merged the positions of Academic Vice President and Dean of the College into one Dean of Academic Affairs, William Mathis, and he appointed a new Director of Business Affairs, Clarence Handlogten. In the long run, neither of these turned out to get good appointments.

Each position had been too long unfilled, and unfortunately, the persons appointed would become failures in the Hope College situation. Mathis was a liberal arts college administrator with experience in at least one small Baptist college in Texas; Hender had professional fundraising experience in small colleges, but none with a school with Hope's oddities, and Handlogten, viewed himself as potential presidential caliber despite having no advanced, or even baccalaureate, degree A few years later when Cal VanderWerf was forced out, Handlogten presented himself to Trustee Chairman Hugh De Pree as the next President but was set straight on that point quickly. In the interim between VanderWerf and Van Wylen though, Handlogten served as part of an administrative two-headed presidential monster... he to handle business affairs, Bill Vander Lugt to handle academic matters.

President Van Wylen moved Clarence onward soon after he became president. Handlogten took a position with one the fast track members of the then Board of Trustees, Howard Sluyter. That did not work either. Soon, Clarence was back in Holland at Haworth Industries but that, too, lasted for a short time. By 1976, Handlogten had disappeared from the Hope College/Holland orbit.

Each of the VanderWerf administrative appointments was necessary but the people were not right: each at one level or other started to signal the end of the VanderWerf administration when they left. Persons chosen from the outside came at market value, and these persons were what could be had for the money Hope had for these positions. None of them lasted.

The growing faculty, however, was to make a difference in Hope College's future: Dick Brockmeier joined the physics department; Dwight Smith and Nancy Tooney joined the chemistry department; and, James Durham, Warren Vander Hill and Robert Peters joined the history department. In many cases, these were one-for-one replacements, causing no growth in the overall faculty size. Cal erred by favoring the humanities so three historians replaced two and Bible/religion added two new positions. Elton Bruins joined the Department of Religion and Bible. Elton made the rest of his career at Hope and, in retirement, began a publishing house funded by donations from the interested. He, and others, called this the Van Raalte Institute.

Great Lakes College Association Interns in Chemistry and Biology

Nancy Tooney came to Hope as a Great Lakes Colleges Association intern in the fall of 1966 and was the first biochemist in the department of chemistry. Nancy Tooney came to Hope because colleague Jerry Mohrig pushed that we should hire a biochemist. Jerry could be very convincing and somehow convinced the biology

275

department to go along with this idea too. At Hope, however, it took a long time for the seed of biochemistry then to grow into a decent sized tree – though eventually it did. Biology departments of the day were having trouble figuring out, internally, who they were. As scientists pushed the molecular envelope at the living versus dead interface, those trained to observe nature without trying to manipulate it, and those trained to try and pin nature down on a molecular level until gave an answer, saw things at the interface quite differently. Biology, the discipline, and biology faculty nationwide as a group, were troubled by their relationships with their colleagues internally in their own departments. They were even more troubled by their relationships to chemistry departments. On a practical level, biologists had little background in chemistry, and most chemists had almost no courses in biology. So the two professional disciplines talked at one another, and in front of one another, and past one another, but they rarely came to quality agreements about mutual interests…until the students coming from high schools forced them to do so. This was, of course, different in the leading graduate schools. Programs set in place in molecular biology at M.I.T., Harvard, Cal Tech, Michigan and other universities in the 1960s have generated enormous economic benefits to their regions in the 21st century because they bridged the molecular interface easily and sooner than others.

At Hope, no marriage was made between biology and chemistry. Togetherness was forced by nature with no concepts being more at the center than the postulates of James Watson and Francis Crick on the structure of D.N.A.—"the double helix."[140]

The G.L.C.A. program was to hire young scientists to work in liberal arts colleges for a year to see if they had a career potential in such places. The program had flaws but it was quite useful. It was explicitly prevented to bring a potential faculty member in, consider their work, and then hire them or not hire them based on the year-long "audition." Operationally, the program specifically precluded that a person be hired to a permanent job at the school at which he/she interned. Nancy Tooney left Hope after one year, and spent her career at the Polytechnic Institute of New York, now the N.Y.U. Polytechnic Institute.

---

[140] The Watson Crick work from Cambridge (1952) developed a huge lore for the study of biology, and high school students became enamored with molecular genetics, as in the eventual cloning of a sheep, Dolly.

John Read, Hope's second G.L.C.A. intern moved from Hope to Mount Allison University in New Brunswick Canada, where he became professor of chemistry in 1979. John served as heads of mathematics and computer science, psychology, commerce and chemistry during his tenure at Mount Allison, was Dean of Science for 21 years, and vice president for Academics and Research from 2004-06. In retirement John took on the problem of literacy in New Brunswick, becoming president of the South East Regional Adult Learning Board that has 25 classes at this time helping adults get their general education diploma (G.E.D.). So, in these two cases of hires, G.L.C.A. interns in the chemistry department at Hope managed because the College made excellent hires.

Politics –September, 1966

In the September 23, 1966, *Anchor,* some attention was given to the impending draft, but even more to an apparent shortage of housing for all the freshman that enrolled that fall. Some male students even had to live in the Warm Friend Tavern. If it was a tavern, it was not much of one, because Holland was a dry town. Many years after the Warm Friend was used as a dorm, it became a retirement home.

The *Anchor* also announced the Cultural Affairs program for the year. Cultural Affairs, the program and later the committee, came to Hope on Morey Rider's initiative. Morey, a professor of music and director of the campus orchestra, managed to get the Board of Trustees to approve a $10 per student surcharge to be used to support a broad range of cultural programming on the campus every year. In this year, the Program included several with high name recognition who would visit the campus because of the Hope Centennial (1866-1966). Anyone alive then had to be impressed with the breadth of the visitors to the campus for that, and their presence in their specific worlds of inquiry/interest. Socialist Norman Thomas, composer John Cage, internationalist Clark Eichelberger, philosopher Mortimer Adler, poet W. D. Snodgrass, cartoonist Jules Feiffer, and the voice of N.A.S.A., John Shorty Powers. I do remember the visit of Sir Tyrone Guthrie the founder of the Stratford Theater.[141]

Graduate education and Hope

When Philip Phelps, Jr. had dreams for Hope in 1871, he saw Hope College as a transitional name. How different Holland would

---

[141] Hope College *Anchor* (Volume 79, Issue 2), September 23, 1966.

have been if Phelps dream for Hope Haven University had become the reality. Even in the 1960s, there were occasional pressures for the school to offer graduate credit. The first department to begin to break with the liberal arts tradition in that era was the music department. It proposed, and the college eventually consented, to offer three degrees – classes would be more frequently in music, with less emphasis in the liberal arts. These included performance degrees, though few Hope alumni became performers.

It was surprising how this approval was managed through the structures at Hope given the pretense of the undergraduate liberal arts focus of the college at this time, but it did. Hope had offered a Bachelor of Music degree in the past and also awarded an occasional Master's degree. Doc Van Zyl indicated on his C.V. that a Master's degree was awarded to one of his coworkers in the 1920s, too. The programs were no threat to the basic liberal arts, and as time went on not only was a 3:2 engineering program added, and so, too, was a four-year program in nursing. Though one has to wonder how undergraduates of that day, or any other, would be insightful enough to offer their opinions. These editorials had to have been spiked with faculty input most likely from the philosophy and religion departments. In a split vote, the faculty agreed to let music add two other advanced degree programs. The curriculum was approved, just in time for Homecoming.

Capital Plans: the Centennial Celebration

Homecoming 1966 also saw the announcement of a ten-year college master plan, the first building was the De Witt Student Cultural and Social Center.[142]

---

[142] Ibid.

Student Center Included

Master Plan Calls for 10-Year Growth; Eight Buildings to Cost $10 Million

Proposed Student Center, Hope College *Anchor* (Volume 79, Issue 2), September 23, 1966. Copyright © 1966, Hope College. Cal's biggest achievement was conceiving of, finding the funding for, and breaking ground for the DeWitt Student Center. Many involved can tell the story better than I. This picture, from the Anchor, is indicative.

At this time, it appeared that the VanderWerf administration was taking shape: the appointment of a Dean of the Faculty, changes in the financial offices, and the appointment of a director of development indicated that. Most of the previous administration had exited: John Hollenbach was on sabbatical leave in Beirut; Bill Vander Lugt was newly appointed College professor at large; Henry Steffins, the treasurer, had been eased out though not without a serious ruckus. The assembly of newcomers was accompanied with consolidation – one vice president replaced several doing parts of one job.

Homecoming events in 1966 celebrated the 100th anniversary of Phelps and Van Raalte's chartering of the College and were exciting, and extraordinary. Norman Thomas, the socialist of the previous generation, was invited to speak. Though it was said he drove the conservatives away from the Chapel, he drew the students with his speech. Thomas said that the American people were becoming "immune to the horrors of war" (as Johnson administration dragged our military more deeply into Southeast Asia). A writer for *Newsweek* countered Thomas' argument, saying the North Vietnamese government in Hanoi was composed of tyrants, that American bombings were reducing their war machine's effectiveness, and that if North Vietnam were to fall to the Communists, all of southeast Asia would not be far behind. Thomas worried that the war would escalate out of control…and it did. The writer, Kenneth Crawford, said the war was turning in our favor and that we had to just stay the course. How often did we hear that from Secretary of Defense Robert McNamara and from Presidents Lyndon Johnson and Richard Nixon? These arguments about the war would resound for almost another decade. I personally have to wonder how many Hope students that would be killed or injured in Vietnam heard these discussions.

The Master Plan for the College unveiled in 1966 seemed outrageously expensive. It offered a $10 million plan that included the student cultural center and a new science facility. Additional dormitories were also in the plan. There were plenty of frustrations and false starts as attempts were made to execute the plan. Looking realistically at the 1966 master plan, in that year the entire alumni of the Institution could only manage contributions of $73,000! [143]

Looking back now, $10 million looks like a drop in the bucket, but it certainly was not then, as schools scraped and scratched for the next operating dollar. Today, Schaap's gifts to the school add up to more, in dollars, than that entire master plan would have cost on a dollar-for-dollar basis, had it been executed in 1966. But, today, our economy is knowledge based, while the American economy of 1966 was manufacturing and agriculture based. The economy today requires creativity and brains. The economy of the 1960s still needed mostly brawn. At that stage, we were still paying for World War II, and the civilian economy that would carry the next generation was barely begun.

It is instructive to look at the financial impact of one scientist/department head on one state university to see how being a faculty member has also changed over that time. My career started at Hope in 1964; and I benefitted enormously from President VanderWerf's role model.

In a retrospective analysis based on my career at Bowling Green, I put together the following in response to articles in an alumni magazine of my other alma mater, Bucknell University complaining about the high cost of the professoriate. As in other cases addressing issues of cost, there were comments about the misuse of tenure, and the notion of tenured faculty. So I penned the following to the president of that school – who also happens to be a scientist.

---

[143] Ibid.

Dear President xxxx:
Re: Rising costs of a college education

All of us can tell horror stories about how much 'college' costs. Few expenses incurred by a parent, or student, rival the cost of an education – particularly an education that includes a residence and meals. Since the professors are the front persons for most college students, it must be their too high salaries that are causing the increased costs of a university education. In Ohio universities, a governor took it upon his office to demonstrate that faculty who taught only a few hours a week were ripping off the state's taxpayers.

"Look at these over-paid professors; they're working only six hours a week."

I invited him, in a private letter, to come spend a day with me. Governor George Voinovich never did.

But though faculty salaries have risen over the last decades, they haven't risen that much. And besides, the professoriate is a college or university's most important 'intellectual property'.

I taught organic chemistry for 47 years at one private college, and two state universities. A recent analysis done of the Foundation funds I generated at the state university at which I spent the last 35 years showed that I left behind, in generous gifts from the supporters of our ideas, and for the education of future chemistry students, more $'s than the school paid me in salary over the course of my career. These endowments were, and are, working capital. And they generated interest sufficient to pay tuition and fees for about 80 Ph. D. student years – a student year represents the stipend and fees for a Ph. D. student at the university. I estimate that tuition and fees generated to be double my total career's worth of salary. In short, in endowment earnings, my school earned more, by a factor of 3, than I cost them in salary.

This says nothing about the millions of dollars of research and development grants my program, my students, and our collective creativity brought to our university – eight figure numbers those. At the small college at which I previously taught, the new science center is named after the first undergraduate research student that worked in my labs, he managed to start a business that generated millions of dollars for him and when he sold his business. He was generous with the proceeds. Both his alma mater, and the university at which he taught, have buildings named for him.

281

It also says nothing about the new businesses spun off my research program. At this time, just one new business in northwest Ohio feeds almost forty mouths every day. And it's growing. This business, in stereolithography and 3D printing, is playing an ever increasing role is moving person to person communication from planar to three-dimensional.

These examples represent merely the economic benefit of full time, and yes, tenured faculty members in a physical science for his university. Not every faculty member has this sort of record; but some scientists and engineers do, and some of Bucknell's finest teachers of chemistry were also its most successful scholars.

Teaching university students is a sacred responsibility. (Recall how moldable you were at 18 or 20, and then realize that every faculty member with whom you came in contact, had some impact, positive, negative or neutral on the   person you became.) We owe it, as the guardians of our society's future, to arm our universities with the most dedicated, hardest working, individuals of our generation. And they owe it to us, to pay society for its confidence in their abilities. But what's wrong with giving them a little job security?

The next time you hear a complaint about college tuition, recall the professors of your youth and their dedication to you and your classmates. In some cases, what your tuition might have paid them in salary was a mere drop in the bucket of what they were actually worth as intellectual property for your university.

Hope's Centennial

Hope's centennial theme was "Education for Responsible Leadership." This was addressed specifically by Governor George Romney in a speech at the Holland Civic Center in October 14, 1966.

In that speech, Governor Romney, himself an active Mormon with no earned academic degrees, addressed the crux of the private educational conundrum, and society's problem as it was evolving in the post-Sputnik era. Private colleges and universities were critical, uniquely American, educational resources. As the North Central Report clearly pointed out, Hope's students were its most important assets. And Hope had excellent students. The question was could Hope pay its bills to remain competitive so that students in the future would find it attractive, too? Romney said that he felt the state would eventually have to provide funds for the private schools. As it turned out they did this by grants in aid that went directly to the students. This occurred in short order.

Romney went on to address the matter of faith and higher education, cautioning that there should be no "breach of the wall between church and state." As a Mormon, he certainly knew and understood religious prejudice and narrowness as he was from a religious minority, speaking of their protection against the religious will of the majority. Though arguably on the same side as those in the college community that would wish, and needed, state financial support, he was clearly aware of its dangers. He was also a governor running for re-election, and the always-present press was hanging on his words.

As alumnus John Piet had pointed out so cogently the prior spring, the search for truth was an objective but faith in God was subjective. In a college or university, the two stood starkly in contrast.

Chapter 19

Major Funding Comes to Hope's Science Programs

Alfred P. Sloan Foundation and Small College Funding

As Cal VanderWerf said in his inaugural address, many in the country were suggesting that Protestant undergraduate-only education in America was finished. It had stood the nation in good stead, taking young people off farms, and into pulpits. But the America emerging after World War II had grown much beyond that, and American young persons needed much more than that. Alfred P. Sloan, C.E.O. of General Motors, was one of the "small handful of men that actually made the automotive industry," according to Henry Ford II, on the occasion of Sloan's death. According to his obituary in the *New York Times*, he was particularly skilled at seeing an oiled organization work, on all cylinders, as though it were of one mind. By 1936, Sloan had made enough money to dabble in philanthropy. With $10 million, he began to turn back the profits via the Alfred P. Sloan Foundation. By the time he died, he'd given more than $300 million to the Foundation that bore his name. This was being given away via the Foundation to various charities – Sloan Kettering Cancer Center in New York, the Sloan School of Management at M.I.T., and Sloan funding for the physical sciences among others.

In mid-summer 1966, President VanderWerf received a letter from Everett Case President of the Alfred P. Sloan Foundation, announcing that the Foundation would award up to $500,000 to each of twenty liberal arts colleges in order to help them maintain their positions in the pre-professional education of scientists. The program director was to be Larkin Farinholt. Farinholt had taught at Washington and Lee before World War II, served as chair of chemistry at Columbia, then served in the Explosives Division for the National Defense Research Council during the war. The Case letter went on to explain what we all knew –small colleges had played a disproportionate role in the education of scientists. The Sloan Foundation wanted to ensure that continued.

Cal called a small group of us to his office, told us about the letter, indicated he was sure the reason it was sent when it was "because most college faculties take summers off, and the Foundation wanted to start the winnowing project with the initial letter." He then appointed a small committee with Norm Norton, the new chair of biology, as chairman. Irwin Brink and I served from chemistry, Jay Folkert from math; and Harry Frissel and Dick Brockmeier from physics. Cal met with us for the first meeting or two, and then sent us on our way to prepare a proposal.

Money was in short supply for Hope College, and most of that which was spent campus wide came from student tuition. This was historical for most small colleges and Hope was a little worse, but not that different, than most. But the game in higher education was becoming you have to pay to play. So we were collectively *thrilled* when Cal told us of the Alfred P. Sloan opportunity. Sloan's history in the sciences was to fund winners; young investigators in many scientific fields known as the highest achievers were named Fellows of the Alfred P. Sloan Foundation. I still carry that honor proudly on my academic curriculum vita.

Sloan Foundation President Everett Needham Case had been president of Colgate University for twenty years before becoming president of Sloan in 1942. During his tenure at Colgate, Case was a staunch supporter of a liberal arts education, instigating at Colgate an initiating year or more for students returning from the Army to acclimate to college and studying. This was necessitated by the large number of veterans returning from military service. They had chance for a college education since it was funded by the G.I. Bill. At the time, higher educational leadership was also addressing so-called "shadow faculties," like Harvard's Conant, who served at Harvard while also working covertly in Washington on National Defense Research Committee matters. Conant must have set foot occasionally in Cambridge, but it is hard to know when and how often. Case worried that because of their wartime obligations, faculty would find government work so compelling and important that they would neglect their professorial duties. Though that was not much the case of liberal arts college faculty members, it surely became the way university faculty in the sciences would behave: in part, this led to the Berkeley riots occurring almost at the same time as Sloan's requests to the small colleges for a proposal.

Preparing the Sloan Proposal

Case and Farinholt understood the value of small colleges and probably were surprised to find Hope with a president like Cal VanderWerf. I am not sure that made any difference, but Sloan gave Hope College a chance and Hope took advantage of it. The Hope committee met a few times in the conference room in the then new physics and math building (now VanderWerf Hall). Then Dick Brockmeier and I took on the hardest work.

Given the energy Hope's sciences were starting to show, we could have written much less specificity into the program than we actually did, and Sloan – following a visit by Dr. Farinholt to the campus – would have funded us. But Dick and I took the charge to do something new seriously, so we spent hours working on ideas that might be volleyed to the Foundation for their support. Dick was fresh from finishing his thesis at Cal Tech. He assured me that Murray Gell-Mann would win a Nobel Prize, which he did in 1969 for his work on quarks as constituents of hadrons. We also talked a little about computers though most of that would come later. Dick also shared his thoughts about the undergraduate curriculum in the sciences of the future. Some of the Cal Tech physics department, and this was later shown to be Gell-Mann as much as anyone, of that day, were fascinated with life, how to understand it and how to do so at the minutest level. Dick took this attitude, too.

Before long, we devised a curriculum that assumed all Hope freshmen students wishing to major in a science would need no additional mathematics. These students could then start physics in the second semester of their freshman year, slide past a little chemistry, and eventually end up majoring in molecular biology. The basic assumptions were logical, though not reasonable. Most Hope freshmen were graduates of high school programs not located in college or university towns, and their mathematics' availability was limited. As I remember, mathematician Folkert told us that only 25% of the enrolling students had the advanced level of math the curriculum required.

Sloan had targeted $10 million for the project saying that it would divide this between 20 liberal arts colleges on the basis of their proposals. They set a limit of $500,000 per grant. Dick and I concocted the idea and started to write how nascent physics students would be turned into biologists by the end of their sophomore year in college brushing chemistry as they went past it.

For such ambitions one needed more faculty and the major budget items of consequence were for additions of faculty to the departments. It was with Sloan funds that Hope started its geology department.

I had to leave Holland for an American Chemical Society meeting in early September, so I left the writing with Brockmeier knowing, well, that the proposal was written in physics speak. I know I told Cal, and my colleagues, particularly Irwin Brink, that the proposal was not in any condition to be submitted at the time and the deadline was nearing. I had also given up trying to make English out of physics speak getting nowhere with Dick so I was also not surprised when Dwight Smith, just hired to join our chemistry department, and I met at the A.C.S. meeting New York and he told me "the Sloan proposal is really badly written." What happened was the Cal called Brockmeier to his office, Cal rewrote it in a day, and the proposal was submitted on time. Farinholt visited us in late fall, and the grants were made just before Christmas. Hope received $375,000. That was what we asked for. At the time this was the largest grant the College had ever received.

Anchor stories were generally quite accurate. Headlines into which misspellings crept and grammatical errors were found more than occasionally were rare. The story about the largest single grant the College had received up to that time, a grant from the Alfred P. Sloan Foundation for

$375,000, is an exception. This story, as singularly significant as the event was, had careless misspellings throughout. Irwin Brink's name was misspelled; I was called Douglas Norton. The grant had been funded and we, the scientists, had cause to celebrate. Through a collected effort that was as much VanderWerf as anyone, we had managed to step up a notch in the competition, nationally, for funds for our programs in a liberal arts college that we were making over. With the recognition of the Sloan Award, Hope had stepped among liberal arts colleges with science programs of quality – Haverford, Smith, Mount Holyoke, Carleton, Washington and Lee were among other winners, and they all had larger presence in the liberal arts community than Hope did. Hope was well known in chemistry, but it was not strong in the other sciences. The Sloan proposal, when funded, started to provide resources to change that.

Foundation Gives Hope $375,000 Grant

The focus of the *Anchor* story was certainly correct because in it was wrapped the reason that Cal VanderWerf returned to Hope College to be its president, and why I agreed to teach there for a time. Small schools, throughout their history, had been feeder schools for Ph.D. study; the sciences had done well but the humanities had not. But that ability—the ability to prepare students for graduate study in the physical sciences—was in danger. In the matter of transitioning America's feeder schools into the post-Sputnik scientific generation, Case was clearly prescient. Hope was honored to be included in that Sloan community. More important though was the initiative that the Hope award by the Sloan Foundation triggered. Another grant of that time, for $30,250 was the continuation of the program to support faculty interns. These funds came from the National Science Foundation.[144]

Clearly America was rebounding its science into peaceful uses, and smaller schools with good leadership and a clear cut vision of what that meant, were prepared to take advantage of the American initiatives. Smaller grants were continuing: the importance of these was that they provided flexibility and local decisions. Shell even gave a few funds to Dean William Mathis and it got more space in the *Anchor* than did any individual's contribution to the $375,000 Sloan grant. The *Anchor* editors could even spell the Dean's name.[145]

VanderWerf Announcement, Hope College *Anchor* (Volume 79, Issue 25), May 5, 1967. Copyright © 1967, Hope College.

The Peters Caper

In the small print of faculty appointments in the fall of 1966 were the appointments of three historians – Warren Vander Hill, James Durham, and Robert Peters. The emerging Dr. Peters, a potential academic jewel, had been captured by the returning (from Vienna Summer school) Paul Fried at a meeting at Heathrow Airport. Peters was a terrific find, we were told. And the campus was pleased to welcome him in

---

[144] Hope College *Anchor* (Volume 79, Issue 18), March 3, 1967.

[145] Hope College *Anchor* (Volume 79, Issue 25), May 5, 1967.

September. Dean Mathis spoke to the *Anchor* on January 13, 1967 about the administration, the students, the relationship with the church, and other critical issues. Mathis, by this time, had been on campus for less than six months but had jumped right in. In many respects it was Mathis' appearance at Hope that, coupled with the Vietnam War, set the campus on edge. Peters began a series of evening lectures set up by the history faculty and the staff at Western Seminary. A subsequent article by Peters, from a late February *Anchor,* dealt with Peters' thoughts about the history of the ecumenical church. One need only skim the article above by Peters about Oxford and Cambridge to conclude that Wikipedia now does a much better job.[146]

Peters Caper, Hope College *Anchor* (Volume 79, Issue 18), March 3, 1967.
Copyright © 1967, Hope College.

Peters made it clear that one can fool most of the people most of the time. It was not that anybody on campus knew what he was talking about, but many in the humanities and the seminary went to Peters talks as though he could provide them with some wisdom. It never dawned on them that he might be a fake. Peters, it turned out, was a charlatan and Mathis, to an extent, became the straight man for his buffoonery. When the immigration authorities came to get Dr.

---

[146] Hope College *Anchor* (Volume 79, Issue 14), January 13, 1967.

Robert Peters, Mathis maintained him to be a scholar without credential.[147]

The Peters Caper

'Dr.' Peters Caught; Has Academic Degree

By John M. Mulder
author-Editor

Robert Peters, scholar, history professor, Anglican priest, husband to several wives, and economist par excellence, was arrested Tuesday, March 7, in Holland by U.S. Immigration officials and is presently awaiting deportation in Detroit.

Contrary to erroneous reports in various newspapers, Peters does have academic training. Until his arrest last week he was successfully leading his double life as an assistant professor of history at Hope.

PETERS HOLDS the M.A. in research from the University of Manchester in England where he studied under Gordon Rupp, professor of ecclesiastical history. However, he does not have a B.A. in history from the University of Liverpool, nor an M.A. or B. Litt. from Magdalen College, Oxford, nor a Ph.D. in Reformation history from the University of Manchester.

Nevertheless, he seems to be the author of various articles written by one Dr. Robert Peters. These include a study of church administration, "Oculus Episcopi," which was in fact written by him and published by the University of Manchester Press in 1963.

His most recent article was published in "Studies in Church His-

tory, Vol. III," which was printed in Leiden, The Netherlands.

ALL OF THIS information was uncovered by one man who played the most instrumental role in exposing Robert Peters, assistant professor of history at Hope, as a fraud.

This man is Dr. James A. Brundage, professor of history at the University of Wisconsin, Milwaukee campus. In mid-December 1966, when Peters was casting about looking for another position for the next academic year, Peters came to Milwaukee to be interviewed for a post in the history department.

Prof. Brundage said he was impressed with Peters at that time and commented that "he made a very good appearance." However, "just as a matter of routine," said Prof. Brundage, he wrote to various friends in England whom he had met while studying there on a Guggenheim fellowship.

ONE OF THE FRIENDS to whom he wrote was a man at Cambridge who had recently reviewed Peters' book, "Oculus Episcopi." His friend sent back a stern warning that Peters was a fraud. Prof. Brundage, investigating with the registrars of the various universities from which Peters claimed degrees, substantiated the charge, and Prof. Brun-

dage proceeded to document most of Peters' activities around the world from 1939 to the present day.

According to Prof. Brundage, Peters has only one legitimate degree, the M.A. in research from the University of Manchester. He is, however, an Anglican priest, although he has been inhibited. This is not the same as defrocking a priest; rather, it is a temporary suspension of a priest from the right to perform services in the Anglican church.

What unfolds below as the life of Robert Peters from 1939 to the present is examined in a very detailed way in order that the record can be set straight and that the rumors surrounding this man may end.

THE FASCINATING HISTORY of Peters and his caper is the story of a professional impostor.

He was born in England on Aug. 11, 1918, and proceeded through various primary and secondary schools until 1939 when he enrolled at St. Aiden's College in Birken Head.

St. Aiden's is a college for men preparing for the Anglican priesthood and is located on the West Coast of Scotland. Colleges such as St. Aiden's do not grant degrees, and thus in 1941 Peters was ordained as a deacon and

in 1942 as a priest in the Church of England.

From 1942-44 he served as curate of two churches, first in Almondsbury and next in Somers Town. During 1944 he married Hilda Brixton, his first wife.

From 1945-46 Peters was assistant chaplain of Oxenham's School, and in 1946 he married bigamously Margaret Gladdish, a 22-year-old nurse. In 1947 he was arrested for bigamy, jumped bail, and fled from England.

THUS BEGAN his double life, and he has been deceiving people around the world.

Peters fled from England and lived for short periods during 1947 and 1948 in Switzerland, Ceylon, Singapore, and Australia. During this time he also changed his name from Robert Parkins to Robert Peters.

In the fall of 1948 he arrived in Canada, seeking employment and claiming an M.A. from Oxford and an M.A. from Melbourne. Unable to obtain a teaching job, he enrolled in Trinity College in Toronto where he read for the Ph.D. in history.

HE WAS EXPOSED and left Toronto for western Canada where he variously portrayed himself as a history scholar, schoolmaster and preacher. He swung down into the United (Continued on Page 6)

Peters Caper, Hope College *Anchor* (Volume 79, Issue 20), March 17, 1967. Copyright © 1967, Hope College.

If that was Mathis' view, he clearly was not that savvy a judge of academic talent. "Robert Peters" held only one of his academic degrees, a Master's in research from Manchester, but he did not receive a BA in History or his Ph.D., as he had claimed. He was found out when he went to interview for another position, at the University of Wisconsin-Milwaukee, and the professor, who interviewed him Dr. James Brundage, did a basic background check by querying friends in the field. They were warned that he was a fraud, and Dr. Brundage exposed the lie to the authorities at Hope. An old axiom of university hiring has it that even the devil has friends. Several of Peters' friends wrote to the *Anchor* to ask that Hope College reward Peters for his teaching ability, negating his fraud.[148]

---

[147] Hope College *Anchor* (Volume 79, Issue 18), March 3, 1967.
[148] Hope College *Anchor* (Volume 79, Issue 20), March 17, 1967.

Chapter 20

1968 Prologue

Few years in its history have been as important, or significant, as 1968... at Hope College and, as it would turn out, in the United States of America.

All was not well with Hope College. President VanderWerf challenged the Board of Trustees through its Chair, Hugh De Pree, in a letter outlining the issues the school faced dated January 6 (see Appendix). Cal was never accusative. It was not in his nature. But in this long, detailed, well-thought out letter, he accuses the extant Hope College community of being unaware of Hope's greatest strengths, and at some level duplicitous in moving to achieve their own objectives.

Hugh De Pree was clearly in the middle between a creative president, bringing Hope College its future, and an entrenched western Michigan body of alumni, mostly clergy, and a Church that neither understood Hope's challenges, or the threats to its existence. Herman Miller, Inc. of which DePree was the President, was one of the most imaginative office furniture corporations in America. Yet, in his College capacity, De Pree was not as strong in supporting growth and development. In his not-for-profit contributions he must reverted, in some ways, to his before business-life.

Cal's presidency was facing a campus community whose attitudes had been poisoned by its own fear for its future. Dean William Mathis really made much of this happen. Whether that fear was with the majority or not was never certain. The Church remained an underground railroad of sorts, not having an agenda of its own save for what had happened in the past. Knowing that the future was often fearsome, the gossip railroad flourished overriding the more thoughtful.

America's morale had been poisoned by the Kennedy assassination, and this invaded every facet of American life. Matters in the country were going from bad to worse. By the end of the year, 1968, America would be a totally different country. Richard Nixon, a man the Nation rejected in 1960 and California dumped in 1962, came back to be President in just six years. One can summarize the western Michigan body politic as being mostly pleased, as native son Norman Vincent Peale originally led the charge against John Kennedy's candidacy out of anti-papist fears. My own reaction at the time was one of disbelief.

Cal should have moved on in 1968. He should have said, "I'm sorry. This was a terrible mistake. I have to leave you to your own devices." But he stayed and Hope was ultimately the better for it, because as he rightly assessed, America needed a liberal arts initiative rooted in Christian values more than ever.

To Hugh DePree

The correspondence between President VanderWerf and Board Chair DePree is rich and full, but mostly one way. Cal was a careful thinker, and precise writer. No small-town business man, no matter how accomplished, could write as clearly as he. It appears DePree did not even attempt it.

In retrospect, Cal's lucid thinking was right on for American higher education then. In his words, President VanderWerf said, "Never before in history has [Hope's] type of liberal arts education, firmly rooted in Christian values, been so desperately needed."

He then provided an optimistic overview:

1. The future is brighter than the negative nabobs of conservatism maintain; we have just lost faith.
2. We owe it to our fine students to give them an education of excellence and integrity.
3. Hope College could achieve this ideal.
4. His career was staked on the enterprise.
5. This ideal would not be easy to achieve or maintain, but would be worth it.

Then he got specific, and dealt with financial issues: in doing so he alluded to a memorandum from a "Finance Control Committee" sent to the Executive Committee of the Board of Trustees that must have listed all the places where the College was in financial trouble. Later in the letter he refers to a nay-saying faculty member, who wished for the VanderWerf administration to fail and who had "walked into his office, and proudly announced to the secretaries that the College would be bankrupt in nine months or less."

Why did Cal stay there? Why did he put up with this?

First: he said that the situation was not as tough as it was being made out to be. His administration was shackled with a large loan, on which 5% interest was due every year, to pay for a hastily conceived mathematics and physics building. This interest on the loan had to be paid from operating funds. In preceding years, when

293

operating surpluses were the norm, these were entirely used – against his better judgment – to pay down that loan.

Second: The College was so backward and behind the times, that funds from the operating accounts had to be used for, what he called, *capital projects*. These projects run from sublime to ridiculous:

1. The school had no electric typewriters in the offices in 1963. By 1968, many offices were equipped with them.

2. Faculty offices were created from available spaces: in the Science building, Jerry Mohrig, Mike Doyle, and I all had offices in former bathrooms. Music, education, the admissions office, and the alumni offices were accommodated, too. Remodeling had to be paid for.

3. Faculty fringe benefits prior to the VanderWerf administration included mostly self-insurances, particularly for the retired. Salaries for the "retired teachers" had to be paid from operating funds. By 1968, the College had joined TIAA and medical insurance benefits were also improved. Hidden debts, as in the payment of retirement benefits, were being eliminated from the current operations.

4. The library holdings, which were hopelessly behind, were being added to.

5. By 1968, a Chicago architect, Charles Stade, was floating around the campus interviewing everyone in sight, as part of the planning process for the Student Cultural and Social Center. The College did not have funds to pay for the Center yet, but this architect – with a Christian connection and recommended by a Board member – was determining how the Center should look and what it should do. Cal questioned how Stade's billings were being monitored.

6. Indigenous to all this was Cal's distrust of the offices of treasurer and financial affairs. Cal felt there to be job duplication and would soon make a change.

7. Then there was a matter that dogs university administrators: set up costs for scientists. When I

arrived in Hope in 1964, I was so naïve that I did not ask for anything for my labs...I felt I could make do. Fortunately, I was coaxed into a different stance by Irwin Brink. But Sheldon Wettack handled his appointment differently. He requested about $14,000 to set up his laboratory, including a Ferrand fluorimeter and a gas chromatograph. Cal's defended his purchases. As it turned out, it was an extremely wise investment for the College.

8. Additional investments in the biology department were less wise. Funds were provided for a certain Robert Fitzsimmons who had no impact on the Institution. Norman Norton, chair of biology, had served as titular head of the Sloan proposal committee and benefitted from this for his department.

9. Cal also listed the upgrades in faculty. He went on to say something that still rings true: "in almost every department, students can hold their heads high and compete on an equal basis with students from other colleges and universities in the country. This is a fundamental type of integrity which we must build into our program if we have any right at all to exist!"

10. "Last year was full of conniving and intrigue that led to overstaffing in a few departments." Cal was not specific, but he likely referred to Dean William Mathis not being a responsible individual. Letters in the file from him to the President had him taking silly positions about what his job was. Overspending also happened because financial responsibility was spread to more than one person. "Had the functions of treasurer and business manager been merged in a single position, a single person would be responsible for preparing the budget *and* presenting it to the Board of Trustees." Cal also pointed out that that because of this disparate structure, he was "losing fiscal control of the College *and* that the legal counsel retained by the College has failed at several junctures during the preceding year." He ended by questioning the

performance of the College auditors as retained by the Board.

11. These were extraordinary accusations for a president to make about the institution of which he was, supposedly, its chief executive officer. In short as a business Hope had too few Indians and too many chiefs. Cal had found that this pertained in business and management, just like it pertained academically "more interest in running the College, than improving their own departments," as Bill Vander Lugt said in the North Central report (April 1964).

12. Finally, there was a question of over-staffing in some departments. Cal pointed out, correctly, that overstaffing was easily corrected by not replacing those who leave. He off-handedly deals with an issue still plaguing the academy – the part time instructor. "If part timers are employed, it makes most budget planning irrelevant." He also says that the Dean of the College should be responsible for the instructional budget and, in him, control of all hiring and firing on the academic side retained. That had not been the case the year before.

In general, and for reasons unclear, the president was viewed more or less as a figurehead. All, or at least, most of the financial decisions were being made by others, likely with direct involvement almost on a day-to-day basis by members of the Board including Hugh DePree.

Next, Cal dealt with campus income. Science faculty, to this day, are expected to pay for most, if not all, of their activities outside the classroom. At Bowling Green, where my income was well beyond my salary each year for thirty-five years, I was referred to by a provost as a "cash cow." So much for my high ideals of teaching and research in that school—I became a money factory. At Hope, that was certainly the case, too. By the end of 1968, the chemistry department alone had managed more than $1,000,000 in outside grants and contracts.

But Cal defended other departments where incomes were less available. Music, Cal said, where much instruction was one on one, could never be expected to pay for itself on the basis of funding from outside the campus. "The fact is a large percentage of Hope students insist on taking music courses and music lessons, adding significantly to the average cost per student at Hope. Yet, in the face of the many contributions of the music department, who is to say we should abolish music in the name of economy?"

He listed the contributions of faculty in the various departments that were not generating income or press. For example, "Intramural athletics are, in the long run, more important than varsity athletics."

Science Instruction in Colleges and Universities in General

The sciences were continually beaten on by Cal's critics. After Bill Mathis arrived, the science/humanities split at the College became seriously unpleasant. But Cal had it right on.

"First of all, to any liberal arts college which hopes to stay alive significantly in the years to come, a science division of genuine integrity is absolutely essential."

1. One of the most important reasons for this is that male students often prefer science, and "we have trouble recruiting qualified male students."
2. 50-80% of highly qualified male high school graduates claim they wish to major in the sciences.
3. Our coeds are brighter than our males; ergo for the male population to keep up, there have to be strong science programs.
4. Many of our finest graduates, across the disciplines, were recruited to our college because of our science programs, and then switched majors.
5. Our chemistry faculty are among our most active recruiters of students in the entire college.
6. Students prefer to go to schools where the textbooks are written – where the authors are. Good students do not want to go to schools where the texts are parroted by professionally inactive teachers.
7. Summer Institutes in chemistry and other subjects were important recruitment devices for the best students. They also partly fill the dormitories during otherwise inactive periods.
8. One has to lead with one's strengths, and until other programs catch up, that means foundations that support the sciences will have to be approached first.
9. Finally, much of what we perceive as the best teaching goes on in one-to-one instruction in the physical sciences.

Research experiences for undergraduates are critical to maintaining the highest quality programs. In Cal's words, "the excitement, the thrill and the adventure of education – as in learning on one's own."

Cal made a comparison later in the letter between the practice of the musical arts, and the sciences. "It is this type of exhilaration and satisfaction which, coupled with a truly meaningful spiritual life, that will afford Hope College distinction that will make her truly worthy of support on a far broader basis than has been yet achieved."

Operation of the Plant

One of the many problems of the liberal arts college is that its physical plant use is inefficient and non-economic, as they run for only nine months each year. One way to combat this was the new chemistry and mathematics summer institutes then beginning which brought in more than 200 people to campus for eight weeks or more.

The chemistry and biology departments were recipients of undergraduate research participation grants from the National Science Foundation in the summer of 1968. These grants provided opportunities for students, but also paid overhead to the College. By 2010, eight campus departments had research participation grants and chemistry was not one of them. More than 150 students were employed during the summer as research assistants.

Cal recognized the contribution of the Sloan Foundation grant, bringing international recognition to the College, as well as gifts of property to the biology department for use as a field station. Mary Jane Gold would give the Gold estate to the College for the Marygold Lodge. Hugh DePree, a few years after Cal left the school, would buy the property for Herman Miller for $1,000,000. Today, that property is conservatively worth $25 million.

Next year's budget

There were hazards ahead, as Cal put it: "There's no doubt we have faced a series of tough breaks." They were as follows:

1. Enrollment was considerably below what we had been given to understand it would be.
2. Because of disappointments in the food service, far fewer students signed up for meal plans than were expected.
3. A church development campaign fell far below its expected outcome and seriously eroded our development efforts.

298

4. The death of a major supporter delayed funding at some level.

"But tough things happen, and we have to manage our way through them," Cal wrote. He also mentioned this would happen in spite of the delight of members of the faculty who resist change as in "a member of the faculty who resists change stated triumphantly in my office that a member of the Board had told him that the College would be bankrupt in nine months."

Cal wrote to DePree: "Nothing can be more devastating to our Hope dreams than this type of attitude. This cuts the heart out of the men who, in faith, with vision, with hope and with courage could actually succeed in doing the job." Cal pointed out that the Board was not living up to his expectations of them and pointed out some ways that this could be done.

Fundraising and Development

By this time, the College had attracted Lee H. Wenke to its development staff. Conversations with Loutit Foundation failed because a member of their (Loutit's) Board was an alumnus of Alma College, but the Charles E. Merrill Trust, in a visit arranged by the President of the New York Stock Exchange, alumnus Robert Haack, and under the leadership of the future Secretary of the Treasury Donald Regan, was more accommodating. Eventually, the Charles Merrill Trust provided some funding. A science grant of more than $200,000, brought by a proposal written by Dwight M. Smith, in the chemistry department, was provided by the Research Corporation on the basis of Cal's work. Wenke, for his part, was working with the Kresge Foundation, with the George F.Baker Trust, and with others.

Cal assured Hugh De Pree that, with all these efforts, the College would be in the black by August, and it was. As far as the next year was concerned, that depended on the fall 1968, enrollment. Hope's rate of growth has been among the lowest of all of the coeducational colleges in the State of Michigan. Even Central College was doing better.

As it turned out, 1968 was an excellent year for the College; major grants were obtained in the sciences, enrollment began to turn around, and other changes began to take effect.

Church

The big gorilla in the room was the church...and the word "church" had several meanings. First it meant the organized Dutch Reformed Church; second, it meant the church, as in the church gossip lines in western Michigan churches; third, it meant the local clergy, and mostly the Holland clergy. Cal pointed out that, in his opinion, the College was closer to the Church now than it was five years earlier. But, this was where he finally got more tough than apologetic.

1. "We can never allow a certain surface piety, formal church connection… become a cover-up, by faculty and staff, for non-performance of [academic] duties, for incompetence or for just plain slothfulness…In matters like this, the Board will either have to back the administration to the hilt or get a new one."

2. "Hope College should assume a true position of leadership in the denomination."

3. "Realistically, I do not believe we can depend on the Church as or our sole source of support. Certain members of the Board believe that we can; I believe this is unrealistic. The time has come when this has to be faced honestly by the leadership."

Cal had thrown down the gauntlet. He then proceeded to outline the many ways in the Church had failed the College. He analyzed, as only a scientist might, the potential "pot of gold" that was the Reformed Church.

Finally, he dealt with an issue that was a realistic issue for a College so dependent on a large number of modest givers: the question of support in the face of controversial political causes. An honorary degree had been bestowed on Michigan Senator Philip Hart, a Roman Catholic. Some members of the Church withdrew support for the College because of that. Then there was the Dick Gregory appearance on the horizon. A black civil rights activist: how could that be at my Hope College? Support was withdrawn by some more because of that. Some Church members opposed dancing. Others opposed some of the fun college students like to have.

All in all, it was a never to be won battle, but Cal called that battle to the Board's attention as something that had to be addressed. "There is a tragic generation gap in our day, and some of the shared forms, rituals, and dogmas of our generation, simple and honestly, don't have much meaning for even the most conscientious and sincere of today's generation."

300

January 6, 1968

Mr. Hugh De Pree, President
Herman Miller, Inc.
Zeeland, Michigan

Dear Hugh:

There can be no doubt that Hope College is in several respects at a cross-
roads in its history at the present moment.

The recent memorandum from the Finance Control Committee to the Executive Committee of the Board of Trustees focuses attention on our current
predicament.

As we study and take action on that document, I feel it extremely important
that we be both fair and realistic in assessing the past and in making projec-
tions for the future. Just as in the past we may have been unduly optimistic
or visionary, so at the moment we may be unwarrantedly pessimistic or
gloomy. Another way of stating it is that we may have lost a good deal of the
faith which once we possessed about our enterprise.

I am totally dedicated to the cause of offering our fine young students a college
education of excellence and integrity, undergirded by a dynamic spiritual climate which has true meaning in the life of each individual student and genuine
relevance to the contemporary world. Never before in history has this type of
liberal arts education, firmly rooted in the unchanging Christian values, been
so desperately needed. I truly feel that we can achieve this ideal at Hope
College; in fact, I have staked my career on faith in our enterprise. But
the task won't be easy, and its accomplishment will certainly depend upon
our being totally realistic about the conditions in which we find ourselves.

First of all, let me say that I do not believe that the situation is as bleak as
the memorandum would have us believe. In the fall of 1963, Hope College had
a non self-li... ...ating debt of approximately $750,000 on the Physics-Math
building, repre ...ced by a loan from the Old Kent Bank at 5% interest. In the
interim, this debt has been reduced to about $270,000, with both the interest
on the loan and, in some cases, part of the principal being paid from our
operating budget. Likewise, at the end of '64, '65, and '66 fiscal years, our
erating budget showed a significant surplus, which in every case was us...
...ieve, to retire debts previously incurred. Each year, when the oper...

From the letter from Cal VanderWerf to Hugh DePree, badly damaged by the
Van Raalte fire. The entire letter is printed in the appendix. Its details though,
are captured in the body of the story here. Like all not for profits, Hope was
having problems with money. This letter deals with much else, but the bottom
line was the cash in was not enough.

Cal said that the College could no longer afford to appease
the trenchant members of the community or they would cease to
function as a college with any integrity. He went back to his often-
cited complaint: while nominally controlled by the Reformed
Church, the monetary drive was lacking. "We differ from a Calvin
which gets all of your support from its Church; a Kalamazoo, which
gets none of its support from a church, or a Wheaton which can
appeal to a whole band of pious supporters from throughout the
Nation while, being non-denominational, it does not have problems
with any specific doctrines or creeds."

Further, if this continued, it would affect the mission that Cal had tried to instill: "We may run into increasing difficulty in expecting any financial support from the Federal government because of Church control." He pointed to the Ford Foundation's rejection of a research grant on the basis of church control, something that would be repeated unless Hope decided once and for all where it stood."

"We expect, and hope, for an enrollment of 2000 students in the fall, but what worries me most is our long-range capital funds picture. Up until now, we as a college and as a Board have not proved we can raise capital funds. To sell promissory notes merely postpones the inevitable."

Derek Bok, a fellow Dutchman, when he was in his last year as Harvard president in 1988, offered a beautiful speech at graduation. Bok apocryphally describes his dream of how to fund the continued development of Harvard University; first he had to sell bonds and obtain a billion dollars, to be paid off in ten years. The payoff – at least of the interests on the notes would come from great athletic prowess, from economic development, then from selling its soul to foreign governments. Finally, when all of this ran out "I woke from my dream."

Hope was far too conservative for any junk bond sale. Its policy had been to pay as it went along. Thus, the Student Center became so vitally important: "It represents more than just a building. It is a symbol of our ability to raise capital funds. This has really become a symbol of Hope for all of us."

Cal pointed out that in four years, slightly over $100,000 had been raised all from outside the church and college. With the Peale science facility on the horizon, where would three million dollars come from? The Board of Trustees was noncommittal and unproven in their ability to raise a successful campaign. He wanted the Executive Committee of the Board of Trustees to work in concert with the administration to develop a strategy going forward, stating "The fires of philanthropy are fueled by the causes it embraces, and stoked by its own great men, the leaders who stimulate others to give and work enthusiastically and voluntarily."

He ended with a veiled threat: "To balance the budget next year is no trick. We lop off faculty, forget about special projects and fire within the limits of the law. But to do that doesn't require a president. A computer would do the trick. There is a better way open to us. Let's not turn our back on it without a valiant and honest struggle."

302

In a subsequent letter dated January 26, 1968, Cal repeats much of the prior analysis: "We have to get the funds together to build this student center. We have excellent students and a developing faculty of some excellence. We have to put the facilities in place to enable the College's future greatness."

The document was recovered badly damaged from the Van Raalte Hall fire, but one can see the issues the school was facing easily from its contents. Tensions were arising mostly from the lack of money, the Finance Control Committee, a bad appointment of Dean Mathis, and an overall lack of faith.

As I say elsewhere, I was chair of the Cultural Affairs committee during the academic year 1967-1968 in what was a more or less sinister attempt by Cal's enemies to put one of his strongest supporters in an awkward academic position. But I handled it. That year is the only year in my academic career that the only thing I published was a book. I went to the Netherlands in September to reignite my research career.

In many respects, this was one of the smartest career decisions I ever made. Because I did reignite it – more, I ignited what was to become my bread and butter for the rest of my scientific life: polymeric reagents in organic photochemistry.

Chapter 21

Campus Unrest

Campus unrest hit Hope in a different way than it did large universities. Administrations and administrators had to lose sleep over student unrest. The *Anchor* went on the attack as young journalists learned the art of "investigative reporting." After The Ohio State University was shut down by the National Guard, students were shot and killed by the National Guard at Kent State University in May of 1970. Hope did not close for any period, but its students were upset nonetheless.

These events were less not the events of students, than of nation coming unglued from a world that had known a righteous peace after its victories in World War II. Americans trusted their government, but the post Kennedy governments, first of Lyndon Johnson and later of Richard Nixon, were not trustworthy.

Many more students were attending college because it became a necessary rite of passage, but not all shared the values of their predecessor collegians. Individuals and groups struck out to make their own way in this new-found educated community. Though the national morale had been deeply impacted by the assassination of John F. Kennedy, later events had a similar effect. Much later, Thomas Friedman argued that the World Trade Center bombing, of September 11, 2001, was, like the Kennedy assassination, a "gift that keeps on giving. Osama Bin Laden really did mess us up. Though we've erased the ruins of the World Trade Center, the foreign policy fear is still very much in us…and we have a national leadership that's offering us insecurity, as in leadership in an age of fear."[149]

Freidman quoted David Rothkopf, Editor of the New Republic: "…9/11 was such an emotional blow to the U.S. that it, in an instant, changed our worldview, creating a heightened sense of vulnerability. Not only did we overstate the threat, we reordered our thinking to make it the central organizing principle in shaping our foreign policy." While the circumstances behind the September 11th terrorist attacks and J.F.K.'s assassination are entirely different, the reverberations that the events wreaked on the generations who experienced them were similar as were the outcomes. America was never the same after either.

---

[149] Thomas Friedman, "The Gift that Keeps Giving." *New York Times*, December 2, 2014.

The 1960s were a difficult time for reasons other than just a change in presidents, or a war in a far-away country. The Supreme Court held that segregated schools were unconstitutional in May 1954, and racial issues were never far from the surface. In the south, when blacks began to exercise power, the outcomes could not help but be confrontational. Black activism was intensified by bus boycotts in Alabama, the emergence of the Rev. Martin Luther King, Jr., Malcolm X in Harlem, Adam Clayton Powell in the Congress, and the push back of southern whites through their political systems.

Printed newspapers were becoming less important and the number of words of newsprint published each day was taking a steady path downward. Papers merged with other papers, and readers were given fewer choices of what to read. A thousand-word story in the local papers, even those in small cities, was gradually being replaced by short evening news sound bites. Television news bytes replaced the morning paper read. This change, more than anything, exacerbated the nation's anxieties over the war in Vietnam. Mothers seeing the evening pictures of body bags returning to Washington area airports knew, for sure, that their son's remains were in one of them. If they were not there that night, they would be the next night.

Television emerged as the way in which information was passed around the country. Though dorm rooms had no television sets nor did the students have access to the iPad, news spread by word of mouth and radio permeated the campus campuses. There were no alternative networks of consequence; no public broadcasting system; no national public radio. Nuremberg-reporting alumni (Walter Cronkite, Eric Sevareid and Howard Case Smith) were more important to the average citizen than was any Senator or Representative. On the evening news programs, one could expect several minutes from/about Vietnam. This usually meant body counts of Vietnamese killed and the number of American casualties, dead and wounded, as listed for some recent period. There would be pictures of battle scenes. Generals in the field would tell viewers that things were going better; one could expect victories soon. William Westmoreland commanded military operations in Vietnam from 1964-68 when the biggest growth in American presence occurred. Professor Thomas Tinsley Tidwell III, then an assistant professor of chemistry at the University of South Carolina, stood at an honorary degree ceremony held by the faculty for Westmoreland on the Columbia campus, and raised a sign that read "No More 'Doctors' of War." Tom, who was a mild, determined, careful scientist, had been involved in integration issues on campus, was given the sign by members of that group. After his action, he was ushered from the hall, and told that though he had tenure at South Carolina to never expect promotion. The University of South Carolina, though potentially an excellent academic institution, lost several chemistry faculty because it was a segregated institution. Tom was forced to apologize and, wanting to keep his job and burgeoning research career, he complied. Eventually, Tom left, became a professor at the University of Toronto, and later Dean of Science at the Scarsdale campus. On retirement, he published a book on the organic chemistry of "ketenes" and is now concentrating efforts on historical aspects of organic chemistry. But his career, though never tainted, took a side road because he stood, publically, against the leadership of the Vietnam War.

The VanderWerfs, both Cal and Rachel, were committed to, and active for, racial equality in Lawrence, and at the University of Kansas. As has been documented in the chapter on the VanderWerfs and racial issues in Kansas in the 1940s and 1950s, they were among the first in their community to make race an issue. What they did not expect was what they found in Holland, Michigan. Mrs. Alta Wilburn had been a part of their family since the girls were little.[150] Alta, born in Wichita was a resident of Lawrence until Cal and Rachel moved her to Holland. Her mother, who lived to be 105, had been left by her father in Lawrence with five children so Alta only finished high school and then had to go to work. She found working with little children fun and fulfilling, though she had no children of her own and her marriage was short-lived. Mrs. Wilburn went to work for the Lawrence League for the Practice of Democracy's preschool nursery. Both Gretchen and Klasina VanderWerf were enrolled. She stayed there for a time, but developed a skin disorder and had to quit. Rachel VanderWerf called her, asked her to do a little baby sitting, and it became nearly a full time job. She became the children's nanny and added some good judgment and common sense to the very busy VanderWerf's household.

In an interview for Holland's sesquicentennial given to the Hope Holland archivist, Mrs. Wilburn said that when she moved to Holland in 1963, the townspeople were inhospitable. One man wrote Cal: "We've never had colored in the town and we don't want any now." Mrs. Wilburn also faced criticism even from Holland's clergy.

---

[150] Larry Waegenaar, "Wilburn Alta Oral History Interview; Sesquicentennial of Holland, "150 Stories for 150 Years" (1997). *Sesquicentennial of Holland, "150 Stories for 150 Years."* Paper 165. http://digitalcommons.hope.edu/ses_holland/165.

One minister told Alta that he was on the committee to "help control her." She was the victim of a lot of gossiping.

In a small act of subversion, Alta participated in a plan to ride in a float sponsored by other minorities in the Tulip Time parade that had the title "Let's Be Friends." The VanderWerf children wanted to ride with her. At other times, the children shielded Mrs. Wilburn from criticism they experienced in the community. She claimed that many of the townspeople experienced changes of their hearts after having associations with her and followed the lead of the VanderWerfs, by making every effort to include Alta in their civic activities.

Alta made pointed observations about Cal's efforts to integrate Hope College, and the community and the reaction he and his family faced. When the grades of black students suffered, it was seen as an indictment against the race and their abilities, even if that was not the full picture. Most of the black students were from poor and uneducated families in the South. They found the cold and the food of Holland a challenge, and when they were refused service downtown restaurants because of the color of their skin they could not understand why one would suffer to get an education in such a place anyway. So they left the school after just one semester.

Mrs. Wilburn was a mature woman who had lived with this racial prejudice all of her life. While she had become accustomed to living within the racial tensions, the younger generation rebelled, and developed a platform for change. That happened in the 1960s. Blacks around the country rebelled, and it was difficult on all campuses; Hope was no exception.

One issue almost all could agree on was their opposition to the war in Vietnam. Lyndon Johnson's Texas twang still rings in my ears when I think the words "My fellow Americans." It was an unbearable sound for those who taught young people or who were around them speaking mostly about Vietnam. It was surely frightening to the young people who could see a stint in Vietnam in their future. The American military was a still oiled, but old, military of World War II, with now those that were junior officers in charge. The problem with the Vietnam War was it was not anything like World War II: those battle tactics that worked, however well or poorly, in Europe or the south Pacific, were not effective against a mostly guerrilla force, fighting for their homeland on their own properties, in Vietnam.

It took more than ten years but in the end, the United States lost the war and over fifty thousand Americans died without result.

In spite of the huge and disruptive criticism on university campuses, and from the intellectual community more generally, the opposition to the war was anything but unanimous. There were some who said "the president is our president, and we'll follow him to the end." There were others who supported the military. But almost without exception, there were no persons that really understood what the war was about or what it was being fought over. The subtlety of the domino theory went beyond all but the few that still harbored the pre-World War II issues that brought about the end to colonialism around the globe.

Student and campus unrest at Hope took several forms. The critical issue that impacted students more than any other age group was the draft. There were continuing protests about it, about the way it was being administered, and, in general, about the way students were being treated by their elders. There was a general unrest developing on the campus, but this had as much the local flavor, as a national one. Compulsory chapel attendance was being studied, and the requirement would eventually be changed. A letter from the American Civil Liberties Union (A.C.L.U.) Western Michigan Chapter to President VanderWerf, dated March 14, 1968, indicated the officers of the organization were to seek a meeting with the Hope administration to talk about the issue. President VanderWerf ignored this letter for, among other reasons, he knew it had been mostly generated by an uneasy student. In the day of protests all over the south over minority rights, and protests on most campuses about the draft, and voting rights for students, that A.C.L.U. would waste money on a minor issue in their eyes at a small college in Michigan was fatuous.

The campus paper was trying to stir up trouble, too. Successive editors of the *Anchor* were feeling their oats in the then new investigative reporting responsibilities of American journalism. Communication methods were improving and changing. Television and other audio/visual media impacted human perceptions. One issue that seemed always prevalent was administrative resignations, and movements. President VanderWerf had dispatched a number of former Hope administrators on taking over the presidency, as was to be expected. But he had trouble finding suitable replacements in a small town in Michigan. So turnover among new, mostly unsatisfactory appointees, at least at the beginning, was brisk. The *Anchor* loved to point this out.

The idea that there was substantial movement of Hope's faculty however, as the cartoon below intimates, had no currency. A letter in the immediately succeeding *Anchor* from the faculties in the humanities and arts pointed out that Hope's faculty attrition was much below the national average. Others would have suggested, had they the time to worry about it, that transition and change in college/university faculty is good and renewing. Hope was a four-year institution, and most professionals that taught there were obviously teaching below their pay grades, if they had Ph. D. degrees. The Ph. D. is a research degree, and holders of it are supposed to teach others the principles of research. Hope did very well with the notion of undergraduate research, which stabilized its faculty. But some of us who got uneasy did so because we wanted to work with more advanced students. The thirst to work at the graduate level went unanswered by Hope College. There was an indication in the letter that a cartoon that had appeared in the *Anchor* was targeted at the announced development of the TIAA CREF retirement program at Hope, which was said to allow a faculty to take his/her retirement funds along when or if the person moved. The value of this much needed faculty fringe benefit far super-ceded any complaints about moving faculty taking Hope's resources with them. I took $2416 in retirement dollars, half of which were from my own contributions, when I left in 1971. At last report, Hope's TIAA-CREF corpus was about $130,000,000 built on contributions from faculty and staff, and from college dollars.

There was a prevailing unrest, about civil rights, about the draft, and about the war in Vietnam, on Hope's campus. However, boys were also being boys. A panty raid energized the males and perhaps the females, too. This caused a letter to all the undergraduate women from the dean of women about how the students should grow up, and another series of letters to the editor, asking "Why?"

In February 1968, Congress began to make noises about changing draft status for graduate students, at a time when there was a publicized shortage of college teachers, according to Nathan Pusey, the President of Harvard. Inner-city congressmen saw blacks more frequently drafted than whites, and the draft law was to be changed. Campus students at Hope, which was among the more conservative undergraduate student bodies in America, were becoming much less supportive of the war. According to the *Anchor* on February 26, 1968, by the end of 1968, 75% of those drafted were to come from the college ranks. Earlier that week, President Johnson had abolished all work and graduate school-related deferments.

The event that emphasized political issues in the country, and that was to impact my life for a long time was in the planning stages: my Cultural Affairs committee was planning to bring Dick Gregory, a comedian turned black activist, to campus, in an organized symposium we called "a Crisis in Our Cities." Detroit had suffered serious racial rioting in the summer of 1967, so our committee was planning a campus wide symposium to deal with the issues that begat these riots. The March 1 *Anchor* detailed numerous criticisms of Gregory, and his speeches. But the campus bought into the events of a "Crisis in Our Cities," all of the fraternities sponsored literature meetings in relation to the symposium. The *Anchor* editorialized that this three-day symposium would attempt to make an in-depth probe of the racial conflagrations that took place in urban centers like Detroit the prior summer. "The symposium is an opportunity every student should seize. It can give each Christian a chance to probe his/her relations with his fellow human beings in the most vital of national concerns."

Gregory's speech blasted the United States government. He called the country insane for the war in Vietnam, and then went on a series of diatribes. The actual speech was much different than the *Anchor* reported. For the most part, American blacks were exercising their first amendment rights to free speech: Gregory represented the situation as it actually had existed in the United States. However strident this might have seemed, these were blacks that had been forced into segregated cars on trains running through the South, and into the South, forced to drink from different fountains, eat at separate tables, and go to different schools than were their white counterparts. They had every right to be mad and they were just that. Things changed but not quickly enough they submitted.

Some blacks made the most of this in events in which they were paid to deliver their message to white audiences. Marion S. Barry, Jr, serves as a good example. Barry was my classmate in chemistry at the University of Kansas entering in the class in the fall of 1960. He was later chosen by civil rights activist, James Farmer who had Kansas City connections (C.O.R.E. – Congress of Racial Equality) to serve the cause in Washington, D.C. Members of the District of Columbia had just gained the right to representation, and could elect its own mayor. According to his obituary in the *Washington Post*, Barry was "the most influential and savvy local politician of his generation. Barry dominated the city's political landscape in the final quarter of the 20th century, also serving for fifteen years on the D.C. Council, whose Ward 8 seat he held until his death. Before his first stint on the council, he was president of the city's old Board of Education. There was a time when his critics, in sarcasm but not entirely in jest, called him "Mayor for Life."[151] Into the first dozen years of the new millennium, Barry remained a highly visible politician on the District of Columbia's political stage, but by then on the periphery, no longer at the center.

Barry came to Kansas following study at Fisk University in Nashville, TN, where I. Wesley Elliott, Ph. D., the first Kansas black to have been awarded an assistantship in the University chemistry department, was a faculty member. Marion Barry was my classmate in the graduate program in chemistry so at one stage in writing this, I tried to reach him and talk with him about his experience at K.U., and his mentor, Sam Massey who was one of America's most significant scientist/civil rights leaders. But Barry was too indisposed to talk with me so I never reached him.

Elliott's colleague, Samuel Massey, was Barry's mentor. A Tuskegee Institute alumnus, Massey had followed George Washington Carver to Iowa State University where he was allowed to finish his degree in organic chemistry by working in a basement laboratory. Sam was later named one of America's 75 most influential chemists on the occasion of the 75th anniversary of Chemical and Engineering News, the news magazine of the American Chemical Society (A.C.S.).

According to Massey, in the summer 1950, the Division of Chemical Education held its biannual meeting of chemical educators at the Ohio State University. Massey was invited to speak at the meeting, but was denied a room at the O.S.U. Union because he was black. So Cal VanderWerf, who was chair of the conference, gave him his room. Cal and his wife were staying in Columbus with her parents. Barry, a student eligible for graduate instruction at Kansas, was sent by Massey to VanderWerf because he trusted Barry would there be evaluated based on his effort and ability. Sam never forgot the VanderWerf's generosity.

Massey was President of North Carolina College at Durham from 1963-1966, and when that went sour, was appointed professor of chemistry at the Naval Academy. A Navy engineer also on that faculty later said that Massey's salary was three times higher than any other Naval Academy faculty member! Marion Barry eventually let Massey, and the department at Kansas that gave him a chance, down. He was, as Barnes said in the *Post*, self-centered and indulgent.

The western Michigan reaction to Gregory was more or less as could have been predicted. Calvin College, being the subservient

---

[151] Bart Barnes, Marion Barry Obituary, *Washington Post*, November 23, 2014.

institution that it was to its church, cancelled a scheduled Gregory appearance. This action was taken by its Board of Trustees who then silenced the campus newspaper, the *Chimes*, for taking a stand against the administration in the Gregory case. Academic freedom was not alive and well at Calvin then, if it ever was. Hope embraced the Gregory appearance proving it much more open to controversial, but needed, political discussion. The events of the speech were heavily covered by the press, but nothing untoward happened because of them. A few weeks later, Martin Luther King was assassinated in Memphis so the whole national attitude took on an air of stridency that took a long time to abate.

In the March 8, 1968 issue of the *Anchor,* the resignation of Larry ter Molen as Director of Development was announced, a significant loss to Hope. Larry moved to Southern Methodist University, a bigger university in a more affluent part of the United States. The lay leadership of Hope was strong-willed but below average in contributions and insights and, even after the Board was reorganized, mostly in-grown. Few had academic experiences other than their undergraduate experiences and those were mostly at Hope College. Many years later, Larry told me that Hugh De Pree, who was chair of the Board for many years, had advised him to leave Hope College. Many of us were trying our damnedest to build an academic institution of substance at Hope and the Board Chair was advising one of most able colleagues to move elsewhere? Hope suffered for many years from De Pree's leadership with this being just the first example of his intervention into campus affairs. One Board member who differed in background from this markedly was Fritz V. Lenel who served on the Board for a number of years during the VanderWerf tenure. Lenel was a German physicist had immigrated to the U.S. in 1931 as the Nazis began to rise. He taught at Rensselaer Polytechnic Institute where he was a professor of metallurgical engineering. He lived to be 96, founding the field of powder engineering. This was based on his work with a division of General Motors immediately after he came to the U.S. Lenel was not a particularly advantageous appointment to the Board, but at least he was an academician.

Boards can be substantial and supportive or interfering and disruptive. The North Central Report (1964) did not deal with that dichotomy at Hope, though it could have. As the VanderWerf administration progressed, it was the Board, more than the faculty, that paralyzed growth, change and development of the school.

Chapter 22

Stridency Rises

After the withdrawal of Lyndon Johnson from the 1968 race for president and the assassination of Martin Luther King in Memphis, the American psyche was on high alert. This alert permeated all parts of society. Students became increasingly strident in class and in campus affairs. The press at all levels became more anti-and aggressive. It took mental toughness to stay the course particularly if one were on a college, or university campus. College presidents had a particularly difficult time: the classic case is that of chemist Kenneth Pitzer who went from being president of Rice to assume the presidency of Stanford to last only nine months as a president despite being familiar with California culture. During the brief time Pitzer was at Stanford, I asked Gene Van Tamelen what he thought of the appointment, thinking I would hear something profound. "If he does his job, I'll do mine," Gene said.

That was the opinion of the research professors in general. They did not care who was president, or what that president did, as long as the administrative behavior of the office, and the administration, let us do chemistry as chemists did it.

I was in San Francisco at the time of the King assassination attending an American Chemical Society meeting. The first edition of *Laboratory Experiments in Organic Chemistry*, the laboratory manual I wrote with Mohrig, had been published. The book's contents were occasionally problematic in its first edition, and neither of us expected the skills of the young users to be as inept as they turned out to be. We over-estimated our audience. But the book included the use of new instrumental techniques for the study of organic compounds and this was a plus. It sold well.

I was exiting a speech by Eugene McCarthy at the San Francisco Hilton on the evening King was shot and killed in Memphis. I moved myself into the back of a ballroom at the Hilton, attending (sneaking into) a fund raising dinner hosted by Congressman Don Edwards for Eugene McCarthy. Sherry Rowland, and I met at the back of the hall as McCarthy criticized J. Edgar Hoover and General Lewis Hershey. A few minutes earlier, King had been assassinated and we now know that King had been a target of Hoover's suspicions. But Rowland and I were in that Hilton ballroom as interested bystanders trying to figure out what McCarthy had to say to the voters at a time when he, almost alone, opposed President Johnson's stand on the war in Vietnam.

The King assassination was another assault to America's dignity. Though death of public figures had been part of the America psyche since the assassination of President Kennedy, none of us though it would happen again. But it did. It was impossible to not be impacted by the King assassination. It directly impacted attitudes and events on Hope's campus. Hope College held a march to Centennial Park in Holland, and many, including President VanderWerf, spoke there. Basketball star Floyd Brady, though not that political, led the march.

A college president at the time, or university president for that matter, was under pressure from all sides. The King assassination and Johnson's ducking responsibility for the mess in Vietnam shot downward. Every institution was impacted at some level, and in different ways. For example, at the instigation of a student, the American Civil Liberties Union (A.C.L.U.) had been contacted about Hope College, a college strongly affiliated with the Reformed Church, for accepting federal funds. The issue at hand was compulsory chapel and how a not-for-profit institution requiring regular church attendance of its students could be eligible for federal funds. The separation of church and state is a fundamental tenet of American democracy that few still argue with. If a grant to a physics department to study the basic properties of the neutrino has some religious implication, I fail to see it. But students of that day, and organizations that challenged for the rights of individuals in a free democracy, were on heightened alert to challenge. So a letter from a local chapter of the A.C.L.U. to a college president of an institution in place for more than 100 years to ask for him to address the question of public support of programs in his institution is presumptuously nonsensical at one level, but in the context maybe expected? The Constitution guarantees due process; and if the A.C.L.U. really felt there was an issue, then the proper action is to sue, carry the case through the courts, and let the matter be decided at the highest constitutional level.

For some colleges, denying federal funds was a savvy business decision. Two institutions that refused federal funds, as a matter of principle, were Hillsdale College in Michigan and Grove City College in Pennsylvania. These schools made a development point out of independently avoiding attachment to the government by not taking its money, and conservative donors were attracted to that. Academic snobs would say "They took no federal money? Why would the taxpayers want to give them any?" But neither Hillsdale nor Grove City took federal money as a matter of choice and their constituencies gave them many more money than the government would have anyway. For these two cases, denying federal aid was good business.

Cal ignored the letter of inquiry by the A.C.L.U. But the *Anchor* devoted ink to it because the A.C.L.U. inquiry indicated another step by those not in control to put pressure on those that were. As in all cases that need time to develop, the semester ended and with that the A.C.L.U. threat went away. It was and is, a practical organization in pursuit of personal rights, and this particular issue did not fit their agenda. It was not big enough.

Not all was morbid and sad those days. Mark Rockley, the *Anchor*'s chemistry major and writer, wrote about panty raids.

PANTY RAID—"Try the door" was the cry from the second floor of Phelps as Hope males raced from dorm to dorm in their second panty raid this year.

Dykstra Hall Is Stormed In Second Panty Raid

By Mark Rockley
anchor Reporter

pushed his way through the window into the dorm, knocked on

Panty Raid, Hope College *Anchor* (Volume 80, Issue 23), April 26, 1968. Copyright © 1968, Hope College.

Even with the stridency, students of the day were still just 18 year olds. So some of the students' attentions were turned to matters of the flesh. Rockley was one of those brilliant youngsters the Sloan program targeted. He came to visit Hope from New Zealand with his father as a fifteen year old, to see if the College fit him. It did and he enrolled, He worked in Sheldon Wettack's chemistry labs. Following graduation Mark took his Ph. D. with friend photoscientist David Phillips at the English university at Southampton. Mark is now emeritus professor at Oklahoma State University.

The Draft

Americans of my day grew up knowing that they would be subject to being drafted by the government to serve their Country. My colleagues, some of them, were happy to get that card because one could "drink alcoholic beverages in New York" when one was eighteen and the draft card served for that identification.

The draft had no impact on me personally. I registered, I got my card, and I received my draft category every year on my birthday. Since I was continuously studying, and we were not at war as a nation, I held the '2S' status until I turned 26. By that time I had been an assistant professor at Hope for two years.

Many years later, Professor Geoffrey Stone, a widely recognized constitutional law scholar from the University of Chicago Law School spoke at Chautauqua Institution giving the first, annual Robert H. Jackson Day lecture. The Iraq war was at a high point of American involvement, and the Bush administration, with hawks Cheney and Rumsfeld doing the heavy lifting, was in the White House. In the question session that followed his lecture, a young person asked Professor Stone, "How do we get university students to protest the war with Iraq?"

"Institute a draft," was Professor Stone's immediate answer. He was aware of what that particular student questioner was not – that it was the military draft that instigated student activism and protests in the 1960s and not the war. The students were against the war because they were being selected to give their lives to fight it.

Stone had Vietnam era experience, and knew of the many protests that marked those years. In the late 1960s, because the military services were demanding more and more personnel, Congress agreed to reinstitute compulsory conscription or the Selective Service System. Though the draft had been part of life for eighteen year olds for more than a generation, the Vietnam war was so unpopular and not supported by the American people, that young persons were protesting it at various levels. Reed College seniors declared that they "would not serve" and did so in large numbers. Their opposition to the draft was not *in vacuo*. It was strongly supported by the Reed College faculty, as well as by faculty at many colleges and universities across the country.[152]

---

[152] Hope College *Anchor* (Volume 80, Issue 7), October 7, 1967.

What campus protests of the draft caused, indirectly, was the draft lottery. There were various tensions because those college graduates headed for graduate school were getting student deferments to continue their studies. This caused Congressional exercise, because it seemed the draft was unfair to the those of low-income, and less-educated Americans were preponderantly in the military service. Johnson stopped this deferment of students headed for graduate school, and professional schools first. Then Congress instituted a draft lottery – the 366 days of the year (including February 29) were treated equally and numbers were drawn 1 through 366 to decide who was first to be drafted. All young men born 1944-1950 were included and lottery numbers were drawn in each of three successive years. So the effective birth dates were 1944-54, and then the lottery and the draft were stopped. In the 1969 draft lottery, men having the first 1944-50 birthdates drawn were later called to serve.[153] Democrats in Congress took the lead in driving this change in how personnel were sent to the military, especially Senator Philip Hart of Michigan, among the first to speak up about the unfairness of the previous Selective Service methods.

Campus Riots Spread

Columbia University burst into flame in the spring 1968. President Grayson Kirk was held in his office under baricade for days. Nick Turro, professor of chemistry at Columbia later said "At the beginning, many faculty sympathized with the students, but as the riot progressed, hardened protestors from all over the country arrived. At that point, we all agreed it was time for the police to break up the protests."

Lynn Koop wrote in the *Anchor* that police did come and break up the protests. Koop said that as Columbia had grown, it had pushed back growing from the inside out on Puerto Rican and black neighborhoods, impacting those communities. Inherent to the way this is written is that nacent concern about such neighborhoods – positive and negative – that is inherently prejudicial.

Today the Columbia area in New York has a good impact on the region. Reformed Church headquarters are nearby the Columbia central campus at 475 Riverside Drive, and the former properties of

---

[153] Much has been written about the lottery. An early *New York Times* story (David E. Rosenbaum, "Statisticians Charge Draft was Not Random, *The New York Times,* January 4, 1970) and a later detailed analysis in *Science*, (S.E. Fienberg, "Randomization and Social Affairs: The 1970 Draft Lottery," *Science, 171,* 1971) address a few of the issues with the first lottery.

Union Seminary, across Broadway from the main University campus, are now part of Columbia, too.

---

anchor editorials

On Columbia Sit-ins

THE LAST TEN days at Columbia   desire to cool off the situation at Columbia

"On Columbia Sit-ins," *Hope Anchor*, May 3, 1968 (Volume 80, Issue 24). Copyright © 1968, Hope College.

Campus riots shook campus communities to the core, but they also impacted the nation as a whole. There were food riots during the depression, and people were continuously marching on Washington for various causes. Nothing as nationally pervasivie as the Vietnam protests had occurred in the lifetimes of any living Americans. National news dealt nightly with the Columbia riots, in part because they were in New York and in part because the events were so shocking. After all, Dwight Eisenhower had been the President of Columbia immediately after leaving the Army. Eisenhower's prior office was now Grayson Kirk's office. Riots on the hallowed ground of an Ivy league school? These were serious protests. The *Anchor* editorialized about Columbia, and allowed for a long opinion pieces. These riots were the bellweather of many more to come.

Hope Graduation 1968

Hope returned alumnus Robert Haack to speak for commencement. Haack, who was President of the New York Stock Exchange, pointed out that he was sure his son knew that his first degree – his bachelor's – must have been honorary. Haack made a most important contribution to his alma mater after that visit. He introduced development officer Lee H. Wenke to the leadership of the George H. Baker Trust. Wenke managed something few other institutions outside the Ivy League had managed – a series of scholarships called the Baker Scholars program, and this still exists at Hope.

Lifestyle Issues

Compulsory chapel attendance was not the only thing on student's minds that spring. The notion of *in loco parentis*, or college administrations acting in the parent's steads, was under pressure as well. Eventually women students would achieve the same freedoms on campuses as male students. But that was to take many years, and Hope's women had just gained the right to wear pants, not skirts. At Stanford, the women wanted, and managed to get, much more.

My professional life was taking new turns too. I decided, after four years at Hope working with just undergraduate students, that I needed to get on with my career goal. I had wanted a research career in organic chemistry when I left Harvard, but felt that I could achieve my early objectives as a faculty member at Hope and be helpful to the College as it transitioned into the instruments age in research while doing so. Living in the liberal arts college enviroment had been extraordinary for my growth as a person and scholar, but it had taken its toll. I was selected to be chair of the Cultural Affairs Committee with the main contribution being the controversial but timely appearance of Dick Gregory. That had a negative impact on me professionally, and I was now focused on my research and teaching. So I opted for an opportunity in the Netherlands, which took little doing on my part. It was formally an exchange program with the University of Groningen because Dick Kellogg, an American doing post-doctoral work in Groningen, came to Hope for the year (arranged by Dick's preceptor in Groningen, Hans Wynberg).

Neckers to Go to Netherlands In Chemistry Faculty Exchange

Neckers to Netherlands, "*Hope Anchor*, May 3, 1968 (Volume 80, Issue 24). Copyright © 1968, Hope College.

Hope College will participate in a faculty exchange program with the University of Groningen in The Netherlands during the 1968-69 academic year.

Dr. Douglas C. Neckers, an associate professor of chemistry, will spend a year in The Netherlands teaching organic chemistry while Dr. Richard Kellogg of the University of Groningen will teach at Hope.

The exchange program is administered by the Petroleum Research Fund of the American Chemical Society.

Dr. Neckers is the author of two textbooks, "Mechanistic Organic Photo-Chemistry" and "Laboratory Experiments in Organic Chemistry."

Dr. Neckers was graduated from Hope in 1960, received his Ph.D. from the University of Kansas and has done post-doctoral work at Harvard University. He has taught at Hope since 1964.

DR. DOUGLAS C. NECKERS

Dick married a Dutch woman, and was planning on immigrating to the Netherlands, but Hans wanted to make sure Dick knew what he was doing. Hans' way to impact this was to ask, or tell, Dick to go to Hope for a year. Hans had taught at Grinell College in Iowa, and thought he was well aware of the values of the small schools in America. So, he was all for Dick coming to Hope and supportive of a position for me in Groningen. I had no thought about staying at Hope for a career so I saw this as very good thing to do. Sue and I, with baby Pam, left in September, and mostly had a great time in the Netherlands.

Chapter 23

Academic Year, 1968-69

The academic year, 1968-1969, began at Hope with a series of complaints by students against the College, the administration, and just about anyone that they deemed in control of their behavior/destiny. These attitudes were commensurate with declining citizen morale nationwide.

So President VanderWerf greeted the returning alumni in mid-October with "lumping all the students of this generation

together in one ball of wax...has replaced baseball as a national pastime. Students change here and mostly for the good, but we change too because we learn from our students."

Cal's insights continue to impact me. This statement is so right. After nearly fifty years of teaching, I honestly know that I learned much more from my students than they learned from me. I said that to the few of them that were assembled at Hope spring, 2015, when I was named a Distinguished Alumnus.

Cal, 1969.

What Cal could do, in speeches that he really thought about, was get to the crux of the issue. "We applaud the sincere majority (of students) that honestly attempt to realize the Christian ideals they've heard our generation proclaim. We try and look at the world through their eyes, though, and would hope they would look at it through ours." He challenged his generation of scholars to "cultivate a culture and climate that fosters and develops their maturity... but do so by, in some ways, listening to them."

He was also firm, telling the alumni, "students are not running the College... neither is the administration, the faculty, the alumni, not even the Board of Trustees. We are all in this together."

VanderWerf enjoyed this Homecoming. It appeared after several years that the Activities Center would finally happen. Tom Renner must have suggested a picture be taken with a spadeful of dirt blotting his face. Cal was alone in this photo, so committed he was to the Student Activities Center. But the job of president was clearly exhausting him.

DeWitt groundbreaking, *Hope Anchor,* September 1968. Copyright © 1968, Hope College.

In loco parentis concepts continued to surface. Later, these issues would more or less pass but Hope College and other colleges were having trouble in 1968 deciding to which generation it belonged. It was not alone. Most colleges and universities were engaging in the same rites of passage. A long *Anchor* article in the spring 1969 discussed changes in the dorm visitation policies for women students at Carleton College – a college similar in size to Hope, but non-sectarian with students who generally had higher test scores. Carleton liberalized visitation policies that spring so that the woman's dorms were open from 8 a.m. through midnight every day. They were closed after midnight, not so much because of *in loco parentis*, but so that the majority of woman could get some sleep!

Superimposed on all of it was the fact that young men could be drafted to serve in the military to fight a war many opposed, and most did not understand. At this time, and not only in Holland, they were neither able to either vote or drink beer.

The election of 1968, and particularly the Democratic convention in Chicago in August were, after the Columbia riots of the spring, the next really large set of national protests. Even though Lyndon Johnson was gone by that time, the war raged, a Democratic president was being held responsible for that, and his party needed to bear the brunt of protests for it. Riots in Grant Park during the convention were accompanied by a multitude of arrests. Chicago's Mayor Daly was in charge, and was a front row convention delegate who national television caught yelling derisive calls of "kike" when disagreeing the positions of Senator Abraham Ribicoff of

321

Connecticut. Specific instigators of the protests were targeted by the police. Eight were arrested. They were later dubbed the Chicago Seven, for their actions in aiding and abetting. More correctly they were charged with conspiracy in inciting riots. The seven included counter culturists, and protestors in general. One organic chemist, John Froines then an assistant professor of chemistry at Oregon State in Corvallis, was among the seven. Froines was targeted for arrest because it was alleged that he had assisted in purchasing riot materials. He and one other defendant, Lee Weiner, were acquitted in a trial that began on September 24, 1969. But Froines career was never the same. Eventually he moved to Vermont, and where served on the faculty at Goddard College, and later he directed O.S.H.A.'s division of toxic substances. He eventually retired from U.C.L.A.'s School of Public Health but even that was not without controversy.

At the convention, Julian Bond, a young Georgia state senator, was a most impressive presence. Bond represented new blacks: articulate, soft-spoken, and determined. He came to Hope later in November and called for the rise of a "New Democratic Coalition," saying in a news conference that "there is no black problem; but there is a fantastic white problem." Bond, who apparently never raised his voice (in contrast to the acerbic Dick Gregory of the spring) said that the election of Nixon would signal a difficult national time for blacks.

The election of 1968, as had been the election of 1964, proceeded with significant controversy. George Wallace, former Governor of Alabama, was running as a third party candidate, put forth by southern segregationists. Wallace's candidacy was controversial, and he brought the segregationists of the South to the national arena. Seven thousand supporters gathered to hear him in the fall in Grand Rapids, though he was heckled during the speech. It was thought by *Anchor*'s writers (Oct. 4, 1968) that he might impact the Republican vote in Ottawa County. The Editors of the *Anchor* endorsed Nixon! Wallace-talk disappeared with the Nixon victory that November.

As the new school year began, the traditional campus concerns still held the attention of students even with the national crises underway. For the first time ever, students had the option of using pass-fail as a grade choice. Sixty-two juniors and seniors elected to use the option. A debate ensued over who should be allowed to smoke cigarettes on campus, with the main debate centering around female students and how lady-like the practice was. Five students were arrested for possessing marijuana in Kollen Hall, with the Anchor reporting that Hope College's drug problems were the same on campuses across the country. With the finances under review, even the band director was forced to eat on the ground at times.

On the positive side, Hope was awarded $276,100 by Research Corporation and $130,300 by the National Science Foundation for departments, and for research. In total, more than $1 million dollars would come to Hope College in 1968 for science programs. Humanities faculty primed the *Anchor* editors to point out how unfair grants to scientists were to professors of English. But two physics faculty countered these assertions by talking about how much work a research proposal, or almost any proposal, required.

Lake's Michigan and Macatawa were on Hope's doorstep. Pollution of both was obvious, smelly and untamed. Hope's attempts to work on the polluting factors in Lake Michigan were stop and start. It was a huge lake and it had many environmental problems. Attempts to work on the lake's problems by scientists were welcomed but they had to be efforts of substance. So some felt the College scientists were negligent of the lake. Sadly, none of Hope's scientists were prepared for lake work of any consequence. People that used Lake Michigan were impatient with additional studies. Infestations of lamprey eels had ruined much of the sport and commercial fishing in the Lake. By this time this had been mostly solved as scientists killed the eels during spawning season in the streams flowing into the Lake. Ocean going vessels had brought everything from alewife infestation to residual fuel wastes to Holland shores. In the hot summer, the smell of dead alewife permeated even downtown Holland. At times, one could not walk along Lake Michigan beaches because of the tar residues that had been washed from tanks of ocean going vessels in the Lake. But no scientists at Hope had seriously identified with these issues; at least there was little being done by researchers at Hope that might impact them.

Hope's geology department was begun in 1967 in part with funding from the Sloan grant. With the appointment of Cotter Tharin, geology had its first department chair. Tharin had been a colleague of chemist Dwight Smith's at Wesleyan University in Middletown, Connecticut, and visited Hope because of that. A Holland businessman gifted a boat to the College on September 27, 1968, so Tharin, with an oceanographer and chemist David Klein, began research on the Lake's waters. The gifting of Marigold Lodge fit with this initiative. Unfortunately, Hope was too small, private, and parochial to have much impact. That belonged to larger, mostly government agencies who funded research scientists in the state universities though they have not done so well. Lake Erie has problems still to this day with pollution and fouled water supplies for the cities that take their water from it.

Grants earned by faculty were one facet of the science humanities split that had developed at Hope. Maybe it was always there, but in Van Zyl's time it did not surface. In any case, President VanderWerf pointed out that it was hard for faculty in the humanities to generate external funding, but with a strong faculty, grants will come. This became true but later when some of the non-science faculty became even more respected in their fields than were the scientists.

Consumerism in higher education

In the late 1960s, student attitudes at most every college or university changed from "Thank for all your help as I seek my role in life" to "Damn it, what have you done for me lately?" This got to the point nearly forty years later in my career, at Bowling Green, that students were treated like customers, and faculty could not tell their customers to go away if they were bad performers or had bad behavior. One kept them around anticipating their next purchase as in next semester's tuition payments.

Dormitories became more user friendly. At Hope, for example, phones were installed in dorm rooms, though one could not call off-campus from them. Clarence Handlogten, treasurer and director of business affairs who later thought he would make a good College president said "We feel it will improve on campus communications." Handlogten clearly could not see how the College could deal with billing problems for each individual phone in every dorm room. So why did he not say that, instead of making something up?

These administrative gaffes were little things, but indicative that the president was not being well served by those whose job it was to make students, of which there were increased numbers, more at home.

On the other side of the President's plate was the need to placate the old-fashioned but restless church. With the appearance of student unrest, it was increasingly wanting more control. Reformed Church ministrers did not think much differently on church matters than had their parents, and that translated into strangely archaic attempts to control the College.

324

A much larger issue caused by the church/gown conundrum was what resulted as Hope became a more vibrant academic community. Academic vibrance brushed against academic freedom, and these issues were identified in the editorial in that *Anchors* of late 1968 – the financial impact of controversial campus speakers: "This college, like all institutions, has a vital and continuing need for friends, including those with large and loose purses. Such support however does not entitle the contributors to have overt influence on internal policies at the school. Well-intentioned, intelligent opinions should be considered, but disregarded when they clash with those of students, faculty, and administrators."

What triggered the editorial was not Dick Gregory or Julian Bond. It came from an apparent attempt by certain financial contributors to get the athletic programs to generate more attention to campus athletes. The (group) even suggested that certain scholarships be awarded for athletic skills. The editorial writers were aghast at this and "proud of athletic prowess that comes without athletic scholarships." And the athletic department was dismayed that the Board was examining athletic policy, because, and this was one of the more compelling arguments for why it needed to be examined: "it had never been evaluated before."

The editorial went on: "Likewise, controversial speakers like Julian Bond and Dick Gregory come under the guise of academic freedom, and their messages must be heard." In a subsequent *Anchor* article, it was reported that a "church had rescinded its gift of $15,000 to the College immediately after Julian Bond spoke on campus."

President VanderWerf said he was not sure that there was a direct causal relationship. Clarence Handlogten, the non-academic, disagreed with the President. He was sure that there was though he could not prove it. I have always liked it that the financial types are sure they know what is best for an academic institution. In Handlogten's case, this was clearly apparent, even though he had no college degree. VanderWerf should have fired Handlogten immediately. It was already evident that he was marching to the beat of his own drummer and brash enough to disagree with the President, for whom he worked, publically. Cal, unfortunately, did not.

Cal stated: "The resistance to outside pressure… should be maintained. College policy should never be dictated by opinions from outside the campus, nor should the campus community sacrifice pricnciples in the name of pragmatism."

Still, on February 14, 1969 the Board, at the President's instigation, appointed a special committee to report on its review of Hope's athletic policies. They found that:

1. Hope's athletes were mostly at the College to get an education.
2. There was overall satisfaction with campus emphasis on sports.
3. The Director of Athletics should report to the head of the Physical Education department.
4. No individual should be head coach of more than one major sport. (Russ De Vette had been head coach of both the football and basketball teams; this was aimed directly at him.)

Eventually the Board committee prevailed, Hope's athletic programs remained primarily without scholarships, and De Vette resigned as head football coach. Woman's sports at that time were few and far between. Title IX, which guaranteed women's athletics a financial par with men's, had not yet occurred and there was surely no equality with MIAA champion teams fielded by the men.

But the athletic department, particularly those comfortable under the 1950s systems, blamed the president. Though he had every good reason to encourage a much higher visibility of sports teams, because it was good for campus recruiting particularly of male students not skilled or big enough to play in the Big Ten, the athletic department opposed him at every point. Male students were often as much attracted by a campus where they could continue to play their sport, or sports, as they were to get a first class education. Cal felt more attention given to sports teams would improve Hope's male student population.

Another issue that began during these VanderWerf years was that of student evaluations of their teachers. Good departments and competent faculty had been using evaluations for a long time. I used student evaluations because I wanted to know what I was doing that students found off-putting, and I wanted their help in becoming a better teacher.

But formal, mandated all-campus evaluations were a different matter. They suggested, as was vigorously stated some years later, that students were consumers and faculty were selling their services to their customers. This I found objectionable as did several more mature Hope faculty at that time. Though I spent many years as a department head in a large state university, I have never changed my mind. Student evaluations that were mandated for administrative control served no purpose. I knew, as a department chair, who the good teachers were and those that were doing a poor job. Mandated evaluations gave too much responsibility to immature young persons.

In the fall, student evaluations of faculty became mandatory at Hope College. Department chairs used them to guide faculty as needed. Some of the poorer chairs used evaluations as sledge hammers to determine raises. Good teachers got an infinitesimally larger raise than poor teachers, and unless that poor teaching was done by a future Nobel Prize winner it was penalized. We had begun the Sloan curriculum by the fall of 1969, when I returned from the Netherlands. I often joked that teaching organic chemistry to 180 freshmen meant this was the only time in my academic career I had seen my name written on the walls above the urinals in the men's room in the old Hope Science building. But it was clear soon that we had over-challenged, and over expected from, Hope's highly motivated science majors. They still needed time to mature. So the Sloan curriculum at least was altered significantly a few years later.

Since I was often teaching undergraduate courses that were required, either for a chemistry major, or as part of college requirements, I would occasionally play with student evaluations. One year, I gave an exam just before the final, and gave extra credit at some level that was tied to a student's evaluation scheme. Neither the extra credit mattered, nor did evaluations change much. I did the same thing occasionally when a seminar speaker (who I wanted to impress with how big, and active, our campus was) came to lecture. "Ten extra points on the exam for those that come to the seminar! And I'll take attendance." That filled the hall for the speaker, but ultimately had no impact on letter grades while making some of the students feel better and making the seminar speaker think our university had huge numbers of budding professional chemists.

Years later when I was teaching a freshman course for non-science majors at Bowling Green, and I was department head, I wrote out my own questionnaire. My questions had "agree or disagree" answers.

1. "Neckers doesn't know anything, but boy can he teach." Agree or disagree?
2. "Neckers really knows the subject, be he can't teach for smotz." Agree or disagree?
   I got 50% agreement on both questions.

This Bowling Green class that I was teaching was so under motivated, and so beneath the level that I felt college-able, that I wrote the president of the university and told him I would never teach another course for undergraduates at his University again, and I never did. Such had become of the consumeristic pandering to unable undergraduates become. Even if students did not attend classes at Bowling Green at least, they might be retained and given additional chances to graduate.

Student evaluations at Hope became a test ground for the use of the new campus computer. The first year, and the first time it was to be used for this, it broke down. No red faces, just "That's the way these things go," said Dick Brockmeier and Clark Borgeson who were managing the evaluations. Dick was having fun playing with the machine. Clark was feeling his oats.

Campus Safety: Maintenance

Hope saved money when it came to keeping its plant in repair. This picture from 10th Street shows part of its lawn mowing force in the late 1960s. The crews that worked for Henry Boersma, Director of Operations were, however, very hard working and dedicated. I never cease to be amazed at how much those good people did for the school with so little. The school had gotten bigger than could be managed on tight budgets, but they gave it the old college try. One janitor that served the science building for part of a year was 80 years old – many who lived in Holland, Michigan, then had not been taken care of in proper retirement contexts.

Clearly there was a lot more wrong with the campus than ice on the hill from Van Vleck Hall to the then Science Building. For one thing, our instruction in chemistry was safety irregular if not safety absent. For some of us, naive though we were, it was just the thrill of this that attracted us to chemistry labs in the first place. Fortunately, we never had an accident of consequence in those Science Building days. We had plenty of small fires, small cuts, and minor explosions. I was quick to grab the fire extinguishers in the organic labs. But nothing happened that was consequential. The hood system in the building came as constructed by some plumbing shop in the pre-war Holland. Doc Van Zyl was very proud of it, though with it one could manage, possibly, to find that water vapors from a chunk of dry ice might be dragged skyward toward the vent at the top of the hood. But as for saving lives from toxic substances? It was perhaps better to work with safer stuff (Or maybe not).

328

The hoods made terrible racket. When young thespians were at practice in the 4th floor little theater in the Science Building, MacDuff could hardly utter a word that MacBeth could hear. So they shut the ventilating systems off. I am not sure how many of us faced near poisoning from second-act- hood-turnoffs by aspiring actors!

Safety in campus housing was another matter. On February 21, 1969, a fire in Voorhees Hall evacuated all the residents in the middle of the night. Fortunately, as was reported in the *Anchor*, no one was hurt. This Hall was home to many women students and the fire caused a large scare. It was a traumatic event for the campus. The problem of safety in that Hall had been known for a long time, but now it had to be addressed immediately.

The fire was caused by an overloaded electrical circuit and it was not itself serious, itself. Nonetheless, the dorm had to be evacuated, and the Holland fire chief called the fire a providential act. More than three years prior, students had protested unsafe conditions in Voorhees Hall, by marching to the President's house late at night and demanding fire escapes for the Hall. Though plans were drawn up, they were never acted on because, as always then it seemed, there were no funds. The *Anchor* editorialized that "Voorhees Hall is/was a fire trap. It has wooden stairways, poor wiring, woefully inadequate fire escapes and warning facilities."

This time, they would not let go and the problem was quickly fixed. The concluding paragraph was a little tough: "Plans to renovate are more than two years old and have yet to be acted on. This instance indicates how low a priority the safety of the students is in Hope's financial scheme." The counter pressure, we all knew, was the never ending idea that one needed to keep tuition low so that students with limited resources, from Church families, could afford the College. And to do that, the College had to keep its expenses at a minimum because still, then, 80% of the College budget came from tuition. What to do for the students displaced from Voorhees Hall in the meantime became a serious issue. Some were moved to the Warm Friend Hotel while others moved to other dormitories.

The Voorhees fire was a wrench for everyone. But the best way to get the attention of the board was to make sure that they knew they could be individually sued if harm was done to a person, or persons, at the place, or in the organization, for which they had fiscal responsibility on their watch. This was a ploy I used more than once later as a department head in a state university; but in the instance of Voorhees Hall it was real and potentially serious.

329

In 1969, when this fire occurred, governmental control was just beginning to be asserted. No government agency could make Hope College assure that its living facilities were safe, but this was changing. No one could make its chemistry faculty assure its teaching labs were safe, but that, too, was changing.

As in all cases it took major incidents to cause changes. Patient consent, as in consent to use experimental treatments by physicians on their patients, was a direct result of the exposes at the Nuremberg Doctor's trial. It was discovered in the course of the trials, that doctors had experimented on human subjects in heinous, criminal ways at Auschwitz and other camps. Brandt was Hitler's physician, but Josef Mengele had escaped to Argentina, it was found years later, or he would have been one of the primary persons on trial in the second Nuremberg trial. The result of these findings was that all physicians and treatment centers in the western world undertook the responsibility of informing their patients about experimental treatments before they could be undertaken. These rules have become ridiculous at some level now, but they are all in a good cause, patient consent and patient information.

Chemistry labs have become much safer too as have facilities that manufacture chemicals on large scale. In the former case, this happened because of student injuries. At U.C.L.A. a few years ago, a young graduate student was killed by an explosion in her research lab. The professor was not present at the time, and eventually charged in a criminal case for negligence. Surely the labs at UCLA and around the country are on heightened alert because of this. Explosions of chemical manufacturing facilities have happened since companies began making things in large quantities. But safety regulations did not reach the common awareness until after World War II. The case of the explosion of an ammonium nitrate manufacturing facility in West Texas is just the most recent of these. Regulations there were lax; and those that did exist were not followed closely.

In the Voorhees fire, what was also the case (and this showed in President VanderWerf's remarks about the fire, and other campus safety issues, later) was that "safety" was and is a relative term. Prior generations were happy to have a dormitory or science building without much thought into the safety or sustainability that dorm or that science building. By 1969, society could be much more concerned about quality – and not just availability. Now they could ask "what kind of dorm or building do we have?" The Holland fire commissioner predicted the Van Raalte Hall fire in 1980.

330

Fortunately, the Science Building, though a fire hazard at some level, never burned down while the Board wrestled with what to do with Voorhees.

Nothing represented the VanderWerf era more succinctly than was its attention to academic matters as the highest priority. The genesis of an application to Phi Beta Kappa was a capstone for that. A graduate's election to Phi Beta Kappa recognizes a high level of academic achievement for the person elected. Rachel VanderWerf had been elected to the Ohio State chapter at the time of her graduation, and it always rankled Cal. "Rachel has a Phi Beta Kappa key and I do not." That was because Hope had no Phi Beta Kappa chapter.

Cal and Rachel undertook to fix that. They made application to Phi Beta Kappa. A visiting team of Phi Beta Kappa evaluators interviewed a number of faculty on a campus visit. I was one of those interviewed because I had been elected to Faculty Honors as a senior student but like Cal, had no chance to be elected to a non-existent chapter. The application for Hope's chapter was successful and a year or so after he was forced out at Hope, Cal was the Hope chapter's first and only legacy elected to that chapter.

Student unrest on campuses continued that spring and though it was mostly universities where the unrest was the greatest, Hope was not immune. Some of the changes demanded by student protestors were in the purview of the campuses to change, and mostly they did just that. One was who taught freshman classes: at major universities after these protests, large freshman sections of most courses were taught by full professors. The late Nick Turro, taught freshman chemistry at Columbia successfully for many, many years.

The realignment of the Board of Trustees that President VanderWerf had worked so hard to achieve was also taking form. Robert Haack joined the Board and as a Hope alumnus, he impacted the finances of the institution often in the years ahead. In March, retired President of John Hannah of Michigan State was to join the Board of Trustees. Howard Sluyter, from Dallas, was a former Grand Rapids used car salesman, and of less consequence to the College financially, though an active member of the Board. John Dinkeloo was a highly visible architect at the time the Board was reorganized; he studied at Hope from 1936-1939, transferred to Michigan in architectural engineering, and began his practice with Eero Saarinen in 1950. He and partner Kevin Roche took over the firm when Saarinen died, and did not miss a commission. He was active and helpful at the time too, but he was located in New Hampton,

331

Connecticut. What became of greater consequence in the next year or more however, was that the entire executive committee of the Board was the same old, same old, and the composition was almost entirely western Michigan. Business persons they now mostly were. Academics they were not. So the diversity even of a potentially able board was missing. This was well before the day of the conference call, so except for full board meetings arranged a few times each year, most of the decisions were taken locally. The reorganization of the Board in the 1960s reduced the cost of Board meetings. Years later, it was apparent that appointments to the Board made for apparent financial reasons only, were mostly negative going. As has always been the case at Hope, these were internal, self-serving and not of serious academic value.

Computers were just starting to be used on colleges campuses in 1969 too, but access to them was awkward. It required complicated 'near typing' of punch cards using a clumsy card readers and key punches. Their keyboards were designed by engineers, not typists so even if a user could type on a typewriter, he/she could not use them easily. Nonetheless, according to Bill Anderson, Hope College treasurer, Ken Vink's use of the computer and accounting software adopted from Big Dutchman in Zeeland, Michigan, was crucial to getting campus accounting running at a professional level. This happened in the early 1970s. Chemistry major Jim Brainard wrote in an *Anchor* in early May that Hope's computer was under-utilized. At that time, IBM did not make user-friendly machines. Apple would change this.

Hope alumnus John Tysse spent several years at the College, first in the admissions office and later in the development office. He served as administrative coordinator of the Centennial celebration in October 1966. But, as was often the case, young Hope alumni had multiple chances to pursue other things even during their tenure on the Hope staff; Tysse was no exception and he left to start his own Holland business in this time frame too. These creative Hope-ites made major contributions to their alma mater even if they only stayed a short time. Hope became a very different school because of the work of people like John Tysse and many others. John was, for several years, President of the Board of Brewton Normal School, an institution run for black high school students in Brewton, Alabama by the Reformed Church.

As always, programs of the Cultural Affairs series required advertising, and that advertising was not always successful. Holland was a small community, and audiences were hard to come by, even for ordinarily sold-out events. The cartoon in the May 2 *Anchor* is a case in point.

"I'm sure I speak for all of those present in welcoming the New York Philharmonic to our campus...."

Hope Anchor, May 2, 1969 (Volume 81, Issue 23). Copyright © 1969, Hope College.

Art Buchwald's columns were in the Anchor every week. This one, about student protests and faculty responses to them, is both poignant and funny.[154] Mark Russell, also a political pundit with a sense of humor, credits Buchwald's humor with helping him get his start.

---

[154] Art Buchwald, "Understanding Faculty." Hope College *Anchor* May 2, 1969.

Art Buchwald

Understanding Faculty

by Art Buchwald

One of the things that impresses people about the student demonstrations is the strong stand that some members of the faculty are taking on the issues.

I was on the campus of Northamnesty University and ran into a professor who was trying to stop his nose from bleeding. His clothes were torn up and he was walking with a pronounced limp.

"WHAT HAPPENED, Professor?" I asked, as I helped him search for his glasses.

"The militant students just took over my office and threw me down the stairs."

"Why, that's terrible," I said.

"From my point of view it is, but I think we have to look at it from their point of view. Why did they throw me down the stairs? Where have we, as faculty, failed them?"

"Are you going to press charges?"

"ON THE CONTRARY. If I pressed charges, I would only be playing into the

"BUT THAT'S A terrible thing to do."

"I don't think we should make judgments until all facts are in. I would say burning down a philosophy building could be interpreted as an unlawful act. At the same time, there are moments when an unlawful act can bring about just reforms."

"But the books, the records, the papers are all going up in smoke. Shouldn't we at least call the fire department?"

"I don't believe the fire department should be called until the faculty has met and voted on what course of action should be taken. There are times when a fire department can only inflame a situation. We should also hear from the students who started the fire and get their side of it. After all, they have as much stake in the university as anyone else, and if they don't want a philosophy building, we should at least listen to their arguments."

Art Buchwald, "Understanding Faculty." Hope College *Anchor* May 2, 1969. Copyright © 1969, Hope College.

As now, the Middle East was a huge problem. President Landrum Bolling of Earlham College, a Middle East scholar, spoke at Hope on the role of the United States during the impending war in the Middle East. For many of our students, this was the first experience to hear the history of tensions in a region fraught with anxieties. Bolling was optimistic that there was achievable peace in the Middle East, and shared his optimism with students that the United States and Soviet Union would reach an easing of tensions in the region.

Kay Oae's father, Sigaru, was the first Japanese organic chemist to study in the United States after World War II. He came, first, to the University of Kansas where he worked with Cal VanderWerf. Subsequently, Sig Oae went to Charles Price's labs; then he returned to Japan where he became professor at Osaka City University and subsequently Professor at Tschuba. Kay came to Hope to study because, obviously, of her father's connections with the VanderWerfs, and brought her father to international night at Hope College.

The May 9 *Anchor* reported I was returning from the Netherlands to reassume my position at Hope in the chemistry department. A small announcement in that issue also pointed out that

Rick Bruggers would begin study in biology at Bowling Green State University. Little did I know then that Rick's path, and mine, would cross again when I became Chair of Chemistry at Bowling Green in 1973. Rick went on to a most distinguished career mostly with the United Nations where he worked in pest control in African nations.

My main contributions to Rick's careers were to talk him out of things he did not do well, play basketball and organic chemistry. I convinced him to focus on those things he did very well, like long-distance running and field biology. The long term proves the old guy gave him some good advice even though it did cost the chemistry department at Hope one FTE student. Rick never gave up the desire to be a basketball player it seems.

The final *Anchor* of that academic year devoted an entire page of news print to the opinion of two *Anchor* editors, George Arwady and Tom Hildebrandt. Their writings were insightful and, especially in retrospect, useful. Though the *Anchor* was forever dwelling on change at the College, Arwady pointed out that most of these changes had really strengthened the faculty, and academic programs, of the College. He opined that keeping the best in Christian values in college charged with educating the young in the broadest possible way, is a tall order. He came down on the side that the College needs to open the doors to more dialogue and dissent. But he was mainly positive in the directions being undertaken by the VanderWerf administration.

Hildebrandt was also strong in his support of the institution's goals, objectives, and directions. However, he also pointed out that the students pay 73% of the College budget in tuition. Contributions from outside the school are meager though growing. And as in all things, they grew enormously in the future but the development causes took years to come to fruition.

The academic year 1968-1969 ended with basically positive input from these two generally aggressive critics. It looked like things were on an upward path for the coming academic year even if critical questions, as always, lingered.

Chapter 24

President VanderWerf's Last Year – 1969-70

The faculty in 1969 was expanded to meet enrollment growth: 600 freshmen, approximately 35 of whom were non-white, enrolled for fall classes, and there were now more than 2000 students. Both numbers were campus records. By this time, Voorhees Hall had been remodeled into make do faculty offices that were accommodating faculty from the departments of history, political science, and religion.

The academic management of a growing faculty was, fortunately, changing. In the prior year, under Dean Morette Rider's leadership, three main campus boards were organized: Administrative Affairs; Academic Affairs; and Campus Life. These boards had the expected responsibilities. Administrative Affairs was to deal with organization, administration, public relations and general faculty/student welfare. Academic Affairs was to examine and act on matters of curriculum, instruction and cultural affairs. This committee was chaired by the Dean. Campus Life dealt with co-curricular and extra-curricular activities. Its responsibility was life on campus.

After this arrangement was finalized, the institution had organized its faculty management and its businesses around logical personnel/inputs. No longer would everyone's nose be in every matter of the College, as opposed to their own departments. Campus life, specifically a lack of available dorm space, plagued the fall. But by spring it had worked its way out. This was a time, in the nation, when college and university enrollments were increasing dramatically.

The draft remained at the front of student's attentions for obvious reasons, and a committee headed by a Reformed Church pastor was accepting draft cards, to be destroyed in protest, from student protestors. President Nixon announced in early September that the year's draft requirements would be for 50,000 less draftees than originally anticipated. Not all saw this change in the best light. Some would have thought the large student enrollment was driven by student deferments. A campus peace march, to include a day's suspension of classes, was being proposed by the students. The administration decided against cancelling classes wisely, because they were placed in the position of taking a political stance if they did so. The students, as was their right, held the event anyway.

Larry deBoer, a member of the class of 1967, composed the letter at the left to the editor of the *Anchor*. His letter showed the dedication of some members of the young crowd in America to the ideals of the administration. Larry was later killed in Vietnam and a campus tree was planted in his memory.[155]

The national war moratorium was held on October 15, 1969. I happened to be away, giving lectures in Texas and New Mexico for the American Chemical Society. On that day I was on the Texas A&M campus in College Station. There was no protest at all that I could discern, though several students seemed to be playing frisbee around the iconic cannon that was the identifier in the middle of the campus in honor of the R.O.T.C. corps of cadets. Such was the difference between those in the south that saw the military as important to their nation when their nation was the confederacy, and much of the rest of the Nation, even if it did have to fight wars. Campuses like Texas A.&M. were the exceptions, however. Most student bodies bonded together around a common, get-out-of-Vietnam initiative. It would be a mistake to conclude this was the opinion of the entire country. It was not, and the military had plenty of support particularly from veterans of two world wars.

The moratorium came together at Hope though, and nearly 1500 students, faculty and staff participated in it. Students, faculty, the student church, and many others wrote letters to the *Anchor* in support of the unofficial moratorium. President VanderWerf felt that an official sanction of the protest, as in a day off from classes, would

---

[155] Hope College *Anchor*, September 19, 1969 (Volume 82, Issue 02). Copyright © 1969, Hope College.

place the College in a political, as opposed to academic, position, and this was untenable and unadvised. So classes were not cancelled. President Nixon was said to play a deaf ear to the moratorium and protests. But he soon announced that he had fired General Lewis Hershey, the Director of the Selective Service, an individual who signified the dysfunctional U.S. policy on Vietnam. Hershey had been a lightning rod for protests. He was a carry-over from World War II This cigar smoking general on the Washington scene was doing his job, and seemed to enjoy it, though it was hard for him to understand how anyone could not want to be inducted into the military.

Students did take time out, or at least the *Anchor* did, to complain some more about the President and the lack of campus activity on the Student Center. What the campus wags did not realize was the President VanderWerf urged the students to become more energized when the student center was first dreamed. It was he who organized the first student protest. The delay came from the cost of the expected construction. When the bids were first opened, they exceeded estimates by more than $300,000. Two weeks later the issues were resolved and bids were let for the construction of the Center.

Student expectations had become front and center in a confrontational way. No single student then at the school expected that a student center would be open when they arrived, so why should they have been so acerbic when none was found when they got there? The answer was that they were different students than those found in 1964. This group was really much more dissatisfied in general and on edge, and for good reasons. If school did not go well, or even if it did, they had a war to fight.

Over the next several months, and the remainder of Dr. VanderWerf's presidency, the *Anchor* was filled with letters about the war: how we could get out of Vietnam; how to stop the draft; how students should have an impact on the national decision making processes. The draft lottery was officially announced in mid-December and before the academic year was out, those students that could expect to be drafted soon knew it. Men whose numbers were in the first 122 picked could expect to be drafted. Those from 123-244 had an average probability of being drafted. And those below 244 had a low probability. Still, local boards were assigned quotas to fill so if one came from a Board with few eligible candidates, one could be drafted even with a very high number. It became difficult, though not impossible, to concentrate on one's interests with the draft always

338

in the background. The news on the national front was disturbing on a daily basis.

Music events were highlights of the year usually. Maurice Durufle conducted his Requiem with the Chapel Choir on campus the *Anchor* said. The campus was saddened when Jim Tallis died tragically from a brain tumor in Dallas in September. He had taken a position at Southern Methodist University that fall, but almost immediately was diagnosed and died rather suddenly. Loren Eisley was given an honorary degree in May. Eisley was a VanderWerf colleague at the University of Kansas, later becoming the Benjamin Franklin Professor of Anthropology and the History of Science at the University of Pennsylvania. His book *Darwin's Century* received the *Phi Beta Kappa* science prize in 1959. But even Duffy Wade's Blue Key bookstore was no longer sacred territory. The students opened their own bookstore to buy and sell used textbooks.

An editorial that was intended as, and was, a slap in the face for President VanderWerf, appeared in the *Anchor*[156]. This editorial was, as much as anything, still another reaction to the failed national policies that were killing young persons, against their wills, in a war they did not want or make. The blows to the gut the then President was forced to take from his alma mater, a putatively Christian college, were unpleasant. This was followed by the joy of the *Ranchor* that, Art Jentz, a Hope alumnus and professor of philosophy reflected in a letter to the *Anchor*, there should be more of. "More Ranchor; Less Anchor," Jentz wrote.

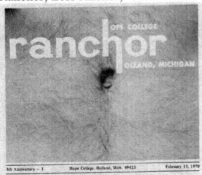

Hope Ranchor, February 4, 1970 (Volume 82, Issue 15). Copyright © 1969 Hope College.

---

[156] Letter to the Editor, Hope College *Anchor*, February 4, 1970 (Volume 82, Issue 15).

Further insult was added to injury when an advertisement was put in the *Anchor*, reading "Wanted: Male. For position as chairman of major university chemistry department. Must have PhD in chemistry. Must have published in the field." Similarly, an ad was placed below it for "Director of Public Relations," advising that applicants be "evasive, slick, and professional." [157]

As a result of all of this, and his reaction to it, President VanderWerf landed in the hospital with double pneumonia. For him, the time to dream was over. When astronaut Frank Borman came to receive an honorary degree on February 20, 1970, Cal was in the hospital so Rachel entertained the faculty on his behalf. He was deathly ill and the Hope to which he had given his complete attention to resurrect it from its potential grave was coming to life. In the words of the North Central committee, he came to do an experiment and discovered he had to invent the entire periodic table. This lovely addition from the *Ranchor* suggests that the Board, some members of the administration, and others were probably feeding the student writers poison.

Students continued to push for more freedoms. There were discussions of open visitations in the woman's dorms. A school that just two years before was celebrating panty raids now wanted woman's dorm rooms open for visitors? Students complained vigorously when a professor in the department of education was terminated. This was typical behavior for faculty failures on the academic scene. When discharged, cry foul and get the students to front for you. The case of Claude Crawford did not come to much though because it was a departmental decision not to continue his employ.

THE VANDERWERF BUST–For the first time in Hope's history, the Bored Trustees won first prize in the College's annual snow sculpture contest. However, it is doubtful that the bust will last through the summer months.

Hope Ranchor, February 13, 1970 (Volume 5, Issue 1). Copyright © 1970, Hope College.

The student church continued to expand and grow. Though first opposed by the churches in Holland it had, by this time, been in operation for four years. It had its own consistory or a reasonable facsimile thereof, and students

---

[157] *Hope College Anchor*, February 4, 1970 (Volume 82, Issue 17).

with no prior church affiliation could affirm their faith after attending the campus chaplain's church membership class. One of the things that puzzled me for years is a narcissistic attention to formal recognition that seemed to be most prevalent in the western Michigan Reformed Church. While pillorying one of their own in the person of the president of Hope College, they were seeking greater recognition from the young people then studying at the school. The *Anchor* editor in 1970, another Donia, wrote about greater cooperation between the colleges "funded" by the Reformed Church as if some must have wanted that. As near as I could tell, the Shultes crew in Pella, and the Van Raalte tribe in Holland separated in 1847 never to ever be remixed. The only issue that was of potential consequence outlined in a long article in the *Anchor* was the issue of black studies at the Reformed Church campuses. The Dutch, as a nation, had had their problems with black minorities over the generations. I remember remarking, in Groningen at lunch in the fall, 1968, about the headlines in the Amsterdam Courant referring to Bob Beamon (long jumper who set the world record at the 1968 Olympics) as a "nigger athlete" in Dutch. Dutch professors at the table, some who had survived the Holocaust, were quick to point out that this headline did not contain the connotations in the Netherlands that it had in the United States.

Hope was having significant issues with its black students at this time because they wanted more recognition, separate housing, and other concessions. Marshall Anstandig, a Jewish student from Detroit would run for, and win, the presidency of the Student Council. His express campaign promise was to work with the Holland community to improve community relations with the College, and with the students toward this end, too. Marshall, according to the special election issue of the *Anchor*, had been

341

appointed to an ad hoc Human Relations Council to increase black participation in the committee structure.[158] Mike Doyle, professor of chemistry, was also on this council. Mike was the first Roman Catholic appointed as a full time faculty member at the school. Cal, in this letter to Doyle after he had left Hope, was seeking news of its activities because he had an interest in civil rights that traced back to his Kansas days.

Marshall Anstandig also knew that the community of which he was part, as a Jewish student, felt unwanted by Hope College and the Reformed Church. This was had started to change under the VanderWerf leadership, but quickly reverted to its old focus after Cal was terminated. Among other things the change in attitudes toward Jews on campus post-VanderWerf were felt in the Holland community. This drove the changed Padnos stance toward Hope in the Van Wylen administration. Padnos' money went to Grand Valley State University where they became major donors, and Don Lubbers took great advantage of Hope's bigotry.

The older faculty continued to retire that year. One was Clarence DeGraaf, a professor in English since 1928. I knew Clarence, and enjoyed riding with him to the campus the first year I taught at Hope. He shared many things that he found challenging, and interesting at that time. After all he lived through two World Wars and the Depression. This was 1964-1965 though, and the hellishness of the late 1960s was still about to hit our communities and the campus. In an interview with the *Anchor*, De Graaf complained about the tendency toward specialization on the campus then, and particularly that specialization of male students. "Liberal arts has

---

[158] Hope College *Anchor,* April 14, 1970 (Volume 82, Special Edition).

become feminized while the men have gone to the multi-university."[159]

In some respects, he was right. But this changed dramatically, for better or worse, with woman's liberation. In any case, I personally believe my liberal arts education as important, or maybe even more so (with the benefit of hindsight), than my pre-professional training. I think the late Mr. DeGraaf indicted the campus community unfairly and incorrectly. What he should have said, was the majoring in English was now, more likely, appealing to female students on campus.

The campus continued to change personnel, even in the most sensitive places. Ray Smith was named new head football coach. I was only sorry that Ray did not get a chance to appreciate Cal VanderWerf at this best because they would have become good friends. Smith's story (below) got to share headlines with John Denver who was to appear on campus then in concert.

Then four students were murdered by the Ohio National Guard at Kent State University in Ohio. Nearly half the Hope student body refused to attend class the next day in protest. If campus protests were to reach a nadir – the Kent State massacre was that. Every campus in the country was impacted. Few remained open for the remainder of the spring semester. Ohio's governor Rhodes was never really sorry that it happened. He felt sorry for the individual students, and their families, but defended as long as he lived the decision to send the National Guard to the campuses during those days of student protest.

In a reminder of the good old days, Professor David Myers is shown babysitting a campus computer as the eternal card reader jams. Main frame computers took a lot of work to understand, and use. I never managed that saying, as part of my final oral exam for a Ph. D. degree at Kansas in 1963 in response to a computer programmer's question: that I would probably not use computers in my work until computers became smart enough to teach me how to do so. Little did I know that, in the hands of Steve Jobs and Steve Wozniak 17 years later, that is exactly what would happen.

---

[159] *Hope College Anchor,* April 24, 1970 (Volume 82, Issue 22).

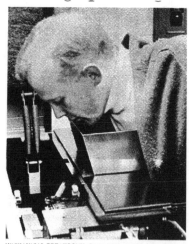

MECHANICAL BREAKDOWN—Dr. David Myers, assistant professor of psychology and director of the Office of Institutional Research, tries to loosen some jammed computer cards. The computer is one of the tools Myers and the OIR uses to survey the College community to obtain information used in making College policy decisions.

Hope *Anchor*, May 15, 1970 (Volume 82, Issue 25). Copyright © 1970 Hope College.

Finally, it became clear that the College was starting to benefit from competent development leadership. The *Anchor* announced a $500,000 challenge grant to the College awarded by the S. S. Kresge Foundation to fund, in part, the laboratory science facility.[160]

Here, thanks to Lee H. Wenke, is the rest of the story.

The Storm and the Sunset
We shall steer safely through every storm, so long as our heart is right, our intention fervent, our courage steadfast, and our trust fixed on God. -St. Francis de Sale

In the long pull across the years, there will be times when we shall need dogged courage to keep us going when the going is hard. And, what is the source of such rugged courage? Surely, that sense of God's presence when we hear him say, "I am with you always." - Norman V. Peale

I awoke in the wee hours of February 25, 1970. A storm was raging outside and I could hear the sleet and hail thrashing against the roof and windows. I was anxious for I had an important assignment

---

[160] *Hope College Anchor,* May 15, 1970 (Volume 82, Issue 25).

to be completed that day in Detroit. Could it be done? What would happen if it could not be done? I paced the floor and pondered.

Apart from the weather, we were part of the Hope College community that was being buffeted by the sweeping winds of change that had begun in the 1960s. Americans, including the community of Holland, and Hope College, grappled with the issues of the war, civil rights, and the counter culture. Hope College was also divided over internal issues such as retaining mandatory chapel, the hiring of non-RCA faculty and staff, student diversity, and the introduction of educational program offerings that in some ways challenged traditional values.

President VanderWerf and Rachel tried to heal the divisions, develop a greater consensus on how to resolve the major issues, and complete the master plan for development that had been adopted by the Board of Trustees in the centennial year of 1966. Beginning in January of 1970, Cal and Rachel hosted a series of dinners in the President's residence for faculty, staff, administration, and students. The purpose of these dinners was to discuss the major issues, and how to address them.

Joan and I attended the last of those dinners on the Saturday evening proceeding February 25th. We were astounded to learn that immediately following the dinner Cal entered Holland Hospital, exhausted and suffering from double pneumonia. When I visited him the next day he asked me to get an appointment with Stanley Kresge and ask Hugh DePree, Chairman of the Board of Trustees, to accompany me in presenting a proposal for a grant of $500,000 for a new science building. We had only a few weeks to meet the federal requirements for a $1,000,000 grant and a $2,000,000 forty-year loan at 3%. The Board was to make the final decision on the project at its May meeting.

I contacted Mr. Kresge and secured an appointment for 11:00 A.M. on February the 25th. Unfortunately, Hugh was scheduled for a Herman Miller Board meeting in New York City. He suggested that I contact Ekdal Buys, emeritus Chair of the Hope College Board of Trustees.

I called Ek and briefed him on the situation. I also felt that I should inform him that a couple of the members of the Board of Trustees, several faculty and administrative officers had warned me that there was much opposition to the project and that I might be putting myself in jeopardy

Ek listened patiently. Then he asked: "Lee who do you work for?" "I said that I worked for the President and reported directly to him". Ek then asked: "Did Cal ask you to do this"? "Oh, yes", I replied. "In fact, he was going with me; however because of his illness he cannot do it." "Then we are going to do it," said Ek. "I will clear my schedule. You meet me at my home at 7:00 A.M. on the 25th and we will go together. Don't worry about what others think. We are depending on you to spearhead the development program and the Board has approved the master plan. No one is authorized to give you direction but the President. I will see you on the 25th and we can talk some more."

Later that day, I called Norman and Ruth Peale, friends of Stanley and Dorothy Kresge, regarding the planned visit to Mr. Kresge. I told them   that Cal was in the hospital seriously ill. "Can you give us any suggestions or advice," I asked. "Yes," they responded. "We should pray for God's guidance and blessing." They indicated that at the time of our meeting with Mr. Kresge, they would both set aside time to be in prayer asking God's help. Ruth indicated that they would call Mr. Kresge prior to our meeting and tell him that they would be in prayer during our visit. They suggested that the faculty and staff of Hope also join in prayer at that time, which they did.

On the morning of the 25th, well before 6:00 A.M, I departed for Ek's home in Caledonia. Driving conditions were terrible. Poor visibility, ice, high winds, freezing rain, and narrow rural roads raised my anxiety to an even a higher level. I reached Ek's home on schedule. He had listened to the weather reports, but showed no concern. "Let's go," he said with a chuckle. "So far we're right on schedule."

Little by little, we made our way across the state. Many cars had been abandoned along I-96; however, at no point was the interstate blocked. There was virtually no traffic. We proceeded from drift to drift, rut to rut. It was slow-going. By 10:30 A.M., we had only reached Pontiac. Ek placed a call to Mr. Kresge and apologized that we would be late. "Don't worry," said Mr. Kresge, "drive carefully. I will stay right here until you arrive." He also said that the streets in Detroit were passable since the intensity of full storm had not yet struck the city. We proceeded. Ek's calm and confident manner relieved my anxiety. He was so strong and determined and I could not help but model after him. Shortly before noon, we arrived at Mr. Kresge's office in the Detroit Bank and Trust Building. He greeted us and showed us into his office. No other employees were present, for most businesses were closed because of the storm.

"Well," said Mr. Kresge, "what is happening at Hope College in Holland, Michigan, that would cause you to travel across the state to see me on such a day?" Ek explained. Mr. Kresge listened intently and asked an occasional question. Ek gave him a copy of our proposal. Mr. Kresge set it aside. "I knew it was very serious," Mr. Kresge said. "Norman and Ruth Peale called me about an hour ago and said that they were in prayer. Let's pray for Cal and Hope College." He then offered a prayer and it was apparent that the Peale's had briefed him thoroughly.

"You haven't had lunch," said Mr. Kresge. "I ordered some sandwiches and made a fresh pot of coffee. Let's go into the conference room and have lunch and talk some more." As we ate together, it became apparent that Ek and Mr. Kresge shared many common interests. Both knew and loved Milton "Bud" Hinga, faculty member, athletic coach, and Dean of Students. Bud played baseball for Hope at the same time Mr. Kresge   played for Albion. "He was such a gentleman, a fine Christian, and very handsome," said Mr. Kresge.

As our meeting drew to a close, Mr. Kresge invited us to spend the night at his home. He indicated that Dorothy could put a dinner together, it would be no trouble to have us as guests, and we could leave in the morning. Ek thought about it. He asked if we could check the road reports. Mr. Kresge called the state police and inquired about road conditions. The report was that the roads were improving since the storm was moving through and the further west you drove the better they were. Ek then indicated that he would love to stay; however, he wanted to call on Cal VanderWerf that evening and indicate that we had delivered the request. He felt that the news would help to pull the President through.

"That is a great idea," said Mr. Kresge. "If you will give me just a few moments I want to write him a personal note". He thumbed through his well-worn family Bible and found just the verses he wanted. Along with a   personal note, he quoted from Isaiah 43:

"You are my servant,
I have chosen you and not cast you off;
Fear not, for I am with you,
Be not dismayed!
For I am your God;
I will strengthen you,
I will help you,
I will uphold you with my victorious hand."

347

"Please give this to Dr. VanderWerf," said Mr. Kresge. "I will remember him in my prayers. Please keep me informed on his progress. I will take your request to my Board very soon. I still have some influence. And, by all means, keep the faith."

Mr. Kresge walked with us to Ek's car. As we emerged from the Kresge home building, we could see blue sky in the west, for the front was passing through. After warm good-byes and the promise to keep in contact, Ek and I headed home. We drove in silence for a time until we reached Pontiac where the roads were plowed and salted. Ek asked, "Did you feel God's presence during our meeting?" "Yes," I replied. "So did I," said Ek.

Soon we could see the sun as it began to set. We silently meditated on the symbolism. My headache subsided, and I felt my body relax. I thought about Ek and how much he meant to Hope, Cal and Rachel as they struggled to bring unity and strengthen the College, and Irwin Lubbers who once said in chapel:

"God does not merely cause the sun to set,
He paints the sky in glorious beauty"

We continued to drive west. I picked up my car at Ek's home and followed him to Holland Hospital. We visited Cal. Ek briefed him on the events of the day and handed Cal the personal note from Stanley Kresge. We had prayer together and returned to our homes.

In a few days, Cal was released from the hospital. He gradually resumed his duties. One day in mid-March, Cal called me into his office. He handed me a letter from The Kresge Foundation signed by Stanley Kresge with a check of $500,000. Cal invited the science faculty to gather at the President's home at 4:00 that afternoon to share the information and indicated that the science building would be built.

For the next few weeks, we concentrated on raising the necessary matching funds. In May, the Board of Trustees authorized the construction of a new science building.

There have been many satisfying moments in my development career, however, this story and the people involved in it will always remain very special. I will always remember the comments expressed by Cal and Rachel - *"Each evening when the sun sets and God paints the sky in glorious beauty will forever be a remainder."*

Hope was on the way and the VanderWerf administration was clearly starting to have an impact. Unfortunately, Cal would not survive to appreciate his work. In September, a new administration would great returning students.

VanderWerf resigns

VanderLugt named interim chancellor

Closing

Cal, Rachel, and the smaller children packed their bags, loaded the College Oldsmobile which had been given them by Hugh DePree, and left Holland in the summer of 1970 just as they had come in 1963. I presume it was hot and dry and for them, very sad. Cal's health had recovered. They were heading for Fort Collins, Colorado. Cal's friend Bill Cook was Dean there, and he had arranged a position for Cal for one year at Colorado State University. For the first time since he was a graduate student at Ohio State, Cal was "looking for a job."

They bought a house in Fort Collins, and one would thought this might be a good place for Cal to stay, but his research career was over. He had made a decision to be an administrator. Three universities offered him deanships during that year, but he found the position at the University of Florida to be his best option. So in the fall of 1971, Cal, Rachel and the four younger children packed again and headed for Florida.

By this time, the children were old enough, so Mrs. Wilburn did not need to accompany the family on the move. She went back to Kansas City where she had a sister. There she lived the rest of her life.

Some thought it incongruous that Cal would take a position in the Deep South. But his friend and colleague from Kansas days, Harry Sisler, was there and in a very strong position on the campus. In autumn 1971, Cal was named Dean of Arts and Sciences at the University of Florida in Gainesville. He finished his administrative career in that position but eventually, like all administrators do if they can, went back to his teaching career. By the time he died, he was one of the most popular instructors at the University.

The VanderWerfs lived the rest of their lives in Gainesville. Cal died suddenly of a heart attack in May 1989, and never regained consciousness. My eulogy was delivered at the Chapel a few weeks later. Rachel died a much more arduous and tragic death. She developed a brain tumor and was treated for almost a year before she passed away. She was also eulogized in Holland at Western Seminary.

As I reflect on Hope College after more than fifty years have passed, I am convinced that the College had some insightful persons to whom it might have listened more. Had it done so, Cal's tenure would have been more appreciated at the time. Some were in positions of leadership; some were professors. John Piet at Western

Seminary warned the College that religious dogma and truth were contradictory. Some were historical. Philip Phelps, Jr. generations earlier knew, that a university would enoble the Dutch in western Michigan to have more than just a college, and he tried to move the new Hope College to be Hope Haven University. Others were loyal alumni. Bruce Van Voorst, for example, was a humanities alumnus who continuously questioned the impact of the then Church on Hope.

Winston Churchill, whose Anglican heritage brought him to the podium of the world's stage to utter the most profound and weighty, the most bellicose, or at some times, the most vicious of human utterances all as though he was singing a psalm from the choir at Westminster Abbey, seem a good place to end this.

Churchill's aspirations, foibles, strengths, and leadership brought the British victorious through World War II when immediately after the invasion of France he uttered the following in the House of Commons: "I speak about the tragedy of Europe, this noble continent, the home of all the great parent races of the Western world, the foundation of Christian faith and ethics, the origin of most of the culture, arts, philosophy and science both of ancient and modern times. If Europe were once united in the sharing of its common inheritance there would be no limit to the happiness, prosperity and glory that its 300 million or 400 million people would enjoy. Yet it is from Europe that has sprung that series of frightful nationalistic quarrels, originated by the Teutonic nations in their rise to power, which we have seen in this 20th century and in our own lifetime wreck the peace and mar the prospects of all mankind."

After World War II, Churchill realized that Britain was unprepared, and in part at fault, for not being able to challenge the "Hun to the east." He did not admit or likely did not know, though science historians knew, that when Lyon Playfair brought Britain German organic chemistry, its schools chose the classics because the British educated only the elite. Perkin made mauvine by accident in London, and was smart enough to turn it into a large business selling dyes eventually as Imperial Chemical Industries. But A. W. Hofmann, the German who came to manage his chemical school, found the support for research in chemistry better in Berlin than London. So he returned there to assist, via his students, in founding AGFA, Bayer and other dye companies. Along the way sprang up Weiler ter Meer, and the explosive organic nitro compounds – trinitrotoluene, T.N.T., trinitrobenzene, and picric acid.

Nor did Churchill understand, when he made his Iron Curtain speech, that *détente* with the new great iron monster to the east would, in the day of nuclear bombs, prevent them from being used. Nor did he realize, as a politician born in the 19th century, that in the world of science and engineering's endless frontier, having as much or more than the other guy is the best defense.

Churchill doffed his cap to Britain's failures in the sciences and engineering in a speech at M.I.T. where he was invited to plan for its future in 1946. Churchill said, "We have suffered in Great Britain by the lack of colleges of University rank in which engineering and the allied subjects are taught. Industrial production depends on technology and it is because the Americans, like the prewar Germans, realized this and created institutions for the advanced training of large numbers of high-grade engineers to translate the advances of pure science into industrial technique, their output and consequent standard of life are so high. It is surprising that England, which was the first country to be industrialized, has nothing of comparable stature. If tonight, I strike other notes than those of material progress, it implies no want of admiration for all the work you have done and are doing. My aim, like yours, is to be guided by balance and proportion."

Churchill's words were uttered at M.I.T. where their techies had taken the Brits "radar" and made it work. But they could have also been said in New York, Peoria, and Indianapolis, where the British Empire's penicillin had been turned into a drug every U.S. G.I. had to carry. America's scientists continued to function even during World War II's horrors in part because Vanevaar Bush, James Bryant Conant and others pushed our nation's universities into scientific research in the Nation's defense through the National Defense Research Committee. The committee was activated almost eighteen months before the U.S. got involved in the Wars in Asia and Europe. Through it a pattern of grants and contracts to faculty in universities, where they would marshal the services of young science students, produced many of the critical inventions that causes America's World War II effort to later be dominant.

On the American side after World War II, Bush and Conant, worked with our politicians to develop a new long term support system for American college and university scientists. Since it took the British scientists a long time to rebound from the war and German science was destroyed, the Bush/Conant steps put American science in the world's lead after the war, a position that it has yet to relinquish.

Gerrit Van Zyl participated at little bit in the U.S. scientific effort during the war, and he understood, when many of his colleagues in colleges did not, that chemistry was an experimental science. Because of that he found funds, even during World War II, for his brightest students to do chemistry as chemists did it – through experimentation into the unknown. That later became known as undergraduate research. N.S.F., for its part, started some programs in this in the 1950s and, because of Hope's scientists following Van Zyl, Mike Doyle, Jerry Mohrig, and me, these programs survive to this day. At last count, Hope offered undergraduates support through N.S.F. funded programs in eight different departments.

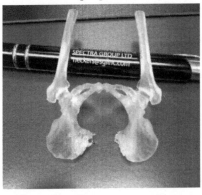

In my current life, I work with some of the brightest three dimensional printers in the world. These are people that can take digital output, and format it so that it can be printed one pixel at a time; thus a C.T. scan of a person's hip becomes a small print in a few hours for limited cost. One of my colleagues in the field describes his business as a place where "everyone works at a pace of frenetic, controlled panic all the time." That business is located within walking distance of Nick Negroponte's Media Lab at M.I.T.

Cal VanderWerf, the President and man, attracted criticism. Creative people always do. Cal was not just an activist, he was not threatened to have people as smart as he around him, and these people were activists, too. Cal relished having like-minded folks around. So what he energized at Hope in the 1960s was a community of competitive hard working young people. As Tom Renner described the Hope of the VanderWerf day, it was a place of very high energy. Hope was never M.I.T. – a place in a frenetic but controlled panic. But when it found itself in the competitive generation of the early 1960s and its older faculty were seriously threatened by it, fortunately, its younger faculty led it forward into the Sputnik generation. N.S.F. grants became the norm at Hope. And, after difficulties convincing the grant givers of the day that good research could be done just with undergraduates, each Hope faculty member found N.S.F. support for his or her expectation. Hope, today, exists because Cal VanderWerf hired faculty that could lead it more insightfully than did the pre-World War II leaders of England lead the British Empire. Cal managed to get Hope to respect science, and doing science. For many years after him, the notion of research with undergraduates was their strongest product.

Ironically, I recently have gotten to know rather well actually, the Dean of the College of Musical Arts at Bowling Green. His father was my first program director at the National Science Foundation. My first grant was awarded in 1972. As I have reflected on how so much has changed, I recall that my first grant of $40,000 was little according to the N.S.F. administrator. But it meant the world to me. As a result of work done under its auspices, I developed two polymeric reagents – one of which was and is the catalyst that Paul Schaap's company, Lumigen, still uses to make chemiluminesent dioxetanes.

When I sought renewal of the program though, the program director had gotten a different set of reviewers and my program was determined to be too small for future support. Fifty years later that 'too small' is generating millions in earnings per year for an American company, and a second invention millions of yen for various Japanese enterprises. Though I did not work just with undergraduates in this program, they certainly helped with it. The critical invention of polymer Rose Bengal, for instance, came right after a time when I had left Hope.

I was recently awarded a Hope distinguished alumnus award (April 25, 2015). I wondered why, actually, because though I taught at Hope, I had left in the early 1970s and rarely had contact save with individuals after that. But as part of the citation, it became clear that the Hope of today is really driven by the energy toward research with undergraduate students that we started then. I have described this in earlier chapters, and pointed out, with great respect, how it ultimately traces to Gerrit Van Zyl, Gene Van Tamelen and George Zuidema. I also recognize, as I said in my speech, that my biggest take-home lessons from Hope came as much from my experiences with the liberal arts as with my experiences as an undergraduate research scholar in Van Zyl's labs.

But it is very true that I pushed research with undergraduate students hard when I was a faculty member there. I was an organic chemist, and there was an intensity about organic chemistry as a discipline that is due to its German heritage. I had been at Harvard with some of the brightest young organic chemists in the then world, and this rubbed off. I knew that the best Hope undergraduate students could compete in that arena, and it was my objective to get them ready to do that. As any academician can tell you, in those days teaching chemistry in a small school without Ph. D. students to mentor could have been the kiss of death. It was easy then to be trapped by the environment, lose one's fire, and settle into C. P. Snow's version of what a college teacher might be. I was too young to let that happen to me, and I went to Hope to start my career. I did not see it as an end in and of itself. Cal knew that and said "Give me five years," which I did.

But while I was there, I was determined to make a Hope career for its young faculty a place where they could thrive scientifically and professionally. Some, like Sheldon Wettack and Dwight Smith, moved successfully into permanent administrative posts. Mike Doyle managed his entire career because Hope gave him the chance. For that I claim a lot of the credit.

I can say, based on the records today of Hope faculty in chemistry, physics, geology, and the related sciences, engineering, and nursing, the possibility of a research career at the school has never been greater than it is at Hope College in the environment of the mid-21st century America.

Dear Doug,   4/22/15

Martie and I wish to congratulate you on your receipt of
the Hope College Distinguished Alumni Award for 2015. This is
a recognition richly deserved and long overdue. Your illustrious
career in teaching and research speaks for itself! The quantity
and quality of your other awards is extraordinary!
Relative to Hope, I want to thank you, together with President VanderWerf,
for giving birth to the undergraduate research program at Hope.
Without question it is one of the pillars in a Hope science education
and is most certainly one of the key factors in Hope's academic stature
and strength in admissions. What foresight you had in the 60's !!

Best wishes always in all you do, with gratitude for all you've
done to build the Hope College we cherish today.   Cordially, Jim

---

After I was awarded the Distinguished Alumnus Award, I
received several letters of congratulations for what I had started. One
was from former President Jim Bultman. I said in my remarks, that
Cal VanderWerf was Hope's greatest president. But that was a
reflection on my take of what had to happen to the Hope from which
I graduated. Cal was a university professor who returned to lead a
small liberal arts college. Without his presidency, Hope would be a
much different place now, if it remained at all. He was too ambitious
perhaps, but so was his young faculty. He was too aggressive for
Hope College I suppose, but so was everyone new that he hired to
teach there. He tried not to mess with those of the older generation
who were made uncomfortable by him, but that could not be helped
either. A professor without a terminal degree would not, in the future,
be a professor at his Hope. Subsequent President Gordon Van Wylen
managed the church, faith of our father's conflict, much better than
did Cal. But Gordon came from a different philosophical position,
it would seem from his record, than did Cal. There was no reason, in
Cal's view, that blacks should be treated differently than whites
whether they were in Lawrence, Kansas or at the University of
Kansas. He wanted to change that. They were also treated differently
in Holland, Michigan. There, as president of Hope, he had to be less
out-spoken. But Mrs. Wilburn did not need to be, and she was not. I
remember Rachel telling me how puffed up Mrs. Wilburn was when
she heard Dick Gregory speak. For the first time in their respective

relationships, Mrs. Wilburn could think she was as important as was Mrs. VanderWerf, and they had, respectful personal relationship. But blacks were generally thought inferior to whites, whether they were or not, and Mrs. Wilburn felt that. Though not an activist, she did manage at some level to change that. And there was no reason, according to Cal, that small college across the disciplines could not be on the same academic plain, if they deserved it, as were their university science colleagues.

After Van Wylen's presidency, as I understand it, the College went through some difficult times at the academic – organized religion interface. Chapel, the program, became separated from the religious experience. One lived for itself as an entertainment for 18-22 year olds. The other was jettisoned. This separation the College still feels. Alumnus Bruce Van Voorst, who I knew in correspondence, speaks to this in his letter.

Hope College, today, is marketed quite well. It is a joy to see that its research strengths, even though they are accomplished with *just undergraduates*, are the best in the United States. To its future, and to a future that remains sprinkled with Cal VanderWerf's legacy, goes all of the best in the world. I respect the school, and cherish the years I spent there.

The job of the historian is to present the facts so that credits are properly assigned as due, and brick bats thrown (perhaps in the absence of objective information) neutralized.   Cal's administration found an institution on the verge of going out of business in its historic, academic form. A sea change was required or, as the North Central Committee evaluating the College said in its report in 1964 "Lavoisier came to do an experiment, and ended up inventing chemistry" (translation mine).

Cal came to work at his historic college and ended up inventing its 21st century future. Hope College, today, advertises that it offers graduate level instruction in a current sense to undergraduate students. And it does so in the liberal arts, Christian, setting. As I say in the book, I could have written its current advertising statements myself. They are extraordinarily perceptive. America needs what Hope says it offers. The one place that comes up short when one is limited to undergraduate students is economic development. I have often wondered how Hope would have differed had Philip Phelps, Jr. succeeded in turning it into Hope Haven University.

Following the VanderWerf administration, Hope turned back to the education of its fathers. The Van Wylen administration had a distinct Christian Reformed Church flavor – something at least

357

Don Lubbers, and probably Irwin, his father, were concerned about. The Van Wylen administration did little to sop Cal or Rachel's personal wounds. She, particularly, felt this discourteous.

But, in the end, time heals wounds and those inflicted during the VanderWerf years have been mostly healed.

In another life, I'm chair of the Board of Directors of the Robert H. Jackson Center in Jamestown, New York. For ten years the Center has hosted, in late August at Chautauqua Institution, its International Criminal Law dialogues. All current international criminal law prosecutors gather to discuss their trades. Jackson, of course, set this path in world history with his landmark work at the International Military Criminal Tribunal at Nuremberg, Germany following World War II. There, the Nazi leadership was identified, and through evidence discovered, tried in a court of law to stand responsible for their crimes against humanity. At one of the Jackson Center events years ago, I asked the international criminal prosecutor for the tribunal in Cambodia why, now, was this trial still going on? His answer— because, at some small level, trials like it bring closure to the victims and their families.

This history presents the facts as they are documented from writings of the times. I was at Hope for most of the VanderWerf period. So I add my personal observations when they are useful, and relevant. I hope, for the family, and for the many who may read what I have written that it will bring closure to a career that was among Hope's most important.

Cal's Hope – Spero in Deo.

Appendices

Compiling this information about Cal VanderWerf, his career, students, and impact has been a labor of love. It has taken me almost forty years to complete it. First, as a student, and then, as a faculty member, I found Hope College a much better school than its external reputation commanded. I opposed the narrow view – this is a College for Reformed Church undergraduates. I saw Cal as an academician whose attitudes about undergraduate education coincided with mine. In retrospect, that is probably a reason I took the Hope job. I also knew that Hope students, in chemistry at least, could compete with the best in the world. My mission, if I had one as a young assistant professor, was to convince a broader audience this was so. Cal had lived in the competitive world of academic research, and he helped us frame our careers in a positive way. What I learned from his career model, I passed to my colleagues and students in the few short years we were together at the College.

I decided to leave out most pictures and images in the hard copy. We have many originals supplied by Tom Renner, and copies from the *Anchor*, the Kansas papers, the *Holland Sentinel* and even the *New York Times*. But I decided, in Van Raalte's image, that words were our currency so we stuck with that. Some of the images from the Archives suffered in the Van Raalte Hall fire (1980), too, so leaving them out was necessary; others suffered in digital translation. Where they conveyed a message, I left them in the hard copy. But there are many images in the on-line version of the book of better quality so please avail yourself of that and by request, if you are so interested.

My academic perspective comes from the school of hard knocks. I know what it takes to compete as a research scientist in America because I became, and am, a research scientist in America. I can attest that my respect for the liberal arts institution, which I mentioned in detail in my Distinguished Alumnus speech in 2015, pervades everything thing I now do. I have long forgotten the details of Gerrit van Zyl's organic chemistry classes, but I remember vividly Paul Fried's discussions about recent European history. So my academic perspective of the liberal arts comes from my encounters with the humanities faculty at Hope.

I know my writings will rankle some; I suspect they already have. I am fully responsible for any mistakes and prepared to argue my cases. I cannot thank the VanderWerf children enough. Both Gretchen and Klasina helped me get this project up and running.

Cal and Rachel had a big family so I hope they will buy a lot of books. Profits, such as they are, will be given to the College.

Contents

Note: There are other items accumulated for the manuscript that simply could not fit into the book. These are all available at DougNeckersExplores.com.

Among them:

The *Report of Review Committee,* the College's response to the North Central Report, put it right on the line to the College – it had to make revolutionary changes or it would not continue as a viable player in higher education in the immediate future.

The *Five Year Report to the Executive Committee of the Board of Trustees, August 9, 1968,* identifies those ways in which the College had been impacted by the Vander Werf presidency.

Cal had a major impact on undergraduate research throughout his lifetime. This was *memorialized* upon his death in a way that captured that impact.

Other documents, presidential speeches, "Kudos and Brickbats," contain little of long-term consequence but insights into the community at Hope College. The *speech by President Bok* of Harvard is both insightful and amusing.

The full list:

Report of the Review Committee – North Central Accrediting Association, 1964- The contents of this review were quite negative.

Letter to Hugh De Pree, Chair of the Board at Hope College, January 6, 1968. Cal was getting very frustrated with the lack of action by Hope's Board of Trustees, even as reorganized.

Five Year Report to the Executive Committee of the Board of Trustees, Hope College, August 9, 1968- summarizing the accomplishments of the VanderWerf administration's first five years.

De Pree remarks – VanderWerf Recognition dinner – President VanderWerf's termination was announced to the campus in June, 1970. De Pree's notes were incomplete, and little detail remains.

Derek Bok speech from Harvard commencement, September 18, 1971.-President Bok with tongue in cheek, spoke about how Harvard University might be funded in the contemporary business/university community.

Gordon Van Wylen – Inauguration speech, 1973.

Convocation Program – Dedication of VanderWerf Hall, October 9, 1981.- This dedication was bittersweet: We all knew that the logical building to have been dedicated to President VanderWerf was the chemistry building.

John Jacobson Inauguration address, 1987.

Memoriam – Calvin A. VanderWerf –Council on Undergraduate Research. Cal was active in undergraduate education for his whole career. This memoriam by a Council of American college faculty recognized his contributions.

Kudos and Brickbats – an assembly of recollections of President VanderWerf.

Freedom of Information Act request (denied)-for a copy of report I believe Cal VanderWerf wrote following World War II on the chirality of the phosphorus atom in sarin. As a student of Mel Newman's and with his insights, I believe it was Cal VanderWerf who first postulated this, but until the report shows up, we will not know for sure.

361

PARTICIPANTS IN THE SERVICE

President Calvin A. VanderWerf, Ph.D., Sc.D., L.L.D., Presiding

Marshals: Professor John W. Hollenbach, Ph.D., Professor A. James Prins, Ed.D.

The Hope College Chapel Choir
Professor Robert W. Cavanaugh, Ed.D. in Music, Director
Mrs. Carol Davis, Soloist (Class of 1970)
Miss Edith Rens, Accompanist (Class of 1970)

Organists: Miss Phyllis Thompson (Class of 1970)
Mr. Kenneth Nienhuis (Class of 1970)

The Brass Ensemble

The processional and recessional hymns are from a service of music
composed by Thomas Canning in 1957 for Hope College.

THE
INAUGURATION
of
Calvin Anthony VanderWerf
as Eighth President of
Hope College
on November 16, 1963

THE INAUGURATION of Dr. Calvin Anthony VanderWerf on November 16 as eighth President of Hope College was an impressive and majestic occasion in Dimnent Memorial Chapel. Even the weather cooperated with unseasonable pleasantness helping to create an atmosphere of warmth and good will as students, trustees, alumni, faculty, community, and delegates from over 200 colleges and universities, learned societies and foundations gathered to witness the installation and congratulate the new president.

More than 300 persons participated in a colorful, academic procession preceding the ceremonies. Ekdal J. Buys '07, chairman of the Board of Trustees gave the charge to the President. The Rev. M. Verne Oggel '11, president of the General Synod of the Reformed Church in America, gave the invocation and the Hope College Chapel Choir, under the direction of Dr. Robert W. Cavanaugh, sang the anthem "Hope Thou in God" written for Hope College during the past summer by Dr. Haydn Morgan of Eastern Michigan University. The Rev. Herman Ridder '49, president of Western Theological Seminary, gave the benediction.

Dr. VanderWerf's inaugural address is printed in full in this magazine.

The inauguration at 2:30 in the afternoon was preceded by an inaugural luncheon in Phelps Hall attended by 400 delegates, faculty and special guests. A reception in Phelps Hall for all who attended the ceremonies closed the official program of the day.

Inauguration CONTINUED

Everyone agreed that the day's program was practically perfect. Everything proceeded on time as scheduled and the feeling was that of good will and faith that the College is in good hands. The only slightly critical comments—and they were made with twinkle in eye, however, concerned the marching during the processional when the delegates, thinking they might be holding up the schedule, took up the slack by hurrying down the aisle at double, almost triple, time. Those who criticized did so on the basis that it was too fast for the eye to scan the faces of all in the procession. Then there was another little misunderstanding in the recessional — the last in, decided on their own to be the first out, which harassed the marshals and annoyed those in charge of this detail. But, as others said, there has to be one "departure from accuracy" to make such an occasion perfect and memorable.

And so it was, a perfect and memorable day on the Hope College campus.

THE CHARGE

*"The leadership
of our college
determines her course"*

Chairman of the Board of Trustees Ekdal J. Buys '27 gave the charge of office to President Calvin A. VanderWerf '37.

Pᴿᴱˢᴵᴰᴱᴺᵀ VᴀɴᴅᴇʀWᴇʀғ, we bring this charge today in behalf of the Trustees of our College, representing our beloved denomination, THE RE-FORMED CHURCH IN AMERICA. Those who now attend Hope, those who support her, and those who lead her, are our deep concern.

Needless to say, those of us who heard or read your convocation address to the student body, thrill with the challenge to personal commitment to God. As you ably said, "We cannot do this for you," but we can set the stage. We cherish for all the men and women here the good things that can be provided for a meaningful four years at Hope. Our first duty is the preparation of lives here that will make an impact on the complexities ahead. The record tells us that this freshman class has the highest academic entrance record in our history. Let this certainly be directed for good.

Those who support us make a warm fellowship to enjoy. It has been said of colleges "Some are community leaders, some withdraw into intellec-tual seclusion." The question arises as to what extent a college should make itself felt in a com-munity. What should be the relationship between the "Gown and the Town?" At this time, may I digress for a moment to publicly express our thanks to this Community for its cooperation and support? The Community Council for Hope worked diligently in our major campaign of re-cent years. This, Sir, should be continued not only at this level, but more areas of influence around us should be activated.

The parents of our students, both at present and in years past, regardless of their prior affilia-tions, are a source of strong support. Contact with this group should be maintained now and in future years.

The Alumni, naturally, make up the major area of support and may I also say, justly so, of criti-cism. However, the give and take of this group is commendable and should be a real inspiration to you. Alumni records tell us that 85% of our graduates are in the service fields; doctors, dentists, nurses, school teachers and administrators, ministers and missionaries in every direction of the compass. The history of this group alone is an impelling heritage.

I hesitate to single out any group, but the women of the church have entered into a relationship of not only financial aid, but a year around effort in our behalf. This is now reflected by the fact that at the last Board of Trustees meeting, five women were present representing various Synods. This is a good trend, but is being watched carefully.

The need for our continued cooperation with inter-college groups, such as the Great Lakes Colleges Association and the Michigan Colleges Foundation is self-evident.

The REFORMED CHURCH IN AMERICA is the foundation of our college. Constantly remember that their prayers and resources are always with you. When all the negatives these days seem to be chipping away at our foundations, thank the Lord for Church related institutions.

The leadership of our college determines her course. The dedicated group of Faculty are our prime asset. Academic excellence is evidenced by the constant high ratings of our students both here and at the graduate level. The areas of their influence demonstrate the emphasis they have received. The Deans, Administrative Officers and all the many people that pursue their daily tasks on this campus are your constant help and support.

As for you personally, Mr. PRESIDENT, this cannot be said without deep personal emotion. Twenty-six years ago, we were on the same platform as classmates. You, because of your academic excellence, gave the valedictory address. Again, because of this same continued achievement in your life, God has providentially returned you as our President. We hope now and then to peer through the windows of your home and see you completely relaxed, either pampered by your six lovely ladies, or pestered by your other man about the house. Your fine family is a God-given blessing, take time to cherish and enjoy them with all your heart.

As we review the Presidential History of Hope from Dr. Phelps to Dr. Lubbers, we find that each made a unique contribution in their own right and this you will also do. The changeless note through these years has been our Motto from the Psalms "HOPE THOU IN GOD," "THIS IS OUR ANCHOR OF HOPE."

May God grant you and your family health and wisdom for the days ahead.

Here they are, bless 'em: Hope's first couple, beaming on Inaugural Day! No wonder alumni are confident that the College is in good hands!

366

by Calvin A. VanderWerf

CHAIRMAN BUYS, Members of the Board of Trustees, Distinguished Guests, Fellow Faculty, and Students,

We are met here this afternoon at the heart of a great and noble college. Her life-beat has, in the past, transfused into thousands of young lives that profound sense of meaning and purpose which leads to the high road of self fulfilling service. We pray that this may ever be her glorious destiny; it is to that purpose, Mr. Buys, that I dedicate myself in accepting the sacred charge which you have laid upon me.

its youth, the stuff of which the future is fashioned.

And I look upon the faculty, the students, the Board of Trustees, the alumni, and the Reformed Church, not, as some would lead me to think of them, as five sets of bosses with demands as irreconcilable as they are insistent, but as friends whose trust, confidence, and steady support makes these weighty responsibilities shared responsibilities, and therefore easy to bear.

Actually, a college president is, indeed, of all men most fortunate. He lives constantly in an electrifying atmosphere, charged with the crea-

The
Inaugural Address

This is a moving occasion, certain to arouse deep emotional responses in all but the most stoic and unfeeling of men. Being neither stoic nor unfeeling, I am deeply moved. I stand before you this afternoon in complete humility as I contemplate the awe-inspiring responsibilities of the chief executive of this college; responsibilities, indeed, that stagger the imagination.

My concern stems not from the warnings of my friends who have been *quick to point out that* "a college president is a person who attempts the impossible while he awaits the inevitable"; — *quick to observe that* "an old college president never dies, he just gradually loses his faculties"; —*quick to philosophize that* "the true feeling of insignificance, that which comes to a man when he makes a mistake and no one notices it, will never overcome a college president";—*and quick to declare that* "a college president is paid to talk, a college faculty to think, and a dean is paid to keep the faculty from talking and the president from thinking."

No, my humility stems, rather, from the fact that society entrusts to our fellowship and tutelage its most precious and priceless possession,

tivity of an imaginative, intense and dedicated faculty, the sweep of whose concern covers the realm of human knowledge and culture. He senses the exuberance of an eager and intellectually restless, yet surprisingly mature and purposeful student body. And he is reminded over and over as he deals with precious lives that a teacher's influence never ceases, but lives to all eternity.

And so it is, in deep humility, to be sure, but also with a high sense of exhilaration, that I dedicate all the talents and energy I possess to the task of providing leadership in fostering the climate which will encourage the faculty and students to grow together as a community of free and consecrated scholars, in the high tradition of Hope College.

THAT TINY BUT HARDY and resolute band of pioneers who founded Hope College believed passionately in education. Striking out from the Netherlands to realize for themselves and their children freedom in Church and State, they arrived on this very spot in 1847. Even as they felled trees to build their homes and churches and

367

coaxed a bare existence from an uncooperative soil, they found time in 1851 to establish an Academy. Twelve years later, in 1862, the first freshman college class became a reality, and on May 14, 1866, Hope College was incorporated and chartered by the State of Michigan.

Parenthetically, I take pride in pointing to Hope's role in the area of co-education. It is as amazing as it is refreshing to note that a pioneer band of hard-headed Dutchmen should found a college open to women, when the spirit of the times was quite otherwise. That spirit was accurately expressed by President Eliot of Harvard when he declared in his inaugural address in 1869: "This corporation will not receive women as students in the college proper, nor into any school where discipline requires residence near the school. The difficulties involved in the common residence of young men and women of immature character and marriageable age are very grave. The necessary police regulations would be exceedingly burdensome."

OUR FOREBEARS established Hope College to provide education of the mind and of the heart, a curriculum for intellect and competence, for character and reverence, that produced God-fearing citizens, diligent in the propagation of knowledge and of faith. In those pioneer days, any college education was an education of quality, and few, indeed, were the privileged who entered college.

For Hope College, in those early days, it was a question of survival. As the immigrants eked out a bare existence, a few of the less visionary questioned the luxury of diverting manpower, energy and means for something as impractical as a college. It was then that their undaunted leader, Reverend Albertus Van Raalte, uttered his prophetic words about this endeavor of faith: "This is my anchor of Hope for this people."

He realized that education was the potent lever by which his group, representing a small minority nationality and denomination, could magnify mightily the force of its influence upon the new nation in which they found themselves.

"I am determined," he said, "that my people will not become the fag end of civilization."

AND SURVIVE THE COLLEGE DID. Still a small college today, almost 100 years later, her accomplishments belie her size. For her graduates, all over the world, are making rich contributions in the arts, the professions, in business, and in religion, far out of proportion to their number.

It would be pleasant to dwell on the inspiring past, to review the College's glorious achievements, to pay tribute to those mighty leaders who with great distinction have brought Hope to her present position of eminence.

"If we can see farther than other men, it is because we stand on the shoulders of giants," such giants in our days as Dr. Dimnent, the profound scholar, who so ably kindled in faculty and students alike, a renewed love of learning and in increased devotion to academic excellence; Dr. Wichers, the noble statesman, who so masterfully guided Hope College to new heights through the most threatening period of her history; Dr. Lubbers, the master architect, who so inspiringly charted her course of unprecedented growth and development. But I will resist the temptation to look back, for as Lloyd George remarked, "No army can march on a retreating mind."

And so today, at this moment, we look to the future. Once again we stand as pioneers. Once again, the question of the survival of Hope College, indeed of all Christian liberal arts colleges, is heard in the land. And why is this so?

The central fact of our age is the explosion of knowledge. We all stand at the frontier of the age of the intellect; we are all immigrants in a new life. Although scientific in its origin, this explosion of knowledge carries economic, political, and cultural by-products which are creating entirely new and uncharted dimensions of human thought and endeavor. From earliest ages, man has, of course, been engaged in mining new knowledge, though only in recent decades with any degree of success, and only in our latter years has he struck a lode of unprecedented length and depth.

THE SPAN OF MAN'S INTELLECTUAL LIFE has been unbelievably brief. Compared to the life of sun, moon, and stars, it is but a fleeting moment. The recorded history of man almost vanishes if we look back 100 life spans. As Dr. J. D. Williams has observed, "You could receive a message from Plato which need not have passed through the mouths of more than 33 men. For Jesus, 28 men would suffice. You could get word from Gutenberg through a chain of 7 men, and from Newton through 4. Two life-spans ago the steam engine and the lathe, keys to the industrial revolution, came into our hands, and the internal-combustion engine was developed only 1 life-span ago." (In fact, the insistent questioning of our 8-year-old son some days ago finally elicited from me the grudging admission that the lowly, everyday zipper, if not invented, has certainly been exploited only in the latter half of my own lifetime.)

"The rate of development during our lives has become fantastic. It took us only a few months to become blasé about man-made satellites. Most of what man knows and has accomplished is incredibly recent." As a chemist, I cannot help but observe that two hundred years ago all men believed in phlogiston; one hundred years ago we had no periodic table and knew nothing about the electron; today we are on the verge of sending an expedition to the moon.

LADIES AND GENTLEMEN, our knowledgeable freshmen in college today know more mathematics than Descartes, more physics than Newton, more chemistry than Madam Curie. In fact, during the past 50 years, less than the lifetime of many in this room, there have been more and greater scientific advances, both pure and applied, than in all the previous ages of man's existence.

But all this is just a fumbling beginning. The real significance of these facts lies in the future which they portend. A famous scientist recently remarked that "if you were to step to the Pacific ocean, dip a tennis ball into the ocean, and then withdraw it, the amount of water retained by the tennis ball would bear the same relationship to the water yet remaining in the Pacific ocean, as our present scientific knowledge bears to the scientific knowledge yet awaiting discovery."

Knowledge in many fields is doubling every decade. Ninety percent, 9 out of 10, of all scientists who ever lived are alive today. Man is just beginning to learn how to learn. Mankind has just begun to sense that knowledge is power. We are all children of the great scientific revolution, which, for better or for worse, has thrust us all into the maelstrom of history.

And as a result, never before in the recorded annals of mankind has such a premium been placed upon the trained and educated mind. At long last, man has come to know that this, the trained and educated mind, is his most priceless, his most significant, his one indispensable commodity. As John Gardner, president of the Carnegie Corporation has observed, "In all the changes that characterize our revolutionary era, this society's new attitude toward educated tal-

Inauguration CONTINUED

ent, may in the long run prove to be the most significant." Everywhere the search is for the uncommon man and woman, for intelligence developed to its highest potential.

AND ALONG WITH the knowledge explosion, the first tide of the population explosion is also just emerging. If we could muster accurate figures we would, I am sure, find that the adults now living constitute a very sizeable fraction of all those who have ever lived. The great mass of humanity is just now beginning to arrive on our planet. By 1970, there will be 10,000,000 students in our institutions of higher learning; by the year 2000, that number will reach 25,000,000, and we can be sure that by far the great bulk of this increase will be found in our tax-supported institutions.

Now, once again, we as pioneers on the frontiers of this new Age of the Intellect, are faced with the question of the survival of the Christian liberal arts college, to be specific, of Hope College. Many critical, honest, and competent observers of the American scene believe that this combination of circumstances I have described spells the demise of the Christian liberal arts college. They are convinced that by the year 2000 it can be written off completely. With the rapid decline in the percentage of the total student population at non-tax-supported colleges, with mounting competition for superior students, with the dollar struggle for competent teachers, with sky-rocketing costs for teaching equipment and apparatus, and with the rising minimum critical staff size required to cover adequately the subject matter in certain fields, the small denominational college, it is said, will no longer be able to offer *quality* education.

AT CONFERENCE AFTER CONFERENCE, we scientists hear that the small liberal arts college can no longer afford, in personnel or equipment, to educate science majors for graduate study. This, in spite of the fact that historically many of our small colleges, including Hope, have been indeed the recognized cradle of our nation's scientists.

A true friend of the small liberal arts college, Dr. W. Max Wise, Associate Director of the Danforth Foundation, just this week warned that Protestant higher education is in danger of becoming "irrelevant to the sweep of history." Our Protestant colleges, he observed, are already in the process of becoming "secure havens" in which students from the nation's middle and upper economic classes are comfortably removed from the winds of social concern that dominate our contemporary lives.

By the year 2000, say other friends, the small denominational college will speak, if at all, with a small, still voice. Many there are who declare that the great challenge to the Christian liberal arts college in the next few years is to keep from passing into oblivion, slowly miring in a morass of mediocrity.

ON BEHALF OF HOPE COLLEGE, I accept that challenge without for a moment minimizing the magnitude of the problems or the severity of the threats that face us. I emphasize them, in fact, because I fear that the hour is already late for us to cast aside self-satisfied smugness. There is only one answer, of course. We must continue to offer, under terms which students can afford, a liberal arts education of distinction and excellence, to the able, and to the ordinary, who come to us from this community, from this state, from this country, and from the far reaches of the world. We must promise that any student, a combination of whose ability, burning desire to learn, and willingness to work, indicates that he can succeed in college, will always find the door at Hope College open to him. And to keep faith with these students and with their parents we must continue to provide truly quality education.

S URELY SMALLNESS, in itself, is by no means a guarantee of excellence. Nor can any amount of pietism gloss over an academic wasteland. In our age of the intellect, piety will constitute a miserable crutch for the academic cripple. Today society reveres and demands excellence and competence. For these there can be no substitutes.

How can we at Hope College meet the challenge of a revolutionary future? How can we fit our academic chain to the sprocket of opportunity?

First of all, by maintaining a competent, creative, professionally alive, and dedicated faculty such as we now possess, largely through Dr. Lubbers' genius for recognizing and attracting teachers and scholars of great promise. Without inspiring teachers, quality education is, we all know, impossible.

Joan VanderWerf Steve '20, Charles Nichols and Anne VanderWerf Nichols '38, Vote about their pride in their brother's big day.

S ECONDLY, we at the small colleges must exploit to the full our unique advantages. Two obvious features of a small undergraduate liberal arts college such as ours are, first, that it is small, and second, that it is strictly undergraduate. This means that our primary and sole concern, in a very personal way, is for the total development of the individual undergraduate student. We can and must shape an understanding of liberal education that leads to the growth and maturity of each student as a real person, that directs and inspires him to initiate a life-long quest toward self-fulfillment through service.

The reception line kept the President and Mrs. VanderWerf, Board Chairman Boys, and Mrs. Boys occupied with congratulations and good wishes for well over two hours.

371

Our size permits, also, a coveted degree of flexibility, of maneuverability, of adaptability. The college which would remain vital must be constantly open to critical self-examination, to experiment, to trial, and to innovation.

Only the foolhardy would venture to predict all the startling changes that lie ahead in higher education in this era of revolution, of orbiting satellites, of gyrating stock markets, of crumbling empires and emerging nations, with men all over the globe crying out for freedom and justice in our day. Eternal vigilance will be the price that every college must pay for survival. I hope shortly to see a joint committee composed of representatives of the Board of Trustees, of the faculty, of the alumni, of the student body, of the administration engaged in an all-out effort at long-range, statesmanlike planning for Hope College. But already we can discern at least the outline and shape of major developments that warrant constant and careful scrutiny.

CERTAINLY ALL THE TOOLS of modern technology will be brought to bear on the problem of increasing the efficiency of the learning process. With television as the major medium for mass instruction and with programmed machines, taped lectures, language audio-tape, and microfilm of all types available for individual study, the complete mastery of the basic substance of any field may, in years to come, be left to the individual student to proceed at his own pace. Master teacher and student, then, would come together individually, or in small seminars, for probing in depth, for critical inquiry and analysis, and for creative scholarship. This implies, of course, a considerable degree of independent study for the student and at least a modest program of significant and stimulating original investigation for every member of the faculty, with honors courses for students at every level. More and more, student and teacher will become inquisitive partners in the eternal quest for truth and wisdom.

CERTAINLY, our liberal arts colleges must, in the future, acquire increasingly an international, indeed a global orientation. As we enter the era of the shrunken and shrivelled Universe, with no part of our globe as far from us in actual time as was Grand Rapids from Holland when this college was founded, we can no longer afford to remain in appalling and abysmal ignorance of the history, the culture, the language, and the religion of nine-tenths of the world's population. We must be able to say with Socrates, as did one of our astronauts upon re-entry, "My country is the World. My countrymen are all mankind."

STUDY ABROAD may soon be considered an essential component of a truly liberal education. Many of these objectives will be best achieved only through the closest type of cooperation with sister institutions through organized groups of colleges, such as the Great Lakes Colleges Association, of which we are proud to be a member.

Although this is an age of specialization (as Mrs. VanderWerf observed when our fifth daughter was born), we must, increasingly, I believe, devise courses and majors that obliterate departmental lines as our new and recently approved curriculum is designed to do. We must strive constantly to close the cultural gap between the natural sciences and the humanities and social sciences. Certainly the liberal arts colleges, particularly, must stand as the great protective bulwark, to keep the arts, the humanities, and the social sciences from being engulfed by the tidal wave of national support for scientific research and development.

Hope College is situated in the heart of a friendly, gracious, and vital community. She will, I hope, accept whole-heartedly the continuous challenge of becoming a more integral part of that community and sharing with it her rich academic and cultural life. We are eager to cooperate with its citizens in making available to the youth of Holland an educational experience from kindergarten through college, that is second to none in the nation.

THERE IS NO DOUBT that the challenges of tomorrow, many of them unforeseen and indeed unforeseeable, are greatly different (though just as stern and demanding) as the challenges which faced our pioneer forebears of a hundred years ago. We are pioneers in an era of breathless

change, when revolutionary upheaval is the order of the day, and we must learn to live in a state of perpetual surprise. But with the change that is the center of the mystery, the drama and tragedy of the world in which we live, we at Hope College hold fast to the changeless, the abiding faith of our fathers which is as real and relevant today as ever. Never before in the life of man have we needed so desperately educated men and women of faith who have recourse to the dimension of the changeless in guiding the direction and course of change.

For change in itself is not progress and knowledge in itself is not goodness.

As MAN HARNESSES the vast energy of the sun and unleashes the almost limitless power of the atom for peacetime uses, he can make possible for all mankind an almost unbelievably high standard of living, or, with the same supply of energy he can in the twinkling of an eye reduce vast portions of the earth to smouldering stacks of radioactive rubbish fairly well annihilating civilization in the process; with his increasing fundamental knowledge of the causes of disease

and his expanding tools and skills for the cure of degenerative diseases, man may soon bring to each of us an average life span of 150 years, disease free, or, with the same knowledge and skills he may set in motion a sweeping biological warfare more devastating and hideous than any nuclear war; with his startling breakthroughs in the understanding of the chemistry of intelligence, of the operation of the human brain, man may soon be able to eliminate most mental illnesses and elevate average human intelligence significantly, or, with his paralyzing nerve gases he may reduce vast segments of the earth's population to a pitiful pulp, to helpless blobs of dehumanized protoplasm.

We live today with the unhappy paradox that civilization needs educated people as never before. Yet modern society has only the educated man to fear because only the educated, or perhaps I should say the technically trained, can wreak total global disaster.

WHAT I AM SAYING is that these are fateful days in which we are living. If our world survives our children or grandchildren may well look upon them and say that they were the turning point in the history of the world. For what we do now may decide the future of the human race.

And the Christian liberal arts college, by its very existence, asserts for all times, and we believe particularly for these times, that education which addresses itself simply to the intellect is not enough, that we as teachers must be concerned "not only with where our students' heads are, but also where their hearts are."

In both our individual lives, and as a civilization, we seem to be tyrannized by change, not directed by the changeless. "Never," as Father Hesburgh, President of Notre Dame, has declared, "is the changeless so important as when change is engulfing us. Our dedication or lack of dedication to values that are changeless will decide the splendor or tragedy of our individual lives, the glory or degradation of our times, the promise or betrayal of our national destiny."

As a college of the Reformed Church, Hope College today, as always, sets at the center of its life the changeless, the eternal verities of the Fatherhood of God and the Brotherhood of man.

These are the central truths which serve to relate one field to another, and to relate all fields to an ultimate sense of values and purpose that makes a vocation a holy calling. Each of us has a place in God's world, a role in His scheme.

KNOWLEDGE, THEN, is not an end in itself, nor do we worship intellectualism for its own sake. As Dr. Buttrick asked in his address last evening, unless we propose to respond, why do we ask to know the truth?

Just to know is not enough, Woodrow Wilson said, "We are not placed on earth to sit still and know, we are placed here to act."

This view of life eliminates cynicism and despair and gives value, aim, meaning, ambition, purpose, and integration. It casts us in the role of committed service; we are co-workers with men everywhere in God's plan. And to search for knowledge and truth takes on new meaning. We would know more, so that we can better serve.

At the reception the President's sister, Lucille VanderWerf Veneklasen '23, is served coffee by Mrs. Hollenbach. Dr. Veneklasen and Mrs. VanderWerf's father, Dr. Harry G. Good (seated), look pleased with the whole thing.

WITH OUR FOREFATHERS we hold that it is God who is the ultimate source of all goodness, all beauty, all justice, all order, all truth. And in this faith our Christian college becomes a free community of scholars, who know the "joy of uncovering truth, the peace of finding truth, and the courage of living truth always." For God is Truth, and therefore the man who faithfully uncovers truth will never find himself alienated from God. As we hold to the changeless, we view man in all his inner dignity as a sacred person. We sense his innate worth in time and his inalienable destiny in eternity.

This is the conviction which constitutes a great fortress and bulwark against all the forces of evil everywhere which would enslave the body or the mind of man, which would rob him of his dignity or freedom.

This is the faith, too, which underscores the divine potential of every individual. It holds the answer to the plaintive and oft-repeated query "What can I do; I am only one." Moses was only one, Socrates, Gandhi, Albert Schweitzer, Martin Luther King, and Christ.

It charges each of God's children with the direct and personal responsibility to be good and to be great. Is not the development of great individuals the task and calling of education?

AND FINALLY, must we not look to the changeless, the eternal verities, to see the ultimate answers to the crucial problems of our civilization? Let us consider just one such problem.

12

374

Experts predict that within five years a dozen nations will be able to equip themselves with enough fission and fusion bombs to blow most of mankind to bits. Which of these nations, then, are the uncommitted peoples of the world to choose? The one which can blow them into the tiniest bits? Or with the battle of force stalemated, will the focus perhaps shift to the battle field of men's minds and spirits?

Was Pasteur right when he said, "I hold the unconquerable belief that science and peace will ultimately triumph over ignorance and war, that nations will come together not to destroy but to construct, that the future indeed belongs to those who accomplish most for humanity"?

Is it possible that mankind everywhere may cling to the banner of that nation which truly recognizes and gives support to the aspirations of men everywhere to be men, to walk free and untrammelled on God's green earth, to feel the wind, and the rain, and the sun on their faces, with shoulders squared and heads held high, in the dignity of true liberty? Can it be that the ultimate victory will be not to the strong but to the good, not to the mighty, but to the just?

The Christian college dares to ask the truly significant questions, to grapple with the truly crucial problems. As never before it must become the leavening influence in higher education, the redemptive force in the academic milieu. The mission of Hope College is sacred. For our community, our Church, our nation, indeed for our civilization, we cannot fail. We must, we will succeed.

WHEN REV. PHILIP PHELPS, the first president of Hope College, delivered his inaugural address he read from the book of Job:

"And unto man he said, Behold, the fear of the Lord, that is wisdom, and to depart from evil, that is understanding."

In this spirit, with the guidance of the sustaining Infinite, and with the reassuring faith and help of all who love Hope College, I humbly dedicate myself, knowing that together we shall strive on to fulfill her destiny, fearing God, and nothing else.

"There'll never be so much chemical knowledge under this roof again," was the President's comment as he posed with three of his former teachers: Dr. Melvin S. Newman, his major professor in graduate school at Ohio State University (left), the President; Dr. Gerrit Van Zyl, his undergraduate chemistry professor at Hope College; Dr. Arthur Davidson, Chairman of the Department of Chemistry at the University of Kansas and prior to Dr. VanderWerf.

The Inauguration of Calvin Anthony VanderWerf as Eighth President of Hope College on November 16, 1963

The inauguration of Dr. Calvin Anthony VanderWerf on November 16 as eighth president of Hope College was an impressive and majestic occasion in Dimnent Memorial Chapel. Even the weather cooperated with unseasonable pleasantness helping to create an atmosphere of warmth and good will as students, trustees, alumni, faculty, community, and delegates from over 200 colleges and universities, learned societies and foundations gathered to witness the installation and congratulate the new president.

More than 300 persons participated in a colorful academic procession preceding the ceremonies. Ekdal Buys, '37, chairman of the Board of Trustees, gave the charge to the President. The Rev. M. Verne Oggel, '11, president of the General Synod of the Reformed Church in America, gave the invocation and the Hope College Chapel Choir, under the direction of Dr. Robert W. Cavanaugh, sang the anthem "Hope Thou Is God," written for Hope College during the past summer by Dr. Haydn Morgan of Eastern Michigan University. The Rev. Herman Ridder, '49, president at Western Theological Seminary, gave the benediction.

Dr. VanderWerf's inaugural address is printed in full.

The inauguration at 2:30 in the afternoon was preceded by an inaugural luncheon in Phelps Hall attended by 400 delegates, faculty, and special guests. A reception in Phelps Hall for all who attended the ceremonies closed the official program of the day. Everyone agreed that the day's program was practically perfect. Everything proceeded on time as scheduled and the feeling was that of good will and faith that the College is in good hands. The only slightly critical comments—and they were made with twinkle in eye, however, concerned the marching during the processional when the delegates, thinking they might be holding up the schedule, took up the slack by hurrying down the aisle at double, almost triple, time. Those who criticized did so on the basis that It was too fast for the eye to scan the faces of all in the procession. Then there was another little misunderstanding in the recessional—the last in, decided on their own to be the first out, which harassed the marshals and annoyed those in charge of the detail. But, as others added, there has to be one "departure from accuracy" to make such an occasion perfect and memorable.

And so it was, a perfect and memorable day on the Hope College campus.

The Charge-"The leadership of our college determines her course"

President VanderWerf, we bring this charge today in behalf of the Trustees of our College, representing our beloved denomination, THE REFORMED CHURCH IN AMERICA. Those who now attend Hope, those who support her, and those who leader her, are our deep concern.

Needless to say, those of us who heard or read your convocation address to the student body, thrill with the challenge to personal commitment to God. As you ably said, "We cannot do this for you," but we can set the stage. We cherish for all the men and women here the good things that can be provided for a meaningful four years at Hope. Our first duty is the preparation of lives here that will make an impact on the complexities ahead. The record tells us that this freshman class has the highest academic entrance record in our history. Let this certainly be directed for good.

Those who support us make a warm fellowship to enjoy. It has been said of colleges, "Some are community leaders, some withdraw into intellectual seclusion." The question arises as to what extent a college should make itself felt in a community. What should be the relationship between the "Gown and the Town?" At this time, may I digress for a moment to publicly express our thanks to this Community for its cooperation and support? The Community Council for Hope worked diligently in our major campaign of recent years. This, Sir, should be continued, not only at this level, but more areas of influence around us should be activated.

The parents of our students, both at present and in years past, regardless of their prior affiliation, are a source of strong support. Contact with this group should be maintained now and in future years.

The Alumni, naturally make up the major area of support and may I also say, justly so, of criticism. However, the give and take of this group is commendable and should be a real inspiration to you. Alumni records tell us that 85% of our graduates are in the service field: doctors, dentists, nurses, school teachers, and administrators, ministers, and missionaries in every direction of the compass. The history of this group alone is an impelling heritage.

I hesitate to single out any group but the women of the church have entered into a relationship of not only financial aid, but

377

a year around effort in our behalf. This is now reflected by the fact that at the last Board of Trustees meeting, five women were present representing various Synods. This is a good trend, but is being watched carefully.

The need for our continued cooperation with inter-college groups, such as the Great Lakes Colleges Association and the Michigan Colleges Foundation is self-evident.

The REFORMED CHURCH IN AMERICA is the foundation of our college. Constantly remember that their prayers and resources are always with you. When all the negatives these days seem to be chipping away at our foundations, thank the Lord for Church related institutions.

The leadership of our college determines her course. The dedicated group of Faculty are our prime asset. Academic excellence is evidenced in the constant high ratings of our students both here and that the graduate level. The areas of their influence demonstrate the emphasis they have received. The Deans, Administrative Officers and all the many people that pursue their daily tasks on this campus are your constant help and support.

As for you personally, Mr. PRESIDENT, this cannot be said without deep personal emotion. Twenty-six years ago, we were on the same platform as classmates. You, because of your academic excellence, gave the valedictory address. Again, because of this same continued achievement in your life, God has providentially returned you as our president. President. We hope now and then to peer through the windows of your home and see you completely relaxed, either pampered by your six lovely ladies, or pestered by your other man about the house. Your fine family is a God-given blessing, take time to cherish and enjoy them with all your heart.

As we review the Presidential History of Hope from Dr. Phelps to Dr. Lubbers, we find that each made a unique contribution in their own right and this you will also do. The changeless note through these years has been our Motto from the Psalms "HOPE THOU IN GOD," "THIS IS OUR ANCHOR OF HOPE."

May god grant you and your family health and wisdom for the years ahead.

The Inaugural Address — Cal VanderWerf

Chairman Buys, Members of the Board of Trustees, Distinguished Guests, Fellow Faculty, and Students,

We are met here this afternoon at the heart of a great and noble college. Her life-beat has, in the past, transfused into

378

thousands of young lives that profound sense of meaning and purpose which leads to the high road of self-fulfilling service. We pray that this may ever be her glorious destiny: it is to that purpose, Mr. Buys, that I dedicate myself in accepting the sacred charge which you have laid upon me.

This is a moving occasion, certain to arouse deep emotional responses in all but the most stoic and unfeeling of men. Being neither stoic, nor unfeeling, I am deeply moved. I stand before you this afternoon in complete humility as I contemplate the awe-inspiring responsibilities of the chief executive of this college; responsibilities, indeed, that stagger the imagination.

My concern stems not from the warnings of my friends who have been quick to point out that "a college president is a person who attempts the impossible while he awaits the inevitable"; —quick to observe that "an old college president never dies; he just gradually loses his faculties"; —quick to philosophize that "the true feeling of insignificance, that which comes to a man when he makes a mistake and no one notices it, will never overcome a college president; –and quick to declare that "a college president is paid to talk, a college faculty to think, and a dean is paid to keep the faculty from talking and the president from thinking."

No, my humility stems, rather, from the fact that society entrusts to our fellowship and tutelage its most precious and priceless possession, its youth, the stuff of which the future is fashioned.

And I look upon the faculty, the students, the Board of Trustees, the alumni, and the Reformed Church, not as some would lead me to think of them, as five sets of bosses with demands as irreconcilable as they are insistent, but as friends whose trust, confidence, and steady support makes these weighty responsibilities shared responsibilities, and therefore easy to bear.

Actually, a college president is, indeed, of all men most fortunate. He lives constantly in an electrifying atmosphere, charged with the creativity of an imaginative, intense and dedicated faculty, the sweep of whose concern covers the realm of human knowledge and culture. He senses the exuberance of an eager and intellectually restless, yet surprisingly mature and purposeful student body. And he is reminded that a teacher's influence never ceases, but lives to all eternity.

And so it is, in deep humility, to be sure, but also with a high sense of exhilaration, that I dedicate all the talents and energy I possess to the task of providing leadership in fostering the climate which will encourage the faculty and students to grow together as a community of free and consecrated scholars, in the high tradition of Hope College.

That tiny but hardy and resolute head of pioneers who founded Hope College believed passionately in education. Striking out from the Netherlands to realize for themselves and their children freedom in Church and State, they arrived on this very spot in 1847. Even as they felled trees to build their homes and churches and coaxed a bare existence from an uncooperative soil, they found time in 1851 to establish an Academy. Twelve years later, in 1862, the first freshman college class became a reality, and on May 14, 1866, Hope College was incorporated and chartered by the State of Michigan.

Parenthetically, I take pride in pointing to Hope's role in the area of co-education. It is as amazing as it is refreshing to note that a pioneer band of hard-headed Dutchmen should found a college open to women, when the spirit of the times was quite otherwise. That spirit was accurately expressed by President Eliot of Harvard when he declared in his inaugural address in 1869: "This corporation will not receive women as students in the college proper, nor into any school where discipline requires residents near the school. The difficulties involved in the common residence of young men and women of immature character and marriageable age are very grave. The necessary police regulations would be exceedingly burdensome."

Our forebears established Hope College to provide education of the mind and of the heart, a curriculum for intellect and competence, for character and reverence, that produced God-fearing citizens, diligent in the propagation of knowledge and of faith. In those pioneer days, any college education was an education of quality, and few, indeed, were the privileged who entered college.

For Hope College, in those early days, it was a question of survival. As the immigrants eked out a bare existence, a few of the less visionary questioned the luxury of diverting manpower, energy, and means for something as impractical as a college. It was then that their undaunted leader, Reverend Albertus Van Raalte, uttered his prophetic words about this endeavor of faith: "This is my anchor of Hope for tis people."

He realized that education was the potent lever by which his grasp, representing a small minority nationally and denomination, could magnify mightily the force of its influence upon the new nation in which they found themselves.

380

"I am determined,' he said, "that my people will not become the fag end of civilization."

And survive the college did. Still a small college today, almost 100 years later, her accomplishments belie her size. For her graduates, all over the world, are making rich contributions in the arts, the professions, in business, and in religion, far out of proportion to their number.

It would be pleasant to dwell on the inspiring past, to review the College's glorious achievements, to pay tribute to those mighty leaders who with great distinction have brought Hope to her present position of eminence.

"If we can see farther than other men, it is because we stand on the shoulders of giants." Such giants in our days as Dr. Dimnent, the profound scholar, who so ably kindled in faculty and students alike, a renewed love of learning and in increased devotion to academic excellence. Dr. Wichers, the noble statesman, who so masterfully guided Hope College to new heights through the most threatening period of her history. Dr. Lubbers, the master architect, who so inspiringly charted her course of unprecedented growth and development. But I resist the temptation to look back, for as Lloyd George remarked, "No army can march on a retreating mind."

And so today, at this moment, we look to the future. Once again we stand as pioneers. Once again, the question of the survival of Hope College, indeed of all Christian liberal arts colleges, is heard in the land. And why is this so?

The central fact of our age is the explosion of knowledge. We all stand at the frontier of the age of the intellect; we are all immigrants in a new life. Although scientific in its origin, this explosion in knowledge carries economic, political, and cultural by-products which are creating entirely new and uncharted dimensions of human thought and endeavor. From earliest ages, man has, of course, been engaged in mining new knowledge, though only in recent decades with any degree of success, and only in our latter years has he struck a lode of unprecedented length and depth.

The span of man's intellectual life has been unbelievably brief. Compared to the life of sun, moon, and stars, it is but a fleeting moment. The recorded history of man almost vanishes if we look back 100 life spans. As Dr. J.D. Williams has observed, "You could receive a message from Plato which need not have passed through the mouths of more than 33 men. For Jesus, 28 men would suffice. You could get word from Gutenberg through a chain of 7 men, and from Newton through 4. Two life-spans ago the steam engine and the lathe, keys to the industrial revolution, came into our hands, and the internal-combustion engine was developed only 1 life-span ago." (In fact, the insistent questioning of our 8-year-old son some days ago finally elicited from me the grudging admission that the lowly, everyday zipper, if not invent, has certainly been exploited only in the latter half of my own lifetime.)

"The rate of development during our lives has become fantastic. It took us only a few months to become blasé about man-made satellites. Most of what man knows and has accomplished is incredibly recent." As a chemist, I cannot help but observe that two hundred years ago, all men believed in phlogiston; one hundred years ago we had no periodic table and knew nothing about the electron; today we are on the verge of sending an expedition to the moon.

Ladies and gentlemen, our knowledgeable freshmen in college today know more mathematics than Descartes, more physics than Newton, more chemistry than Madam Curie. In fact, during the past 50 years, less than the lifetime of many in this room, there have been more and greater scientific advances, both pure and applied, than in all the previous ages of man's existence.

But all this is just a fumbling beginning. The real significance of these facts lies in the future which they portend. A famous scientist recently remarked that "if you were to step to the Pacific Ocean, dip a tennis ball into the ocean, and then withdraw it, the amount of water retained by the tennis ball would be the same relationship to the water yet remaining in the Pacific Ocean, as our present scientific knowledge bears to the scientific knowledge yet awaiting discovery."

Knowledge in many fields is doubling every decade. Ninety percent, 9 out of 10, of all scientists who ever lived are alive today. Man is just beginning to learn how to learn. Mankind has just begun to sense that knowledge is power. We are all children of the great scientific revolution, which, for better or for worse, has thrust us all into the maelstrom of history.

And as a result, never before in the recorded annals of mankind has such a premium been placed upon the trained and educated mind. At long last, man has come to know that this, the trained and educated mind, is his most priceless, his most significant, his one indispensable commodity. As John Gardner, president of the Carnegie Corporation has observed, "In all the changes that characterize our revolutionary era, this, society's new attitude toward educated talent, may in the long run prove to be the most significant." Everywhere the search is for the uncommon man and woman, for intelligence developed to its highest potential.

And along with the knowledge explosion, the first tide of the population explosion is also just emerging. If we could muster accurate figures we would, I am sure, find that the adults now living constitute a very sizable fraction of all those who have ever lived. The great mass of humanity is just now beginning to arrive on our planet. By 1970, there will be 10,000,000 students in our institutions of higher learning; by the year 2000, that number will reach 25,000,000, and we can be sure that by far the great bulk of this increase will be found in our tax-supported institutions.

Now, once again, we as pioneers on the frontiers of this new Age of the Intellect, are faced with the question of survival of the Christian liberal arts college, to be specific, of Hope College. Many critical, honest, and competent observers of the American scene believe that this combination of circumstances I have described spells the demise of the Christian liberal arts college. They are convinced that by the year 2000 it can be written off completely. With the rapid decline in the percentage of the total student population at non-tax-supported college, with mounting competition for superior students, with the dollar struggle for competent teachers, with sky-rocketing costs for teaching equipment and apparatus, and with the rising minimum critical staff size required to cover adequately the subject matter in certain fields, the small denominational college, it is said, will no longer be able to offer quality education.

At conference after conference, we scientists hear that the small liberal arts college can no longer afford, in personnel or equipment, to educate science majors for graduate study. This, in spite of the fact that historically many of our small colleges, including Hope, have been indeed the recognized cradle of our nation's scientists.

A true friend of the small liberal arts college, Dr. W. Max Wise, Associate Director of the Danforth Foundation, just this week warned that Protestant higher education is in danger of becoming "irrelevant to the sweep of history." Our Protestant colleges, he observed, are already in the process of becoming "secure havens in which students from the nation's middle and upper economic classes are comfortably removed from the winds of social concern that dominate our contemporary lives.

By the year 2000, say other friends, the small denominational college will speak, if at all, with a small, still voice.

Many there are who declare that the great challenge to the Christian liberal arts college in the next few years is to keep from passing into oblivion, slowly miring in a morass of mediocrity.

On behalf of Hope College, I accept that challenge without for a moment minimizing the magnitude of the problems or the severity of the threats that face us. I emphasize them, in fact, because I fear that the hour is already late for us to cast aside self-satisfied smugness. There is only one answer, of course. We must continue to offer, under terms which students can afford, a liberal arts education of distinction and excellence, to the able, and to the ordinary, who come to us from this community, from this state, from this country, and from the far reaches of the world. We must promise that any student, a combination of whose ability, burning desire to learn, and willingness to work, indicates that he can succeed in college, will always find the door at Hope College open to him. And to keep faith with these students and with their parents we must continue to provide truly quality education.

Surely smallness, in itself, is by no means a guarantee of excellence. Nor can any amount of pietism gloss over an academic wasteland. In our age of the intellect, piety will constitute a miserable crutch for the academic cripple. Today, society reveres and demands excellence and competence. For these there can be no substitutes.

How can we at Hope College meet the challenge of a revolutionary future? How can we fit our academic chain to the sprocket of opportunity?

First of all, by maintaining a competent, creative, professionally alive, and dedicated faculty such as we now possess, largely through Dr. Lubbers' genius for recognizing and attracting teachers and scholars of great promise. Without inspiring teachers, quality education is, as we all known, impossible.

Secondly, we at the small colleges must exploit to the full our unique advantages. Two obvious features of a small undergraduate liberal arts college such as ours are, first, that it is small, and second, that it is strictly undergraduate. This means that our primary and sole concern, in a very personal way, is for the total development of the individual undergraduate student. We can and must shape an understanding of liberal education that leads to the growth and maturity of each student as a real person, that directs and inspires him to initiate a life-long quest toward self-fulfillment through service.

384

Our size permits, also a coveted degree of flexibility, of maneuverability, of adaptability. The college which would remain vital must be constantly open to critical self-examination, to experiment, to trial, and to innovation.

Only the foolhardy would venture to predict all the startling changes that lie ahead in higher education in this era of revolution, of orbiting satellites, of gyrating stock markets, of crumbling empires and emerging nations, with men all over the globe crying out for freedom and justice in our day. Eternal vigilance will be the price that every college must pay for survival. I hope shortly to see a joint committee composed of representatives from the Board of Trustees, of the faculty, of the alumni, of the student body, of the administration engaged in an all-out effort at long-range, statesmanlike planning for Hope College. But already we can discern at least the outline and shape of major developments that warrant constant and careful scrutiny.

Certainly all the tools of modern technology will be brought to bear on the problem of increasing the efficiency of the learning process. With television as the major medium for mass instruction and with programmed machines, taped lectures, language audio-tape, and microfilm of all types available for individual study, the complete mastery of the basic substance of any field may, in years to come, be left to the individual student to proceed at his own pace. Master teacher and student, then, would come together individually, or in small seminars, for probing in depth, for critical inquiry and analysis, and for creative scholarship. This implies, of course, a considerable degree of independent study for the student and at least a modest program of significant and stimulating original investigation for every member for the faculty, with honors courses for students at every level. More and more, student and teacher will become inquisitive partners in the eternal quest for truth and wisdom.

Certainly, our liberal arts colleges must, in the future, acquire increasingly an international, indeed a global orientation. As we enter the era of the shrunken and shriveled Universe, with no part of our globe as far from us in actual time as was Grand Rapids from Holland when this college was founded, we can no longer afford to remain in appalling and abysmal ignorance of the history, the culture, the language, and the religion of nine-tenths of the world's population. We must be able to say with Socrates, as did one of our astronauts upon re-entry, "My country is the World. My countrymen are all mankind."

Study abroad may soon be considered an essential component of a truly liberal education. Many of these objectives will be best achieved only through the closest type of cooperation with sister institutions through organized groups of colleges, such as the Great Lakes Colleges Association, of which we are proud to be a member.

Although this is an age of speculation (as Mrs. VanderWerf observed when our fifth daughter was born), we must, increasingly, I believe, devise courses and majors that obliterate departmental lines as our new and recently improved curriculum is designed to do. We must strive constantly to close the cultural gap between the natural sciences and the humanities and social sciences. Certainly the liberal arts colleges, particularly, must stand as the great protective bulwark, to keep the arts, the humanities, and the social sciences from being engulfed by the tidal wave of national support for scientific research and development.

Hope College is situated in the heart of a friendly, gracious, and vital community. She will, I hope, accept whole-heartedly the continuous challenge of becoming a more integral part of that community and sharing with it her rich academic and cultural life. We are eager to cooperate with its citizens in making available to the youth of Holland an educational experience from kindergarten through college, that is second to none in the nation.

There is no doubt that the challenges of tomorrow, many of them unforeseen and indeed unforeseeable, are greatly different (though just as stern and demanding) as the challenges which faced our pioneer forebears of a hundred years ago. We are pioneers in an era of breathless change, when revolutionary upheaval is the order of the day, and we must learn to live in a state of perpetual surprise. But with the change that is the center of the mystery, the drama and tragedy of the world in which we live, we at Hope College hold fast to the changeless, the abiding faith of our fathers which is as real and relevant today as ever. Never before in the life of man have we needed so desperately educated men and women of faith who have recourse to the dimension of the changeless in guiding the direction and course of change.

For change in itself is not progress and knowledge in itself is not goodness.

As man harnesses the vast energy of the sun and unleashes the almost limitless power of the atom for peacetime uses, he can make possible for all mankind an almost unbelievable high standard of living, or, with the same supply of energy he can in the twinkling

386

of an eye reduce vast portions of the earth to smoldering stacks of radioactive rubbish fairly well annihilating civilization in the process; with his increasing fundamental knowledge of the causes and his expanding tools and skills for the cure of degenerative disease, man may soon bring to each of us an average life span of 150 years, disease free, or with the same knowledge and skills he may set in motion a sweeping biological warfare more devastating and hideous than any nuclear war; with his startling breakthroughs in the understanding of the human brain, man may soon be able to eliminate most mental illnesses and elevate average human intelligence significantly, or, with his paralyzing nerve gases he may reduce vast segments of the earth's population to a pitiful pulp, to helpless blobs of dehumanized protoplasm.

We live today with the unhappy paradox that civilization needs educated people as never before. Yet modern society has only the educated man to fear because only the educated, or perhaps I should say the technically trained, can wreak total global disaster.

What I am saying is that these are fateful days in which we are living. If our world survives our children or grandchildren may well look upon them and say that they were the turning point in the history of the world. For what we do now may decide the future of the human race.

And the Christian liberal arts college, by its very existence, asserts for all times, and we believe particularly for these times, that education which addresses itself simply to the intellect is not enough, that we as teachers must be concerned "not only with where our students' heads are, but also where their hearts are."

In both our individual lives, and as a civilization, we seem to be tyrannized by change, not directed by the changeless. "Never," as Father Hesburgh, President of Notre Dame, has declared, "is the changeless so important as when change is engulfing us. Our dedication or lack of dedication to values that are changeless will decide the splendor or tragedy of our individual lives, the glory or degradation of our times, the promise or betrayal of our national destiny."

As a college of the Reformed Church, Hope College today, as always, sets at the center of its life the changeless, the eternal verities of the Fatherhood of God, and the Brotherhood of man.

These are the central truths which serve to relate one field to another, and to relate all fields to an ultimate sense of values and purpose that makes a vocation a holy calling. Each of us has a place in God's world, a role in His scheme.

Knowledge, then, is not an end in itself, nor do we worship intellectualism for its own sake. As Dr. Buttrick asked in his address last evening, unless we propose to respond, why do we ask to know the truth?

Just to know is not enough, Woodrow Wilson said, "We are not placed on earth to sit still and know, we are placed here to act."

This view eliminates cynicism and despair and gives value, aim, meaning, ambition, purpose, and integration. It casts us in the role of committed service; we are co-workers with men everywhere in God's plan. And to search for knowledge and truth takes on new meaning. We would know more, so that we can better serve.

With our forefathers we hold that it is God who is the ultimate source of all goodness, all beauty, all justice, all order, all truth. And in this faith our Christian college becomes a free community of scholars, who know the "joy of uncovering truth, the peace of finding truth, and the courage of living truth always." For God is Truth, and therefore the man who faithfully uncovers truth will never find himself alienated from God. As we hold to the changeless, we view man in all his inner dignity as a sacred person. We sense his innate worth in time and his inalienable destiny in eternity.

This is the conviction which constitutes a great fortress and bulwark against all the forces of evil everywhere which would enslave the body or the mind of man, which would rob him of his dignity or freedom.

This is the faith, too, which underscores the divine potential of every individual. It holds the answer to the plaintive and oft-repeated query "What can I do: I am only one." Moses was only one, Socrates, Gandhi, Albert Schweitzer, Martin Luther King, and Christ.

It charges each of God's children with the direct and personal responsibility to be good and to be great. Is no the development of great individuals the task and calling of education?

And finally, must we not look to the changeless, the eternal verities, to see the ultimate answers to the crucial problems of our civilization? Let us consider just one such problem.

Experts predict that within five years a dozen nations will be able to equip themselves with enough fission and fusion bombs to blow most of mankind to bits. Which of these nations, then, are the uncommitted peoples of the world to choose? The one which can blow them into the tiniest bits? Or with the battle of force stalemated, will the focus perhaps shift to the battle field of men's minds and spirits?

Was Pasteur right when he said, "I hold the unconquerable belief that science and peace will ultimately triumph over ignorance and war, that nations will come together not to destroy but to construct, that the future indeed belongs to those who accomplish most for humanity"?

Is it possible that mankind everywhere may cling to the banner of that nation which truly recognizes and gives support to the aspirations of men everywhere to be men, to walk free and untrammeled on God's green earth, to feel the wind, and the rain, and the sun on their faces, with shoulders squared and heads held high, in the dignity of true liberty? Can it be that the ultimate victory will be not to the strong but to the good, not to the mighty, but to the just?

The Christian college dares to ask the truly significant questions, to grapple with the truly crucial problems. As never before it must become the leavening influence in higher education, the redemptive force in the academic milieu. The mission of Hope College is sacred. For our community, our Church, our nation, indeed for our civilization, we cannot fail. We must, we will succeed.

When Rev. Philip Phelps, the first President of Hope College, delivered his inaugural address, he read from the book of Job: *"And unto man he said, Behold the fear of the Lord, that is wisdom, and to depart from evil, that is understanding."*

In this spirt, with the guidance of the sustaining infinite, and with the reassuring faith and help of all who love Hope College, I humbly dedicate myself, knowing that together we shall strive on to fulfill her destiny, fearing God, and nothing else.

HOPE COLLEGE

SELF-STUDY REPORT

For the

NORTH CENTRAL ASSOCIATION

REVIEW VISIT

April 27-28, 1964

Submitted:

April 1, 1964

ART DEPARTMENT REPORT
James Loveless, Chairman

Curriculum

Art History: The Art Department offers a history of art program designed
to complement subject matter in other disciplines. These courses,
furthermore, provide a substantial historical and theoretical base
for students majoring in studio art.

We have need of an additional staff member with more
extensive background in art history. We are presently reviewing the
credentials of applicants to fill such a position.

Studio Art: The strength of our studio program lies in the fact that
the prevailing philosophy of our department is fine-arts oriented.
We place particular importance on the areas of sculpture, painting,
printmaking and basic design. It is in these areas that the student
can realize the essential educational experience of creating significant artistic objects. Such a program is instrumental in developing
the creative faculties in general of both viewer and artist. In
a liberal arts college, this seems to be the only tenable purpose
of an art department.

We are weak in life-drawing. Studio sessions using clothed
student models have been started to supplement the work of students
enrolled in studio courses. Another problem we face is providing
for the major student sufficient exposure to the studio area of his
choice. This deficiency is partially made up by the addition of a
course called special problems which provides an additional semester
in either studio or art history.

A more substantial experience in art education is needed beyond our present two-hour requirement for future elementary school teachers. State and school requirements influence us in this matter. Since we are a liberal arts college, a more extensive art education program seems inappropriate at this time. The general college requirement, mentioned later, will greatly help in meeting this problem.

Student Enrollment and Staffing

Perhaps because the major in art is only in its second year, the total enrollment of students taking art courses is low for a school of our size. There are other reasons for this. The facilities are remote, barely adequate and somewhat decentralized; no general requirement in the fine arts has existed up to the present; and a well balanced staff teaching in the areas of their primary interests and competencies is needed.

The possibilities of realizing significant improvement in facilities (gallery, studios and lecture halls all centrally located) is still at the discussion level. A three-hour general college requirement in music or art will be started in the fall of 1964. The art course will be a lecture course utilizing slides. We are presently reviewing candidates for a new staff position in art. It is expected at this time that this new member will have the necessary qualifications to complement rather than duplicate the present staff's preparation and experience. This new staff member will in all probability have a large role to play in the development of this new general course. The art department will need the services of an art historian and two studio people in the near future.

<u>Art Department Report</u> (cont.)

<u>Budget</u>

The present budget allocations for maintaining our present program, in materials, equipment and library expenses, is adequate. We will need some major equipment expenses in the areas of printmaking, sculpture and art history. The nature of these expenditures will depend on the future staff's background and the physical plant in the next few years. Additional money is needed to defray the costs of handling exhibits and the purchase of works of art for a permanent collection, which we have only just started.

BIOLOGY DEPARTMENT REPORT
Philip G. Crook, Chairman

Staff: Ten years ago the Biology Department had a staff of three, one of whom had a Ph.D. A total of nine sections in six courses were offered. At the present time, there is a staff of six, three with Ph.D.'s and another to receive her Ph.D. in August. We are offering 21 sections in 13 courses. In this same ten-year period, there has been a complete turn-over in staff, with four of the six current staff in their first or second year at Hope College.

Research: Six years ago, an effort was made to establish a research program within the department. A research laboratory was outfitted, and five grants were obtained. Undergraduate students were employed as assistants. During this period, interest was stimulated in a problems course which allows students to work on projects of their own for one or two hours credit. The faculty research ebbed after four years, but the problems course continues to grow and at the present time has 21 students enrolled. The equipment obtained from the grants has been most useful. With lighter contact loads, made possible in part through increasing the size of lecture sections in beginning courses, it is hoped that new faculty research projects will be undertaken. There is now a firm policy of providing lighter teaching loads if a person becomes engaged in significant research.

Program: Two years ago a tripartite program was evolved that seeks to offer undergraduate training in (1) laboratory (pre-med) zoology, (2) field biology and (3) botany. The present faculty reflects this program with a cellular physiologist, an anatomist and a geneticist for laboratory zoology, a mammologist for field biology, and

Biology Department Report (cont.)

two botanists. There is a high probability that at least four of this six will remain at Hope. If replacement becomes necessary, it will be made with people in similar areas of specialization. The courses offered by the department have increased slightly in number but an effort has been made to prevent excessive proliferation. Two separate introductory courses have had to be offered, one for science majors and one for non-science majors. In the latter, one of two laboratories has been replaced by a lecture because of pressure on laboratory space. Courses in ecology, ornithology and botany have been introduced. We would like to see an elective laboratory in genetics and a course in cytology added. This would give each faculty member two, or at the most three, advanced courses to teach.

Equipment: Teaching equipment has improved steadily, but we have not been able to replace much of the old equipment. A television microscope has seen a lot of use. To keep abreast of modern developments, we have acquired a Warburg apparatus, electrophoresis and chromatographic gear and a scaler for isotope work. We are deficient in electronic devices for physiology but are making some of our own. Of major importance has been the construction of a greenhouse, one end of which will temporarily be used as an animal room. A natural history museum has received some attention, and there is an active program underway to collect local plants and animals.

Administration: Improvements are to be made at the administrative level next year when we will have a half-time secretary added to our staff. A more methodical program of counseling and placement will be undertaken and an effort made to follow-up on graduates. We are going

<u>Biology Department Report</u> (cont.)

to apply for N.S.F. funds for a summer institute and for new teaching equipment.

<u>The Future</u>: There are specific matters now being considered. We would like more of our students to benefit from summer opportunities in marine biology, ecology, and research participation. We will have some help here from the G.L.C.A. We are curious about the possibility of using programmed materials to supplement regular course work or to be used in remedial areas. Independent study in biochemistry has been tried, but with poor success, perhaps due to poor organization. This will be further explored. While most of our faculty are relatively fresh out of graduate school, we must keep up to date. Our library collection of books and especially of journals is inagequate and is not being used enough. We would like to have a departmental library and are exploring this possibility. The concerns that we are least able to do anything about are inadequate space and the need to update and replace equipment.

<u>Observations</u>: The number of biology majors has doubled in three years. Our pre-medical and pre-graduate programs are improving. We are getting students interested in some of the new areas. Faculty and student morale seems to be excellent. My impression is that this department is in good shape.

CHEMISTRY DEPARTMENT REPORT
Gerrit Van Zyl, Chairman

The Hope College Chemistry Department is accredited by the American Chemical Society. The affiliate chapter has a membership of 35 sophomores, juniors and seniors who plan to make chemistry their profession. Last year thirteen entered graduate schools on scholarships or assistantships. Many students who are accepted by medical schools also have a major in chemistry.

The department has received substantial grants for fifteen years from the Research Corporation or the P.R.F. (or both). Most of the funds are used to pay superior students working in our summer research programs. This has resulted in the publication of twenty papers in the chemical journals. From five to eight students are involved in this program each year.

The college has received N.S.F. grants for summer institutes for high school science and mathematics teachers during the past three years. For next summer we are offering an N.S.F. institute for high school teachers who have taught a second year course in chemistry. A large number have made application for admission to the institute. Several nationally known chemistry professors, from various universities, will be here for one week each to lecture.

The department has a 1962-1964 N.S.F. matching grant for $27,000 for the purchase of items of new equipment, such as UV, IR, and polarographic instruments.

Upon invitation of the Research Corporation, the college has submitted a proposal which, if approved, will add another Ph.D. man to our five member staff, four of whom now have the Ph.D. degree. The grant

Chemistry Department Report (cont.)

will also include funds for travel to scientific meetings by staff members, student scholarships, visiting scientists, and student research.

A new building is under construction for the departments of mathematics and physics. This will provide much needed room for chemistry in the present building for offices, research rooms, and additional laboratories.

The library, located within the chemistry building, includes the journals published by the American Chemical Society including all of the chemical indexes. It also includes a complete set of Beilstein as well as many other reference works, such as Organic Synthesis, Heilbron, Organic Reactions, etc.

Efforts have been, and are being , made to strengthen and broaden the scope of our undergraduate program of instruction. This is due to the better preparation of the high school students who come to the college, and also due to the new course requirements set up by the accrediting committee of the American Chemical Society.

First year chemistry and qualitative analysis have been combined into a single one-year course. The addition of another Ph.D. member to our faculty staff will enable us to add a course in instrumental analysis. More instrumental methods will be included in our advanced analytical, physical chemistry, and special problems (honors) courses.

Our staff will have more time for independent research with many more of the better students taking part. Each staff member is encouraged to obtain a Research Corporation, Petroleum Research Fund, National Science Foundation, or other grant. Although we are essentially an under-

<u>Chemistry Department Report</u> (cont.)

graduate liberal arts college, and teaching is our main job, we believe

that some research by staff members is necessary.

CLASSICAL LANGUAGE DEPARTMENT REPORT
Edward Wolters, Chairman

There are two faculty members in the field of Classics: Dr. Joseph
Zsiros, Associate Professor of Greek, Ph.D., University of Debrecen, Hungary;
and Edward J. Wolters, Professor of Latin, A.B., Hope College, A.M., University of Michigan, subsequent study at University of Michigan and in Naples
and Rome under the auspices of the Vergilian Society of America.

Within the last couple of years, the Greek Department has expanded
its offerings so as to make it possible to offer a classics major with a
strong core in Greek. At the present time, most of the students who take
Greek are those who plan to enter the seminary. It is hoped that more and
more liberal arts students will be attracted to the courses in Greek.

The Latin Department has for some years made use of the language
laboratory. There are approximately 100 different tapes available, most of
them made by Mr. Wolters. Currently the use of the laboratory is optional
with students. This is partly because of opinion on the part of the instructor, partly because of crowded schedules, and partly because of the difficulty of experimenting when there is only one member on the staff of a
department.

During recent years, the college administration has allocated
rather generous funds for library purposes. For a small library, the number
of books in the classics has always been relatively large. Just now we are
in the process of building a substantial curriculum library for prospective
teachers of Latin in the high schools.

A chapter of Eta Sigma Phi, national honorary classical fraternity,
was established at Hope College under the auspices of the departments in 1958.
At the present time the chapter has 20 active members and nine honorary members

Possibly a special weakness of the Greek and Latin departments is the fact that students working for majors must do so much course work under the direction of a single individual. Another would be that the number of courses is necessarily quite limited. A student who comes with four years of high school Latin finds it rather difficult to obtain the required number of credit hours to secure a major. Greek has a total of 20 hours offered; Latin has 43, 27 of which are above the intermediate level.

ECONOMICS AND BUSINESS ADMINISTRATION DEPARTMENT REPORT
Dwight Yntema, Chairman

The Department of Economics and Business Administration provides academic coverage in the liberal arts tradition in one of the college's social studies areas. In meeting its responsibility, the usual two-fold obligation applies. On the one hand, the department must supply to the college community generally a meaningful introduction to the field along with such additional offerings as students majoring in other areas may require. It must also provide the necessary sequences of study for its own majors, be they in economics or in business administration. Its majors are of two types: those who upon graduation will pursue further study in professional or graduate schools, and those who are terminally oriented and will, for the most part, locate in business jobs after leaving college. At present, more than one-third of the students at Hope College take at least one course in economics or business at some time during their college careers -- about one-half of the men and roughly one-eighth of the women. The majority of these take the first semester course in Principles of Economics. Department majors account for about one-eighth of the upper-class men; very few women major in the field.

In carrying out its duties, the department has proceeded seriously and conservatively. Only minor changes in course offerings seemed needed in the past decade; attention turned more to enrichment and up-dating of existing offerings. The relative number of departmental majors has moved upward rather slowly as is also true of the non-major group electing courses in the area. This stability points to a well-structured foundation that should prove useful in dealing with demands of the future. Thus,

Economics and Business Administration Department Report (cont.)

course enrollments seem likely to rise considerably, a result of overall growth in college admissions in combination with some rise in relative numbers taking department courses. A factor that may contribute significantly to the latter is the imminent shift to the new "core curriculum."

Coping with larger enrollments occasioned by more departmental majors should not prove especially troublesome. It is the introductory offering for the general student that may cause trouble. Larger numbers here will complicate staffing problems because of the special qualifications required of teachers. Larger numbers may also give rise to perplexing questions of course content, especially if the newcomer group proves essentially different in interests and abilities. For example, will we end up with one principles course suited for prospective majors and for others something different yet suitable for "core" purposes? The question may demand careful attention in the future.

Likely to prove more recalcitrant is a different type of problem. This is concerned with serving certain non-major students, often terminally oriented, who may neglect to prepare significantly for job taking after college. There is need for ready access to, and adequate counseling concerning, sequences of economics and business courses that are designed to meet the special needs of these students. Interdepartmental cooperation, and preferably, considerable experimentation will be required. Much remains to be done here.

The past decade has witnessed substantial professional accomplishment on the part of department staff members. Two of the department's three regular staff members completed their doctoral studies and were awarded

403

their Ph.D. Degrees (in business administration). In addition, staff
members conducted two research studies of considerable magnitude for the
Michigan State Legislature, one dealing with unemployment compensation in
the state, and the other with the comparative burden of Michigan's taxes
on business. At the present time, however, there is no comparable major
undertaking in process, so that full attention is now being accorded teaching
duties. The shifting to and from professional research has been accomplished
with substantial success.

EDUCATION DEPARTMENT REPORT
John Ver Beek, Chairman

Strengths in Our Present Program.

Much of what is good in our present department and program stems from the self-study made prior to the visit of N.C.A.T.E. in 1961. As a result, we were granted provisional accreditation for three years. Recently we were revisited, after having made additional changes in our program, and we now have good reason to think that we will receive full accreditation.

We have a better system of teacher selection, including improved standards for admission to teacher education, and a more complete system of record keeping for these students.

With the addition in 1962 of Dr. Douglas Duffy to our Education faculty, we have substantially strengthened our staff and have improved our supervision of the student teaching program. In the fall of 1963, we increased the number of school systems in which student teaching is done to six, whereas a few years ago practically all of this occurred in Holland.

Improvements in the general education requirements at Hope now make it possible for us to send out teachers with a better liberal arts background, and teaching majors have been reviewed in the process.

A Teacher Education Committee, composed of representatives from all the disciplines, plus two senior Education students, now serves as an agency to acquaint the total college with the teacher education program, as well as initiate changes in policy and program.

Teacher placement has been combined with other college placement, and is now administered mostly by one person, with somewhat better facilities for interview than formerly.

405

The department continues to attract a good quality of students into teaching, as is evidenced by the number of senior Education students that graduate with honors and do well in the profession.

Limitations of the Department

Though our placement services have improved somewhat, we have difficulties in coordination, because the department is now located in three areas. Our main office has secretarial help mornings only, and the placement office is too far removed for easy interchange of secretarial services, or contact with department personnel.

The increased number of students to be certified and placed has caused us to ask whether the time has not come to think of revising our materials and procedures, so that electronic equipment can be used to facilitate the duplication of materials.

Increased enrollments and improved programs likewise call for additional staffing, so that more adequate supervision of student teachers may be possible, and so that there can be some follow-up of our graduates to evaluate our present program. Some adjustment may also be necessary in the other disciplines, so that their personnel will have time to assist in the supervision of student teachers. To maintain good working relationships with the public schools in this area, it is important that adequate time be allotted to the staff for these purposes.

As we face the future, the department is studying such problems as:

1. Shall we enter into the pattern of a professional semester, including a full-time student teaching internship, locally or in distant places?

2. Does the department have something to contribute to those who are planning to teach on a college level?

Education Department Report (cont.)

3. In view of the stress on a five-year program in teacher preparation, what
   is our responsibility for the fifth year on our campus?

4. What changes in the Education curriculum and staff are advisable to equip
   our beginning teachers more adequately for the tasks that will be expected
   by school administrators of tomorrow?

ENGLISH DEPARTMENT REPORT
Clarence DeGraaf, Chairman

Strengths:

1. The English Department is staffed with ten full-time and five part-time teachers. Six of the ten full-time people hold doctorate degrees and two more are expecting theirs soon.

2. There is enough variety of special fields among these fifteen teachers to provide a well-balanced program in literature study.

3. The department has two required courses in the core curriculum, six hours of composition and six hours of world literature. This gives us an opportunity to lay a solid foundation for a continued interest in literature and is one of the causes for the heavy run of English majors on our campus.

4. Special attention is given to slow learners and to superior students. For slow learners, we have a non-credit course in Fundamentals of Writing; and for the superior students, we have a program of honors sections and seminars running up through the Senior year.

5. The English major is required to have 30 hours of English including the required hours, but those planning on graduate work will often run 35 to 40 hours.

6. In 1963 and 1964 we required the Graduate Record Exam for all Senior English majors.

7. All English majors are required to take the Survey of English Literature in their last year. This serves to tie together the entire program of period type, and single author courses.

English Department Report (cont.)

8. The department provides for the training of 30 to 40 secondary English teachers each year both in content and in methods of teaching English.

9. The department is well supplied with library materials for the study of literature.

10. The department requires 16 hours of a foreign language for all majors. This is still an excellent discipline in language.

Weaknesses:

1. The department needs to expand the writing program, and for this purpose we have engaged a full-time man specially trained in creative writing.

2. The department is running too many large classes, above 40, to do the best job of teaching. This can be remedied by finding more classrooms and more staff.

3. The department is trying to find a more meaningful body of knowledge around which to center the composition program. This has been wide open to teacher's choice, but now we will select major questions involved in the liberal studies and all concentrate on these with our writing assignments.

4. The department feels the need, especially for secondary teachers, to do more with the history of language (i.e., morphology and linguistic change The course in "The History of the English Language" will be expanded to include more grammar with the emphasis on the teaching of language.

409

GERMAN DEPARTMENT REPORT
E.F. Gearhart, Chairman

Strengths

Personnel

There are four faculty members in the German Department, two of whom are native speakers. One has a Ph.D. degree; the others have the M.A. All of them are enthusiastically committed to teaching. The department is also staffed each year with a native female student who works as an assistant in the language laboratory and department and resides in the German House (for women).

Curriculum

In addition to basic language courses, the department offers a major in German and has had its students accepted in various graduate schools throughout the country. Many have been recipients of assistant-ships and scholarships.

The course offerings beyond intermediate level are limited to six courses in literature for a total of nineteen hours. Three additional courses afford six hours of advanced language study. The courses are so arranged that three basic courses are given every year in a sequence which prepares students for more advanced study. The remaining courses are offered alternate years. Provision is made for independent study, and the Vienna Summer School and Fall Semester Program present German majors the opportunity for European travel and the stimulation of different instructors and courses.

The advanced work is taught by three members of the department, and all three collaborate in the teaching of two of the courses. Thus, students encounter a difference of opinion and method within their own department.

All majors planning to teach or go on to graduate study are urged to work as departmental and language laboratory assistants in order to gain experience. Enrollment in three advanced courses last semester was seventy-five. This semester it is fifty.

Facilities - Language Laboratory

The beginning language instruction is based on an integrated audio/lingual approach combined with emphasis on rapid reading. A modern language laboratory under the supervision of a native speaker and major students provides excellent programmed material for all.

Library

The German holdings are excellent in basic reference works, literary histories, criticism, and periodicals. There is a good selection of works by individual authors, and the collection is growing rapidly as the budget allows.

German House

This housing unit cares for ten girls who are interested in speaking German. German newspapers, periodicals and other reading are available. The house serves as the meeting place for the German Club, and the residents provide leadership for the German Table in the dining hall and inspiration and stimulation for other students of German.

Limitations

Curriculum

Most students enter our program without previous language study. As a result, the number of advanced courses offered must be restricted to enable the student to take them during his course of study. Contacts

411

German Department Report (cont.)

are being made with high school German departments in an effort to recruit
students qualified to take advanced courses earlier in their work at Hope.
Present offerings could then be expanded.

In conjunction with the Department of History, efforts are being
made to develop a more challenging program of independent study for students of German enrolled in the Vienna Summer School.

Requirement

The reduction in the language requirement to one year may adversely
affect enrollment in advanced courses. Students normally become interested
through the reading done in second year work. The department has no control over this requirement.

HISTORY DEPARTMENT REPORT
Paul G. Fried, Chairman

The function of the department is a dual one. It must provide instruction to a large number of students who need at least one introductory course in the field as part of their liberal arts requirement. For the last several years, this has meant enrollment of 400 to 500 students per semester. Under the new curriculum just adopted, this number will increase considerably. At the same time, the department must offer a solid majors program for an average of twenty to thirty students on both the junior and senior level.

On the whole, the department has done an adequate job in meeting both objectives. Until recently the bulk of the non-major students were enrolled in the Freshman European Survey and the Sophomore American History courses. Honors sections in both courses are available for potential majors and exceptionally able students. In the last two or three years, however, there has been increased registration for our middle level courses -- between 50 and 60 enrollees in a number of them.

Although the department welcomes the opportunity to be of greater service to the college community as a whole, we recognize that we need to give majors enrolled in these courses additional training either through upper division honors work, pro-seminars, or some provision for independent study and research. In particular, we are aware that our history majors need to do much more writing than can fairly be required of them in large lecture courses which preclude careful reading and comments by the instructor.

History Department Report (cont.)

The present four man history staff is well balanced in regard
to fields of special competence. The range of courses in American,
European, and Russian history is well taught. In addition, Mr. Wolters,
Chairman of the Latin Department, offers courses in ancient and Roman history, and Dr. Van Putten, Chairman of the Political Science Department,
teaches three courses dealing with the history and politics of Asia.
Unfortunately, both men are approaching retirement.

The most obvious gap in our course offerings is in the field
of non-Western history. "Readings in Non-Western History", introduced
last year, provides a partial solution to the problem. So does the opportunity to participate in the newly developed non-West study programs
(Tokyo, Bogota, and Beirut) available through the Great Lakes Colleges
Association. The department urges majors to take advantage of these and
other foreign study opportunities. The visiting scholar programs planned
by the Great Lakes Colleges Association will also be valuable. Nonetheless, the addition of staff persons well trained in non-Western areas
is almost mandatory.

Another major need of the department relates to the limited
library resources, especially of materials which can be used for seminars
and independent research. This may, in part, be met by a more aggressive
search for local history materials begun during the past two years. At
the same time, the library and the department need to be more systematic
in seeking to acquire special collections of original documents, printed
source materials and microfilms.

Despite the problems indicated above, an increasing number of
high calibre students have been attracted to the department in the last

several years. The most significant change has been in the development
of new climate of opinion in favor of serious scholarship and graduate
study.

MATHEMATICS DEPARTMENT REPORT
Jay Folkert, Chairman

During recent years there has been rapid change in the mathematical world. The large increase in mathematical development and the expansion in mathematical applications during the twentieth century have both affected mathematical education. At the secondary level, programs aiming to improve instruction have had the effect of raising the level of preparation of students entering college. At the graduate level, more preparation is being demanded of students who begin graduate study in the field.

In light of the above developments, the Department of Mathematics has been challenged to keep its program up to date and to meet the needs of the students. To accomplish this, curriculum changes have been made. Intermediate Algebra and Solid Geometry are no longer taught for college credit. The traditional programs of College Algebra, Plane Trigonometry, Analytical Geometry and Calculus have been modified, so that more material is completed at an earlier point in the student's academic program. Courses in Differential Equations and Mathematical Statistics have been enlarged. At the upper level, Modern Algebra and Linear Algebra have been introduced and Advanced Studies in Mathematics has been expanded.

For the liberal arts student, a two-semester course, Fundamentals of Mathematics, has been added. This course emphasizes the nature and structure of mathematics and its purpose is to give the non-major an understanding of the field. Because mathematics is so fundamental to all areas of life, one semester of this course will be a requirement for graduation for all students who start college next fall if they do not include the study of mathematics as part of their major program.

416

<u>Mathematics Department Report</u> (cont.)

In order to effectively teach material of greatest mathematical importance, Surveying, Aerial Navigation, Mathematics of Business, and Spherical Trigonometry, have been dropped and Astronomy has been transferred to the Department of Physics. Furthermore, the department has endeavored to place beginning students properly by means of proficiency tests given during orientation week. Thus students who are well prepared can continue to work in challenging areas of study. Moreover, in-course honors work was conducted this year for students who showed unusual ability. This took the form of a weekly seminar in which students discussed solutions to problems that were more difficult than those in routine assignments.

While the department's program is a good one, some weaknesses remain. Since preparation for graduate study in mathematics is becoming increasingly demanding, more work in geometry, topology, and complex variables should be offered. Presently a semester course, College Geometry, is being offered. Topology and complex variables are introduced in Advanced Studies in Mathematics. Moreover, the offerings in algebra and real analysis are minimal. Expansion in the above areas will need to be considered in the near future.

The staff of the department is diligent, very cooperative and dedicated to teaching. Nevertheless, there is a need for help at the upper levels. Diligent efforts are being made to add a person with a Ph.D. in September, 1964. An additional person of like caliber should be added a year or two later.

Facilities for the department will be greatly improved next year. The new physics-mathematics hall will be occupied this summer. Thus, individual offices and additional classrooms will be available.

<u>Mathematics Department Report</u> (cont.)

Discussions are underway to provide electric desk calculators and a high-speed computer for instructional purposes. To date no decision on these has been made.

In summary, the department has modified its program considerably during the last ten years in order to meet the needs and levels of our students. It will continue to examine its program for further improvement. Since a broad spectrum of students is served by the department, the problem of meeting individual needs will remain important.

MUSIC DEPARTMENT REPORT
Robert Cavanaugh, Chairman

One of the outstanding strengths of the department is that it offers, both through certain courses and participation in performance, opportunities for musical training and involvement to all students on the campus. This is as it should be at a liberal arts college of this size. The high quality of the present staff (and of James Tallis, to be added next year) contributes markedly to this feature.

Another strength is the momentum of the past, in that the qualitative and quantitative growth of the department and its activities has constantly increased interest and support among the constituency (both of and out of the Reformed Church in America). The other point here is that this could become a major weakness if not constantly encouraged.

A third strength is the fact that the college administration has had the foresight and interest to enable the department to offer a sizeable music major program under the A.B. degree -- a program larger than that offered by most colleges of similar character.

While this third strength generally has met the needs of the student who wishes to major in music, increasing enrollment suggests a growing problem in the music major area. The number of students who wish semi-professional training in music to prepare for graduate schools of music and a career in performance, while rather small in the past, is certain to grow. If the college has a commitment to these students, the dual training (theory and performance) in music must be increased. The music department at present is engaged in a study of several possible solutions to offer as suggestions.

Music Department Report (cont.)

Lack of physical space for the work of the department is a growing problem. Present listening facilities are inadequate, and with a music introductory course now in the general curriculum, a large room with laboratory equipment (turntables, earphones, etc.) is needed urgently. The problem of storage space for equipment, especially instruments, stands, etc., is acute. One, or preferably two, additional studios will be needed as staff needs increase within a few years, and more office and library space will be required in that time.

To sum up, the department is keenly, sincerely, and devotedly dedicated to the liberal arts tradition at Hope College, but it is also interested acutely in its responsibility to the student who wishes to make music his life's work.

PHILOSOPHY DEPARTMENT REPORT
D. Ivan Dykstra, Chairman

The following steps have been taken to up-grade the work of the Philosophy Department:

1. Curriculum. A curriculum revision has been approved by the Educational Policies Committee to permit the department to fulfill more acceptably its triple responsibility, for providing an undergraduate program to prepare students for eventual professional work in Philosophy, for contributing to the development of the liberally educated person, and as a service department for the other major disciplines of the school.

   By this revision, the course in Contemporary Philosophy becomes two courses: 19th Century Philosophy, and 20th Century Philosophy, to provide opportunity for study in greater depth for contemporary movements of thought. Courses in Epistemology and Metaphysics have been added. A course in Oriental Thought has been added. The seminar program has been listed in a different fashion to permit greater flexibility in approach, particularly with a view to making possible more individual work by major students. The existing Philosophy of Science course has been split into Philosophy of the Natural Sciences and Philosophy of the Social Sciences, to permit more intensive work in both areas and to give the department's offering greater relevance to the social sciences. A course previously in the Sociology Department, History of Social and Political Philosophy, will be now offered in the Philosophy Department. Courses in Philosophy of History and Philosophy of Language are to be requested for 1965-1966.

421

2. Due to the late resignation of a staff member in the spring of 1963,
   and a failure to find an adequate replacement, the department has
   been seriously understaffed. For 1964-1965 there will be three staff
   people: one with his responsibilities centering in History of Philosophy
   and Introduction to Philosophy; a second (with partial responsibility
   in the Religion Department) having his special concentration in
   Philosophy of Religion and Ethics; and a third, with primary responsibility for Logic, Philosophy of Language, Epistemology and Metaphysics. With this staff, we will be able to provide more opportunity
   for individual work. The three will offer each semester a joint
   seminar for major students.

3. Improvement of Instruction. More use will be made of qualified
   student assistants in Introduction and Logic, to permit more intensive
   assistance to the individual student. We are committed to the principle that wherever students do assist (e.g. discussion group leaders),
   this will be in addition to the announced contact hours between the
   student and a professional staff member.

   Examination procedures are in process of revision and intensification
   in the expectation that these will both provide clearer information
   about the student performance and a stronger incentive for the student
   to be more exact in his pursuit of philosophical inquiry. This is
   to exert a greater degree of direct discipline on that part of the
   student population which still refuses to do what he expects he will
   not be examined on.

   For each course in the department, there will be a spelling out of
   the exact purposes at whose achievement the course will be aimed,

422

and the examining will be done always with a view to determining how
well the course has achieved what its taxonomy of objectives has said
it would try. These purposes will be made the subject of departmental
discussion and where relevant (e.g. in the Philosophy of the Natural
Sciences), of interdepartmental discussion with the departments who
have an interest here, and these will be regularly reviewed.

Major students will be required to perform satisfactorily in a comprehensive examination in the History of Philosophy, to be given
at the beginning of the final semester. This is to be on the basis
of a selected list of major philosophical writings, only some of
which will be included in the regular course work. The student
will be provided such a list at the time he declares his intent to
become a major, with the expectation that he will make use of vacation times to continue his reading in the field.

Where students undertake individual research projects, these will be
henceforth required to be of a very specific sort, to permit the
student to do a very exact job within the limits of a single semester.
This is in contrast to past procedures, where on occasion, students
simply worked at achievement of a general understanding of some
individual's philosophy.

4. Miscellanies. Particularly in the Introduction course, experiments
will be undertaken with a taping of lectures. The course is a large
lecture course with added student-led discussion sections, so that
students will be able to listen for a second time if need be. Also,
the lectures, now given in an informal fashion, can be carefully
analyzed to determine the extent to which they may be expected to

accomplish what they undertake to accomplish.

We may move to a point in the next two years of requiring each major student to "adopt" a major philosopher, with the idea that he will throughout his upper college years make this philosopher's work his special project.

PHYSICAL EDUCATION DEPARTMENT REPORT
Laurence Green, Chairman

The major objectives of the Hope College Physical Education program
are:

1. To improve the physical condition of students to a level which will
   enable them to do a reasonable amount of work per day and still have an
   adequate reserve left at night.

2. To teach students how to maintain good physical condition and how to
   regain it if they should lose it. (This is partly a matter of knowledge
   and partly a learning of techniques.)

3. To impart a reasonable knowledge and adequate command of pertinent basic
   physical skills in selected sports to the end that students may have the
   tools for participation in conditioning activities for after school years.

4. To help develop a desireable attitude toward daily living in our society.

     Several factors have limited the achievement of the above objectives
in our required physical education, intramural activities and intercollegiate
athletic program. These limitations are:

1. Only one gymnasium is available on the campus for use by the entire
   student body.

2. No space or equipment is available for the teaching of apparatus and
   gymnastic activities.

3. Space for outdoor activities in required physical education is limited
   to one field on the central campus.

4. Swimming facilities are not available for use on the campus.

     The real strength of our program is largely enhanced by a dedicated
and well qualified staff. The staff is strong because:

Physical Education Department Report (cont.)

1. It recognizes the importance of participation as the real opportunity for students to learn.

2. It believes that where conflict occurs between athletic and academic pursuits, the academic interests shall prevail, short of complete elimination of the athletic program.

3. It has excellent training for teaching physical education.

During the coming academic year, the program will be improved in the following ways:

1. Academic credit will be given for required physical education activities. Texts and syllabi will be used. A more thorough testing of physical fitness and knowledge of activities will be employed. Expanded opportunity for participation in the program of activities for upper-class students will be offered.

2. The limited use of the National Guard Armory as a second facility will be expanded to include opportunities for more intramural competition.

3. Tumbling and wrestling will be added as activities available to students because we will have more equipment and more space.

4. Through the cooperation of the West Ottawa Public Schools and the Y.M.C.A. we hope to expand the opportunity for student participation in swimming and life saving activities.

5. Limited intercollegiate competition in swimming and wrestling will be offered next year if student interest warrants it.

6. The new Van Raalte campus will be available for extensive outdoor intramural and intercollegiate activities.

Physical Education Department Report (cont.)

We will continue to offer a strong minor in physical education
for students wishing to teach physical education activities after graduation
from college. This minor will include a minimum of twenty semester hours
of required courses. If the student desires to go into full-time physical
education teaching, we will encourage them to take all the courses we offer
and will recommend that they strive for a masters degree program in physical
education after graduation from Hope College.

PHYSICS DEPARTMENT REPORT
Harry Frissel, Chairman

Relative to the size of the school, the Physics Department of Hope College is small. For the present school year, the staff was raised to three full-time members. Since only half of one floor of the Science Hall is available to the Physics Department, the department finds itself seriously handicapped in facilities for laboratory work and project work at the upper-class level.

A comparison of our curriculum content with the recommendations of the Ann Arbor Conference on Undergraduate Curriculum indicates that about 95% of their recommendations on course content are included in our curriculum. However, our advanced courses are given only on alternate years. This makes it impossible to work out a definite sequence of material and students enter these courses with varying backgrounds and varying instructional needs. A rearrangement in course structure, when permitted by student enrollment and staff, would make for more thorough and efficient curriculum.

The department offers forty semester hours towards its major program. At present this is all crowded into the last three years. This situation also makes it more difficult to follow a sequential pattern.

A number of changes are imminent. If an invited proposal to the Research Corporation is successful, the physics staff will be increased to four. The purpose of this increase is to reduce loads so that faculty can keep abreast in their field and can also become engaged in research activities involving students.

A new physics-mathematics building is now being erected and will be available next fall. This building includes, besides the usual complement

<u>Physics Department Report</u> (cont.)

of laboratories and classrooms, four project rooms for students, four laboratories for faculty, and an adequate workshop. Equipment for use in this new structure will be greatly enhanced if a National Science Foundation proposal for a matching grant of $34,000 is successful.

Changes in curriculum that go into effect next fall are (1) an increase by one semester hour each in the Thermodynamics course and the Electricity and Magnetism course, (2) an extension of the general Physical Science course for non-majors to two semesters and the inclusion of a laboratory into the first semester. In addition the department expects to recommend (1) that the General Physics course now offered in the sophomore year, be made available in the freshman year, (2) that a buffer course, incorporating mathematical tools and physical descriptions common to several advanced courses, be injected between the General Physics course and all advanced courses; this buffer course would replace the present Physical Optics course.

As soon as the transition to the new building has been accomplished, the department looks forward to incorporating more independent project work, research participation, and honors programs into the curriculum and to the organization of a journal club.

429

POLITICAL SCIENCE DEPARTMENT REPORT
James Dyke Van Putten, Chairman

The Political Science Department at Hope College, the newest of
the college's departments, has as its major task the teaching of American
National Government to approximately one hundred students each semester.
These students are from all of the departments on the campus. This course,
we think, is well handled here, since there are no restrictions politically
on the course content. The instructors seek to get the students to read
from a variety of sources embracing all points of view -- from liberal to
conservative -- in order that they may obtain a more rounded political
background as a part of their liberal education.

The number of students who major in political science approximate
ten each year. A series of courses that alternate every other year makes
available to the majoring students a well-rounded study of politics from
the local level to the international level. The two instructors handling
these areas have their training one in the local and state and in theoretical
aspects of politics, the other in the administrative and international aspects.
They complement each other well. Both instructors have also had practical
experience in local, state, national and international aspects of politics,
from holding local political positions to serving in the Foreign Service and
working for other governments around the world. The tie-in with the Washington
Semester Program of the American University provides another opportunity
to majors for breadth and depth.

The Political Science Department has had good success since becoming
a separate department in its placement of its majors. They have been accepted
in graduate schools around the country, in the foreign service, and in
appointive positions in local, state, and national positions.

     The limitation that is most evident in the department is in its paucity of instructors. This means that each of two instructors must teach three different courses each semester, with most of the courses not repeated until two years later. As a result, an instructor must spread himself quite thin.

     In order to have a better department, it should be made possible for an instructor to specialize more than at present. This means that at least one additional person should be added to the department in the near future. If research is to be expected, and well it should be, the teaching in so many different areas each semester must be curtailed.

PSYCHOLOGY DEPARTMENT REPORT
Robert De Haan, Chairman

Strengths of the Psychology Department

The past decade has been for the Psychology Department a period of steady, solid growth. Growth has occurred on a number of dimensions. These are as follows:

1. Staff has increased from one and one half (1953-1954) to three (plus occasional part-time help) planned for 1964-1965.

2. Library holdings have doubled in the past five years and probably tripled in the last decade. The department subscribes to 29 journals in psychology.

3. Limited laboratory facilities have been developed. These are for teaching and research pruposes. They consist of standard laboratory facilities for teaching Experimental Psychology and a small group-research laboratory.

4. Limited research has been conducted both under grants from the government and on the initiative of the psychology staff.

5. The number of majors has increased steadily to its present level of 20 to 30 seniors.

6. We have acquired a psychology building that, up to the present, has been adequate for housing both offices and laboratory facilities.

7. The course offerings have heavy enrollment.

A second major strength of the Psychology Department is its basically sound curriculum. The curriculum is grounded in the liberal arts tradition. Psychology courses are taught with both a humanistic and a scientific emphasis. Moreover, in keeping with the stance of the college, they are taught within the context of the historical Christian tradition.

432

Psychology Department Report (cont.)

The curriculum strives for balance in that it is presented
to a broad range of students from those who take only one course in
psychology to those who have a large number of courses and are headed
for graduate study in psychology.

Limitations of the Psychology Department

One major weakness in the curriculum of the Psychology Department
grows out of a shift of emphasis in psychology over the past ten years.
There has been a rebirth of interest in physiological psychology as a
result of some outstanding scientific breakthroughs made in the past
decade. This re-emphasis on physiological psychology has not as yet
been reflected in the curriculum.

A second weakness is relative inadequacy of departmental facilities.
A major growth spurt has been achieved these past four years with the
acquisition of a staff person trained in Experimental Psychology and with
the laboratory facilities that he developed. These facilities are becoming more inadequate each year, however, particularly in that they
have no provision for Physiological Psychology.

The third weakness is also relative. There is a growing need
to provide opportunities to study psychology in depth for good students
going on to graduate school.

Current and Projective Plans for Overcoming Limitations

The Psychology Department is engaged at the present time in
a study of its curriculum with a view to presenting recommendations
in the year 1964-1965. The revisions will point primarily toward strengthening the offerings in the scientific aspects of psychology, particularly

with the addition of courses in physiological psychology, and with the addition of a senior thesis or opportunity to do an honors paper for superior students who plan to go on to graduate work. Plans also include provision for individual faculty members to develop honors level seminars for selected psychology majors. These seminars provide a means to encourage in-service growth -- possibly leading to publication -- for faculty; and study in depth for the student. Emphasis will be placed upon student initiative and independent study.

A second step being projected is the acquiring of a second person in Experimental Psychology with special competences in Psysiological Psychology.

A third projected step, depending upon a number of decisions made at the administration level, is the acquisition of expanded facilities for doing experimental work and teaching experimental psychology. Again, the emphasis would be upon Physiological Psychology, but other areas of experimental psychology would be expanded as well.

RELIGION AND BIBLE DEPARTMENT REPORT
Henry Voogd, Chairman

Studies in the Department of Religion and Bible are regarded as an integral part of the college curriculum, and a significant part of the efforts of the staff is devoted to teaching courses that form the required program for all students. The requirement includes Biblical studies and a course in the Philosophy of Religion.

Strengths of the Department.

A. The major program in Religion is a comprehensive one that embodies a minimum of thirty hours in the department plus substantial blocks in all areas of study recommended by the American Association of Theological Schools. These include Language, English, Philosophy, Psychology, History and Speech. The strength of the major program is revealed in that a significant number of students majoring in the department have gone on to achieve distinctive marks in graduate schools of theology.

B. The teaching staff consists of four men who are ably qualified and prepared for their teaching professions. Two possess doctorates (Princeton and Edinburgh), one a Master of Theology degree (Oberlin), and the fourth is a candidate for the Th.D. degree (Columbia). The specialized studies of the faculty members complement eachother. These areas include Old Testament and Christian Ethics, Historical Theology and the Philosophy and History of Religion.

C. The course offerings represent breadth and depth, embracing Biblical, historical, philosophical, theological and ethical content areas.

Weaknesses of the Department.

A. Non-western Studies. In terms of the challenge to education in our day, insufficient opportunity is afforded students to receive in the department

435

Religion and Bible Department Report (cont.)

an adequate knowledge and understanding of the religious life and thought of the non-Western world. To overcome this weakness, the department has prepared a prospectus for two courses in non-Western religions:

1. Religions of the Far East. (Animism, Confucianism, Taoism, Shintoism, Hinduism and Buddhism.)

2. Religions of the Near and Middle East. (Religions thought in Sumer, Akkad, Canaan, Egypt, Arabia, Iran, Greece and Rome.)

B. Teacher-Student Ratio. The members of the religion staff have a keen interest in students as persons. The collective aim at all times is:

a. to develop in the individual student a mind sensitive to values -- aesthetic, scientific, and historical.

b. to develop a degree of inner consistency in the individual student's total value structure.

c. to prepare the individual student to arrive at cogent decisions on crucial issues of contemporary life.

The average student-teacher ratio at the present time in the department courses is so high that it is difficult to deal with students as persons. The department has recommended the securing of an additional staff member and the augmenting of the library collection of primary and secondary sources on non-Western religions to alleviate the two weaknesses cited above.

ROMANCE LANGUAGE DEPARTMENT REPORT
Ralph Perry, Chairman

Present Strengths of the Department

Curriculum.

1. We have recently inaugurated a "two-track" system in the second semester of intermediate Spanish and French: One track provides rather advanced readings in literature, for prospective language majors and other interested and serious students; the other track gives readings in the history and civilization of the foreign countries involved, basically for those who consider this as a terminal course in the language.

2. There is a great deal of emphasis on oral work at all levels of instruction. Most intermediate and advanced courses are conducted entirely in French or Spanish.

Students.

1. The number of majors, especially those in Spanish, has been increasing rather rapidly.

2. Every year selected native French and Spanish-speaking students are given scholarship-assistantships to serve as resource persons for the French House and the Spanish Corridor on campus.

Teaching Materials and Equipment.

1. Up-to-date textbooks in use in our elementary and intermediate classes in French, Spanish, and Russian especially stress the latest audiolingual methods.

2. Magnetic tapes are used extensively in the language laboratory with most French, Spanish and Russian elementary and intermediate textbooks.

3. Library holdings in several areas of French and Spanish -- for example, in contemporary French literature -- are more than adequate.

<u>Romance Language Department Report</u> (cont.)

<u>Extra-Curricular Enrichment.</u>

1. We have on campus a French House and Spanish Corridor, where selected women students reside and make active use of both of these languages in every-day conversation.

2. There is a local chapter (Gamma Mu) of the French national honorary society -- Pi Delta Phi -- at Hope College.

<u>Present Weaknesses and Proposed Remedial Action</u>

<u>Staff.</u>

The distribution or ranks in the department -- two professors, one visit-ing professor, and three instructors -- is rather uneven, with no assistant or associate professors. It is hoped that this situation may be remedied before long through promotions or additional appointments.

<u>Curriculum.</u>

1. Our present course offerings in advanced Spanish are not as complete as those in advanced French. To remedy this deficiency, we expect to offer, beginning in 1965-1966, a two-semester survey of peninsular Spanish literature, a course in Spanish and Hispano-American culture and civili-zation, and an <u>additional</u> semester course in advanced Spanish grammar and composition.

2. A possible future innovation (within two years) is a course in general linguistics to be taught by a staff member of our department. This course would be open not only to students in Romance languages, but also to those majoring in German, Classical languages and English.

<u>Students.</u>

We believe that our present requirement for a French or Spanish major (25 hours above the <u>elementary</u> course) should be raised a good deal --

438

Romance Language Department Report (cont.)

perhaps to 25 hours above the intermediate course -- in order to give
our students a more thorough instruction in the language and allow them
to compete more favorably in graduate schools with those from other undergraduate institutions.

Teaching Methods and Equipment.

1. We have not been satisfied with the validity of the proficiency (placement) tests in French and Spanish which are currently being used to assign
students to classes at a level suitable for them. This fall we intend
to use for the first time the new Modern Language Association proficiency
tests, which measure achievement in four different language skills
(understanding, speaking, reading and writing).

2. At the present time we are making a conscientious endeavor to fill in
significant gaps in library holdings of books in both French and Spanish
(including Hispano-American) literature.

SOCIOLOGY DEPARTMENT REPORT
Earl Hall, Chairman

Strengths:

1. Good coverage in terms of comparison with a wide variety of other
   schools' offerings in sociology and as evaluated by the chairman
   of the sociology department at the University of Michigan.

2. Small (ten students or less) upper division classes where intense
   communication and extensive seminar discussions are possible.

3. Opportunity for majors to learn about the discipline as assistant
   graders and through small-scale, course-oriented, library research.

4. Two-man department with interests balanced between theory and research,
   very effectively supported by a part-time instructor in social work
   who has his MSW and years of experience as an agency head.

5. Excellent communications with and support from psychology and philosophy
   departments (philosophy of social science is to be offered). The
   department is also able to draw on a good statistics course in the
   mathematics department.

6. Good communication channels for faculty with others outside Hope in
   the field of sociology.

   a. Hope and Calvin College sociology departments meet together once
      a month for an evening's unstructured discussion of sociology
      and college teaching.

   b. Participation with the University of Michigan, Kalamazoo, Albion,
      and Alma in the Michigan Scholars in Sociology program which in-
      volves several meetings of faculty representatives each year and
      one combined faculty-student seminar each year.

7. Access to data of Detroit Area Survey and Cross Cultural Index at U. of M.

440

Sociology Department Report (cont.)

8. Fully equipped small group laboratory.

9. Opportunity to send students to the University of Michigan for graduate work in sociology on a preferential basis.

Limitations:

1. Difficulty in making abstract sociological concepts real for lower level students. (not peculiar to Hope).

2. No anthropoligist on the staff.

3. Not sufficient integration of statistics course and sociology department offerings.

Prospectus to Overcome Limitations:

Limitation #1.

    a. More use of lower level student participation in descriptive research.

    b. The establishment of a summer school program in sociology by the summer of 1965, in which students will live with culturally deprived families in the inner city of a metropolitan area and attend classes in temporary quarters in the same neighborhood. This program envisions using cultural shock and an unfamiliar environment to underline the validity and significance of sociological conceptualizations.

Limitation #2

    A program is being worked out so that the anthropologist on the Calvin College staff will come to Hope twice a week to teach an anthropology course while a Hope College staff member journeys to Calvin College twice a week to teach his speciality in sociology.

<u>Sociology Department Report</u> (cont.)

<u>Limitation #3</u>.

     The chairman of the sociology department plans to sit in
on the statistics course which sociology majors are required to take,
so that he may better know the content knowledge he can expect of his
majors. Then he can more meaningfully make statistical material a
part of sociology courses, and recommend sociological material which
might be used as problem material by the mathematicians.

SPEECH DEPARTMENT REPORT
William Schrier, Chairman

A committee now has under study how the speech requirement for
graduation shall this fall be integrated into the eight-hour block of
"Introduction to Liberal Studies." The speech courses as listed in the
catalog provide offerings for an adequate speech major. Other new offerings in prospect as soon as additional staff becomes available are courses
in acting, voice and diction, and radio.

Our majors fluctuate in numbers -- 16 in 1961-1962, 21 in
1962-1963, and 11 this current school year. All who go into secondary
teaching are required to take Speech 88, the Teaching of Speech, offered
alternate years. In the interpretation-drama offerings and activities,
our present courses provide adequate minimum preparation either for teaching in secondary schools or for entering immediately into graduate work.
The same is true of our public address offerings. Within the past two
years, some eight graduates have gone on into graduate work, many on
scholarships and assistantships. The play program is conducted under
the auspices of the drama group, Palette and Masque, under the supervision of the Director of Theater. The cultural and practical values
of dramatic courses and play participation are open to all students.
A four-play program is usual, often supplemented by one-act plays, a
children's theater production, a religious drama company, and numerous
appearances of individuals and groups in plays and recitals before church
and other community organizations. We have a chapter of National Collegiate
Players for those who meet its minimum standards.

The public address area not only provides helpful courses for
the general college population with varying vocational interests -- law,

443

Speech Department Report (cont.)

3. New Full-Time Departmental Secretary. The tremendous correspondence required in arranging for attendance at tournaments and public meetings, and the mailing lists for town clientele for dramatics productions, make this need urgent. Student stenographic help just does not fill the bill.

4. Radio-Television Programmed Instruction. For many years it has been my feeling that radio should be under the auspices of the Speech Department. Whether or not educational television lends itself to the needs and resources of a small liberal arts college requires an administrative decision since it affects all subject areas and is not confined to the work of the Speech Department.

There is a Great Lakes Colleges Association project under way relating to programmed instruction. We as a department are cooperating with its work, keeping abreast of developments.

Speech Department Report (cont.)

teaching, ministry -- but likewise provides a minimum background for graduate work or secondary teaching. In addition to classroom work, Hope College takes an active part in all of the many activities sponsored by the fifteen-member school Michigan Intercollegiate Speech League, and by Pi Kappa Delta.

Recognizing the need for an audience in any real speaking situation, we are expanding as rapidly as possible town-gown debates, orations and public appearances before various civic organizations and will seek to expand these opportunities in the years to come.

NEEDS

1. New Speech Building. All of us are looking forward to the time when all speech work, with adequate office space and secretarial help, can be housed in one building rather than being separated between third and first floor Van Raalte Hall and fourth floor Science Building. The present dramatics facility is a fire-trap and wholly inadequate to our needs. The work in the dramatics area has of recent years been so outstanding that, given an adequate auditorium, I'm certain it would attract huge audiences among the town's people and be a real cultural attraction.

2. New Staff Member. The addition of planned new courses, the need for field work in Speech 88 in supervising student teaching, the normal expected enrollment increases, the expansion of work in debate and other speech and dramatic activities before groups and organizations, which in some cases amounts to directing a Speaker's Bureau, all these activities point to the need for an additional staff member in the public address field.

REPORT ON CURRICULUM AND STAFF
William Vander Lugt, Dean of the Faculty

Each department chairman has already reported on the strength and
weakness of his department. What I shall attempt is a brief general statement
of these evaluations as seen by one whose main concern is for the welfare of the
institution as a whole. Some departments feel that their greatest contribution
to the college as a whole is made when their efforts are almost wholly given
to departmental matters. They do not wish to become involved in policies or
issues outside their immediate field of interest. Their philosophy is that
if all the departments are strong, the college as a whole will show strength.
A great deal can be said for this position, and in general I am inclined to support
it. Yet there must be, on the part of several members of a faculty, an enthusiastic endorsement of extensive involvement in the development of the academic
policies of the college as a college. This takes time and energy, but this kind
of commitment is necessary if departments are to make their maximum contributions,
and if there is to be a smooth functional relationship between faculty and
administration.

Hope College fortunately has faculty members who fit these two descriptions. I think it is fair to say that in the past the latter emphasis
has been stronger than the former. Department chairmen were extremely sensitive to the development of the college as a college and did not push ahead as
vigorously as they should have with the development of their own departments.
There were exceptions -- notably the Department of Chemistry. In all honesty,
we must admit that a great deal of the excellent reputation the college enjoys
in the academic world has come through the fine reputation of the Chemistry
Department. No other department can approach the number of majors who have

Report on Curriculum and Staff (cont.)

gone on to advanced degrees. It would be false, however, to conclude from this that all departments should have followed its example. The framework provided by the institution as a whole certainly contributed to the effectiveness of this particular department and credit should be given to both the department and the institution.

Mindful of the age in which we live, we must continue to view our task as both specific and general. Our age is an age of specialization and this places great emphasis on sustained purpose. As knowledge multiplies, it must also be directed to personally appropriated worthy ends.

Policy relating to curriculum has always been determined by the general faculty. Our organizational machinery is such that all proposals for curriculum changes originate in the councils. If courses are approved by the appropriate council, they are forwarded to the Educational Policies Committee which is composed of two members of each council, the vice president and the dean as chairman. The actions of the Educational Policies Committee are circulated among the faculty and unless these actions are questioned, they become faculty action. If questioned, they are brought to the faculty for discussion.

This procedure saves much faculty time and still it keeps open the communication line to each individual member. The course offerings of each semester are determined by the department chairman in consultation with the dean. Most courses are offered annually -- some each semester and others on alternate years. There are no courses listed which have not been offered the past two years. The offerings are adequate for a major in the department with some choice as to direction of major. In this respect, we cannot,

Report on Curriculum and Staff (cont.)

of course, compete with the universities, but we also feel that the undergraduate concept of major involves a body of knowledge that is somewhat fixed and determined. Choice must be limited.

In addition to our organizational structure, there is a good deal of communication with faculty members at all levels of rank. There are organizational lines, but academic administration is, to a large extent, part of personality and situation. The effective operation of line organization is frequently enhanced by informal faculty avenues of communication. I do not think I am guilty of overstatement when I say that our faculty uses the informal approach to good advantage.

Some department chairmen should be more concerned about ordering books for the library. The budget for books is not as large as we would like to see it, and yet some chairmen do not spend all the money we budget for them.

It may be a minor matter in that it doesn't apply to too many cases, but sometimes faculty members do not counsel students very wisely in academic affairs. It is impossible to get every student in the courses appropriate to his interest and abilities, and it is too much to expect every faculty member to be fully acquainted with the total college program, but occasionally I wish for more careful counseling academically.

STUDENT PERSONNEL PROGRAM
L. James Harvey, Dean of Students

The following are comments regarding the areas of major responsibility of the Dean of Students.

Counseling

During the past year an attempt has been made to strengthen the college counseling program. An additional trained personal counselor was added (half-time) to the counseling staff. The responsibility for disciplinary problems has been removed from the offices of the Dean of Men and Dean of Women in the hope that these two persons could devote more time to counseling. Up to this time theory has outstripped practice. Our current Acting Dean of Women is not professionally prepared for counseling and our Dean of Men has not had adequate time with the press of other duties. Our future plans are to secure a Dean of Women who does have the counseling competency and to make some adjustments to allow the Dean of Men to work more freely in this area. Our main needs right now are for a woman counselor and for a specialist in vocational counseling area.

Discipline

Discipline is now handled by the Dean of Students along with the Student Court. The Court is a new innovation this year. Normally all cases of discipline are first dealt with in the Dean of Students' office. Except in cases where strict confidences are necessary, cases are referred to the Student Court for disposition. The president of the college must approve all decisions and if a student is dissatisfied with a verdict, he may appeal to a higher court (Nexus Committee) made up of five students and five administrators. The procedures, appropriate hearings, and protection of rights,

449

Student Personnel Program (cont.)

more than meet the minimum recommendations of the American Civil Liberties Union. They have worked successfully during this current college year.

Residence Halls

Five new men's residence units were opened this year. There are 50 men living in each. Each unit is assigned to one of our five fraternities (local) and their members live there. Each unit has a faculty member living within the hall. Their responsibility is to develop an educational and cultural program within the unit. The physical facilities are good. The educational and cultural emphasis has not yet met expectations. We hope to start a program this next semester called the "Interfraternity Bull Session" which may help move us further in a positive direction.

In our large men's hall (300 men) we have divided the hall into six separate units and have developed a social, athletic and hall-government program around the units. This pattern, as developed this year, is a marked improvement over the former more general pattern.

Fraternities

Fraternity life continues to flourish. All fraternities are local. This year the fraternities were persuaded to change to a delayed rush, so that no freshmen may join a fraternity until the middle or latter part of the second semester on campus.

Student Government

The Hope Student Council was completely revamped this year. We now have a Student Senate with expanded powers. We also have a system of representation based on living units (approximately one member for 50 students) plus eight at-large members. For the first time, we also have a Student

Student Personnel Program (cont.)

Court for the handling of certain disciplinary problems. A Union Board
was added last year to the student governmental organization and this year
they opened a temporary Student Union in one of the women's residence halls.
This is open weekends and is run entirely by students.

Chapel Attendance

The Dean of Students' office has also been responsible for adminis-
tering the new chapel attendance system started at the college this year.
The system has worked very well and has the support of the student body.
Minor modifications have been made during the year, but all in all it seems
to be the best system we have had in some time.

REPORT OF THE DEAN OF WOMEN'S OFFICE
Isla Van Eenenaam, Dean of Women

Of the Dean of Women's two major duties, work with individuals and work with groups, the first is of central importance. While working with groups such as the A.W.S., Mortar Board, Panhellenic Board, etc., occupies a good share of the Dean's time, it does allow for a profitable interaction with the student, making possible a sharing of ideas, consideration of values and educational goals -- sandwiching in on occasion some "painless" counseling.

Because it is in these two areas of work that the Dean of Women can realize to the fullest her obligation of service to the student, and because they are the most time consuming, it is extremely helpful when some of the pressures of the office are eased. In this connection, the Association of Women Students, an organization which administers a representative form of government in women's residences, stands out as the strongest single asset in the department. It functions smoothly and efficiently, and its government is accepted because it is an organization that has earned through wise and constructive measures the respect of fellow students. In the area of social and cultural activity as planned by the Activities Board of A.W.S., one of the finest contributions toward genuine acceptance of, and friendship between, nationals on the campus is "International Night", in which the cultures of all nations represented are displayed through exhibits and performances.

One of the most important tasks of the Dean's Office is the selection of roommates among the freshmen women. The time and thought which is given in pairing them as to areas of similarity in taste, habits, and background has been justified and has proven to be one of the strengths of the department. As a result, there is a minimum of trouble and unhappiness arising from incompatibility among roommates.

The Dean of Women's Office is fortunate in the high caliber of Head Residents who supervise the six residences and comprise the Dean's staff. The Head Resident's strength lies in her understanding of young people, her acknowledgement of basic Christian values and the restraint in which she practices her role as House "Mother". She is not a trained counselor, but has had college training and is encouraged to become familiar with administrative policies and curriculum requirements.

The program of using carefully screened upperclass women as student advisors in the residences is highly successful and certainly one of the strengths of the department. The resident advisor is generally the person to whom the troubled student first turns and with whom the well-adjusted one shares her joys and concerns.

Pointing up the strengths in one's department brings into focus clearly also a few limitations. One of them is a need for an in-training program by the college counseling service. Such a program would help the Head Resident in recognizing early danger signals in a troubled student's behavior.

Another limitation in this department is the lack of a voice in the planning of housing facilities. The Dean of Women's Office is in a position to give valuable advice regarding the function of a housing facility and current trends in dormitory living. Hence, there should be closer cooperation with the architects from the very earliest planning stage.

A third limitation is the lack of a continuing "conditioning" program beyond orientation for the freshman women. Because of the many

understandable adjustments facing the freshmen, and because of the view
generally held by them that college is a never ending series of romantic
contacts and social events, guidance through group discussion with upper-
classwomen, through informal talks with the Dean with a view toward broadening
of intellectual goals and the stabilizing of personality is being planned.

VIENNA SUMMER SCHOOL REPORT
Paul G. Fried, Director

Problems and Potential

The Hope College Vienna Summer School, begun in 1956, is primarily
designed to add an international dimension to the liberal arts training available for Hope College students. For this reason, courses offered in Vienna
are mainly in areas in which the European location can effectively add to the
value of instruction. Art, History, Music, and European Literature, taught
in English, and courses in the German language are the basic curriculum.
Other courses have been, and will be, added whenever student demand or the
availability of an outstanding instructor makes this desirable.

The instructional staff employed by Hope College for the Vienna program is predominantly European. The value of exposing students to the different view points and methods of European professors is obvious. This policy,
however, also poses two problems. Since, at present, we can offer employment
only for a period of six weeks in the summer, staff changes are fairly frequent
and the amount of guidance which can be given to European faculty members
in the academic administration is very slight. The policy also precludes
active participation of regular Hope College faculty, except in administrative
or supervisory capacity. Some program of providing faculty fellowships will
have to be developed before we can expect Hope faculty to be fully aware of,
and interested in, the possibilities the program offers for our students.

The physical facilities used for the Hope program in Vienna are
rented from the Institute of European Studies. They include lecture rooms,
library, office space, dining facilities, and the services of the I.E.S.
housing office. Given the present size of our program the arrangement clearly

has advantages for both institutions. There is some danger, however, that other schools will follow the example of Hope, Oberlin, and Wooster, who now share these facililities. This would intensify the already noticeable competition for staff, office and classroom space, and good housing. As a safety device, Hope will need to explore other possible affiliations and perhaps the rental or purchase of adequate facilities. This would be particularly important if we should want to expand our program in numbers or in the length of time students spend in Vienna.

While they are in Vienna, students seem to be fully aware of the new dimension which taking a course like "Art of the Baroque" in a city which abounds in 17th and 18th century palaces and churches can give to their academic work. All too frequently, however, these insights gained abroad seem to be linked more concretely to the excitement and fun of travel, and there does not appear too much transfer of learning when the student returns to the home campus. Undoubtedly the impact of the foreign study experience could be heightened by well-developed pre-departure and follow-up programs at the college. Questionnaires and experimental seminar-discussion sessions used during the last three years, while useful, have not been as effective as we had hoped. A more analytical approach to the problem must be developed.

Perhaps the most serious handicap of the Vienna Summer School as it relates to our home campus is that it has been a "one way street." More than two hundred Hope students have studied in Vienna during the past eight years. In that same period we had one Austrian student on campus for a year and one for a semester. It seems obvious that the presence of one or more students, and, if possible, a guest lecturer from Austria each year could

456

<u>Vienna Summer School Report</u> (cont.)

contribute significantly to the development of a closer identification between the academic opportunities available to the Hope College student in Vienna and at home.

LIBRARY REPORT
John May, Librarian

The Van Zoeren Library thinks of itself as a service organization whose responsibility is to aid students and faculty members in their endeavor to achieve the educational objectives of the college. The library, then, does not have a separate or distinctive program of its own, but trying to be service-minded, it feels that its basic function is that of all libraries -- "getting books and readers together."

The circulation and budget statistics, indicate a very substantial increase in use of, and development of, the library collection. The new Van Zoeren Library building, which was first used in September of 1961, has changed and improved the whole pattern of library use by both students and faculty.

Adequate library service and support at Hope College came comparatively late in the history of the college. Even now, the basic organization of the library materials into a cataloged collection is not complete, but the employment this year of an additional professional cataloger is resolving this situation.

The present book collection of 75,000 is far below the standard set by the American Library Association for undergraduate college libraries, but with the money provided in recent years and the good buying habits of the present faculty, this picture is improving rapidly.

Although no definite plans have been made as yet, the library staff is interested in a training program for public school librarians and a more adequate audio-visual program as future projects.

<u>Library Report</u> (cont.)

Even though it operates as a service organization, the library staff does feel that the library has one distinctive teaching function, and that is to prepare students to continue their education throughout life by being able to find new information on old subjects and material on new subjects after their formal education is ended. The student who learns how to use one well-organized library properly can use any well-organized library in the United States, if not in the world, with reasonable ease. The student who learns how to evaluate books and other printed materials -- to separate the good from the bad -- by the reference and readers' advisory services of the library is also prepared to select from the great mass coming from the country's printing presses when he no longer has a kindly professor to put the "right" books on closed reserve for him. How successful such a system is can only be determined by a study years hence, but in its planning, this ideal seems most important to the library staff.

TREASURER'S REPORT
Henry Steffens, Treasurer

A. Strengths in the Financial Structure of Hope College.

      The Reformed Church in America has supported the college
very well. Throughout the past several years, it has contributed well
in excess of $125,000.00 a year for current operations. In fact, this
is so predictable that if capitalized, it would in effect be the income
from endowments in excess of $2,500,000.00 at present rates.

      During the past several years, the college has made appeals to
the alumni, and last year, as a result of a rather vigorous campaign,
the alumni have evidenced their increasing interest in the operating
affairs of the college to a point where the amount they contributed
was in excess of $100,000.00.

      Businessmen, too, have been approached with increased intensity
during the past few years, and the response has been encouraging. Through
the Michigan Colleges Foundation, businessmen in Michigan have been
taking greater interest in private educational affairs in Michigan and
have provided moneys somewhat in excess of $45,000.00 a year which is
equal to the income on $1,000,000.00 of endowment.

      An endowment of $2,000,000.00, though not large as college
endowments go, has provided a steadily increasing income for operational
purposes. It is well invested and has been growing slowly through the
years.

      In spite of recognized weaknesses, the treasury of Hope College
has been strong enough to provide for the excellent maintenance of
equipment, buildings and grounds, to purchase adequate instructional

460

supplies and equipment, and to pay salaries to the faculty which up to
now have been sufficiently high, so as to attract teachers of academic
standing, quality and competence. Further it has added to student
services, principally in the areas of library and counseling without
having to rely on deficit financing of any consequence during the past
five years.

B. Weaknesses in the Financial Structure of Hope College.

1. Lack of sufficient endowment funds so as to provide for:
   a. Adequate scholarships both as to amount and number.
   b. Adequate income for the maintenance of educational buildings
      as they are constructed and added to the campus.
   c. Additional and increasingly important faculty benefits.

2. The lack of sufficient funds results in our being too dependent
   on student income for general operating purposes. While it is true
   in our case that our student fees are not high in comparison to
   other private colleges with whom we like to compare ourselves, it
   is likewise a fact that our student fees comprise a relatively
   high percentage of our current income.

3. An insufficient income from outside sources such as business and
   industrial gifts and alumni participation in a giving program for
   operational, educational and scholarship purposes.

4. The general lack of income from endowment and from gifts means that
   we are forced to operate at a student-teacher ratio which is greater
   than we desire. The faculty, therefore, can devote less time to
   individual instruction, counseling, and research than is desirable,

<u>Treasurer's Report</u> (cont.)

with the result that we are unable to achieve one of our major educational goals -- that of giving adequate individual attention to
some students.

BUSINESS MANAGER'S REPORT
Rein Visscher, Business Manager

Strengths:

1. Over the past decade, the physical plant has not only grown in size but
   has become far more adequate for instructional and housing, even of
   our rapidly increasing student body. (See next sheet.)

2. The administration has supported a program of constant repair and maintenance to the older buildings, so that in general they are in first-
   class condition.

3. The maintenance and building and grounds staff are a loyal and hard
   working group, and despite the age of many of them, do an amazing amount
   of work on a low budget.

4. The food service the college employs has worked cooperatively with the
   college, and there is reasonable satisfaction among students with the
   board. (This is no mean feat.)

Limitations:

1. Because of budget limitations, we are always somewhat understaffed and
   are pushed to keep abreast. This understaffing begins in the Office
   of the Business Manager and extends to the maintenance crew.

       Consideration is now being given to adding a person at night
   to check the boilers and perhaps other functions.

2. The location of the campus -- in the middle of town, surrounded by well
   established residences and businesses -- makes it difficult and extremely
   costly to add to the campus. As we face further enrollment increases,
   this remains a real problem, as it has in the past ten years. The recent
   addition to the east campus (Van Raalte) may alleviate this problem, but
   presents another, in that we then have a divided campus.

463

Plant Development 1954-1964

1. Renovation of Carnegie Gymnasium. $60,000 (1954).

2. Construction of Music Hall. $225,000 (1956).
   (Auditorium, 2 classrooms, 7 studios, 13 practice rooms, record library, and booths.)

3. Erection of Kollen Hall. $1,000,000 (1956). Residence hall for 300 men.

4. Acquisition of Van Raalte Campus and subsequent development of playing fields and track. $350,000 (1960-1963).

5. Erection of Phelps Hall. $1,000,000 (1960). (Residence Hall for 160 women and dining hall seating 600 students.)

6. Construction of Van Zoeren Library. $900,000 (1961).

7. Renovation of Graves Hall into a Foreign Language Center and Student Office Center. $125,000 (1962).

8. Erection of Gilmore Hall and the five men's halls. $1,500,000. (1963).

9. Addition to Central Heating Plant. $150,000. (1963).

10. Construction of greenhouse for botany program. $12,000. (1964).

11. Construction of Physics-Mathematics Building. $750,000. (1964).

12. Purchase of numerous cottages and renovation for various uses: Psychology and counseling offices, Alumni Office and Guest Hall, German House, French House, and residence units for women and men students. (1954-1963).

464

OFFICE OF ADMISSIONS REPORT
Albert H. Timmer, Director of Admissions

A. Two committees related to the Office of Admissions dealing basically with admission policy and applications.

    1. Admissions Committee, with its personnel consisting of college vice-president (chairman), director of admissions, associate director of admissions, and six members of teaching faculty. This committee holds meetings on occasions to treat the matter of policy as to admission practices, scholarships, foreign student admission candidates, relation with College Entrance Examination Board, etc.

    2. Committee on Admission Applications. This committee, consisting of director of admissions (chairman), vice president, associate director of admissions, two members of teaching faculty, and assistant to the admissions director (secretary), meets periodically to review admission applications on a "rolling admissions basis."

B. Admissions Office Staff.

    1. The director of admissions whose basic duties involve the following: In charge of admissions staff, chairman of committee on admission applications, chairman of committee on educational grants for new students, foreign student adviser, college representative to College Board meetings, admission recruiting duties.

    2. The associate director of admissions whose duties involve the following: member of admission committees, assistant to fireign student adviser, college student adviser, admission recruiter.

    3. The assistant director of admissions, whose duties involve admissions recruiting and contact person for our church constituency.

465

4. The assistant to director of admissions, whose main duty is management of admissions office services.

5. Two admission secretaries dealing with correspondence.

C. Admissions staff until recently consisted of director and associate director for admission contact service. On occasions members of teaching staff were also used. The third full-time person as assistant director of admissions permits fuller contact service with high school counselors and churches of the college constituency.

D. Projected plans to strengthen services by Admissions Office.

1. A new film-strip is presently being prepared treating the basic characteristics of Hope College. Copies of this film-strip will be made available to the constituency at large which the college serves.

2. A fourth full-time admissions counselor is being considered with the intent that the enlarged admissions staff will be able to make rather complete personal contacts throughout our national constituency.

HOPE COLLEGE COUNSELING SERVICE REPORT
Lars Granberg, Head Counselor

The changes in the operation of the Counseling Service during this school year stem mainly from the addition of Robert Brown to the counseling staff. Between the beginning of the school year and the Christmas recess, his appointment calendar was kept light in order to give him added time to complete the requirements for his Ed.D. Following conferral of his doctorate Bob has more than made up the time, as the statistics below will demonstrate. His work has been highly satisfactory. I regard him as a valuable colleague. His presence makes it possible to confer on difficult cases and to refer back and forth as a given student appears potentially more responsive to one or the other of us.

Last year (1962-63) fifty-six men and forty-eight women used the service, or approximately seven percent of the student body. This year a total of 122 men and 72 women used the service, comprising approximately thirteen percent of the student body. A total of 506 hours of counseling time was given these students (Granberg: 91 students for 273 hours; Brown 103 students for 233 hours). This underscores the fact that demand for service increases with its availability.

More vocational testing and counseling was carried on this year by virtue of Brown's special proficiency in this area.

Three students were encouraged to withdraw from school in order to receive psychiatric treatment. Two of these were young ladies who experienced depression with suicidal trends; the other a young man who was referred to Dr. Klaire Kuiper for out-patient treatment for depression, but whose condition worsened and he was sent home.

467

Counseling Service Report (Cont.)

While the waiting period to see a particular preferred counselor on occasion extended to three weeks, the two-man staff has made it possible for a student to see a counselor within two or three days during most of the year; and on an emergency basis usually the same day the service was requested, or at most, the next day.

Approximately 90% of the students who come to us can be helped in five counseling sessions or less, in fact 80% come three times or less. However, there are the remaining ten percent who needed help over a longer period of time. Most of these could be helped within ten sessions, but six students needed more extensive help than this. These are people with long standing personality problems of a sort that seriously impair their academic work, but who can be kept functioning as they work their way through the problem. Most of these cases responded well to therapy and turned the year academically to good account.

I do not know how to estimate how effective our work is. The largest number seem appreciably improved, but there are always a few we cannot help. The growing number of students who seek help suggests growing confidence in us, especially since many students are being referred by other students who have received help.

OVERALL APPRAISAL OF HOPE COLLEGE
John Hollenbach, Vice President

Strengths

Hope College is one of a number of church-related liberal arts colleges that over the years has provided a substantial and solidly competent program of undergraduate education quite out of proportion to its physical facilities and financial resources. Perhaps its very limitations in these resources has been one of its strengths. Furthermore, the sense of purpose undergirding the institution through the Christian faith of its faculty, board, and the vast majority of the students, has led to indirect learnings which in many ways are more significant than the direct learnings. This quality is the most distinguished feature of Hope College. Over the ninety-eight years of its history it has acquired a tradition of concern for developing of Christian values and for solid preparation for entering one of the service professions which is its most valuable asset. The primary strengths of the Hope College of 1963-64 are:

1. A student body of strong moral fiber, religious heritage, seriousness of purpose, and reasonably good academic aptitude. The students come from a broad enough variation of background -- economic, geographic and religious -- to provide the necessary tension and challenge for personal growth. At the same time, there is a strong enough orientation to religious values by the majority to provide a community of concern. In this context the continuing search for and refining of values becomes a meaningful, even at times a painful, part of the college program.

Overall Appraisal of Hope College (cont.)

2. A relatively stable faculty to provide continuity to the program. The faculty, by and large, are genuinely interested in teaching and carry out their tasks with a high sense of responsibility, and with a genuine concern for students as persons. They consistently expect a high level of performance from students.

3. A physical plant that, with a few significant exceptions, is quite adequate for the present program. These facilities have been well maintained and fairly well equipped.

Limitations

1. The experimental cutting-edge of the college is not as sharp as it should be. Along with a stable faculty there can be and has been an over-conservatism and reluctance to experiment and change. Not enough of the faculty are actively exploring on the frontier in an area of their own discipline. The teaching loads have tended to be high and this further has discouraged the exercise of imagination and experimenting in the instructional process itself. On the whole, the faculty are not as sophisticated in their scholarship as they might be. This is changing, but it needs to be changed further.

2. The financial limitations under which the college has operated have prevented the college from taking rapid steps for remedying the limitation cited above in the past. For example it is difficult for the college to proceed as quickly as it should to a more effective leave of absence program, to bringing in top young scholars in new staff appointments, or to lighten staff loads on occasion to permit research activities.

<u>Overall Appraisal of Hope College</u> (cont.)

3. The old and very happy approach to student admissions -- counseling
   with those directed by alumni, friends, and church, who applied on their
   own initiative -- is fast becoming obsolete. The college has been re-
   luctant to move to the newer more aggressive recruiting approach, and
   especially to enter the scholarship bidding contest. As a result, it
   may face an uphill battle to attract <u>good</u> students in the years immediately
   ahead. The present scholarship funds are definitely inadequate.

4. The staff in a number of areas needs to be strengthened over the coming
   years. Active steps are presently being taken in most of these areas.

5. The library holdings are barely adequate for a college of our size and
   programs.

6. As the Profile Committee of 1961 pointed out, the facilities for art
   and theater are poor, as also those for student recreation. A student
   center with art and theater facilities is becoming a pressing need.
   Furthermore, physical education facilities are being taxed beyond capacity. The need is real here too.

7. There is need for new and more effective ways of securing church and
   individual financial support for endowment funds and for current operations.

8. The program of independent studies is strong in some areas and weak in
   others. More needs to be done here. The freshman-sophomore honors
   project needs serious reexamination in view of the new general requirements.

MEMORIAL SERVICE
Calvin A. VanderWerf
Tribute by Douglas C. Neckers
July 24, 1988

A few days after Cal was stricken with the illness which eventually took his life, I called Julie on the phone in Gainesville to find out how her mother was doing. I asked Julie if either she or her mother had any idea that Cal might be ill. She said that they didn't, and then she quickly added "We weren't ready for this." Cal's passing is too soon and none of us were ready. He will be sorely missed by his family, by his friends, by this college, by his students, and by his university. Few men have had the impact on others that Cal has had and few have engendered the devoted affection of so many. Let us all say that we were lucky to have known him and for so long.

Like most of us, I'm sure, I reflect this day on a long professional and personal relationship with Cal VanderWerf. It goes back to a time when at my friend Vic Heesley's continual pestering, I finally wrote then Dr. VanderWerf a letter inquiring about graduate study at the University of Kansas. Granted, I had heard of Ray Evans and Clyde Lavelic and Wilt Chamberlain and Gwen Cunningham, but little did I know that there was anything in Kansas except a few athletes and long distance runners. One letter was all that Cal needed and before I knew it I was a K.U. Jayhawk and, aside, I think I'm sure that I knew that he had a very good time during the NCAA Championships. This year, the Jayhawks are the NCAA Champions. I'm sure Doc Van Zyl was miffed at this sudden turn of events, because as I recall, Cal took me at the University of Kansas without a letter from Doc—a Hope degree was good enough.

Traveling to Lawrence from the shores of Lake Michigan was trauma. Sue and I felt almost like we were going to drop off the end of the earth as we approached the top of every hill across Missouri on old Route 40. Then we got to Columbia and were sure we had. It was with some apprehension after getting there that fall that Sue and I, on a Saturday afternoon, went to the VanderWerfs to meet this professor who had dragged us from Nirvana's gate to a place beyond site and shore. It was a typically hot and humid summer Lawrence day, but little did I

472

expect a boyish, barefoot gardener with a 2-year-old daughter in tow would greet me at the door. No chemist I'd met ever looked like that; most chemists took things much more seriously. Before I knew it, I had a Ph.D. from Kansas, Cal was Hope's president, and I was an assistant professor.

Cal and I became very good friends and over the years he's had a very important influence on my life. He's remained my confidant, my advisor, my academic mentor. We talked often on the phone and just a few days before he was stricken had still another conversation about a certain career opportunity which seemed possible in the near future. On the day Cal died, Sue and I were coming home from a meeting in Northern Italy. Almost on impulse, we got off the train in Parma. Near Parma stands the village of Busetto where Giuseppe Verdi was born and spent the majority of his life. It was a thrill to walk where Verdi walked, to feel what Verdi felt, to see what Verdi saw and just to think about the legacy of this man who left so much for each of us. Though Verdi has passed, his art remains for all generations to enjoy. It's said that Verdi was not a particularly religious man, and he wrote only his requiem mass, and very late in life, wrote small sacred pieces for the liturgy. Written in honor of an Italian author Allesandro Mitsoni, who in Verdi's mind represented the real Italian spirit, the requiem was essentially the only spiritual statement of Verdi's career. It was to the requiem, rather than chemistry that my thoughts turned on the plane ride home, after I found out that my dear friend Cal VanderWerf was no longer living. Verdi's mass represents the ultimate of creative genius. In it, he ties the awesome humility of the Roman Catholic prayer of the living for the spirits of those who have passed to the brilliant artistry of a 19th century operatic composer. The texts are the texts of other masses, written at different times and in different places—Requiem Aeternam, Kyrie Re Eleison, Dies Irae, Sanctus Benidictus, Angus Dei, Lux Aeterna, Libera Me. And as I was riding on that plane, my mind turned to the first time I'd heard the mass. It was in Toledo's marvelously beautiful, gothic, Rosary Cathedral and it was there that I had sung Verdi for the first time, some years ago. I'd studied the score for years alone at the piano with my recordings, and for this particular performance, in rehearsals. When the time for performance came, however, I had laryngitis, so I had to sit in the almost empty cathedral and listen while my colleagues rehearsed. I remember the night of the dress rehearsal vividly, the night when soloists, orchestra, conductor and choir all came together as one together to put Verdi's requiem together one more time for still another performance. Though I knew the music intimately, part by part, Latin word for Latin word, the reality of its beauty only struck me when the soprano in the Libera Me soared down on that high C, "Libera me de morte aeternam requiem aeternam et lux perpetua"—Lord grant them eternal rest and let the light forever shine upon them. Because she stood above all the rest, over the choir of 200 voices, over the orchestra of brass and full throttle and over the rest of the soloists in the quartet, she provided the capstone, the lock and the key which made all the parts seem as one. In that singularly beautiful moment, she revealed Verdi's art form in a way which says to all who could hear, in the adoration of God the creator and Christ his son, this is the very best which man can do. And as I rode in the quiet of that jet plane, I heard the voice of many who have known Cal VanderWerf saying "Cal VanderWerf was the embodiment of that climactic moment in the Verdi requiem in the lives of all of us." As he lived, he soared like that soprano to that high C over all. Through his influence it was that we could bring just a little bit more of ourselves to the world. For others he lived, and for them his life was the capstone, the lock and the key. His presence made all of us, through his intellect, his skill, his enthusiasm, his decency, his spirit, his zest for life into that finest of art forms, the quintessential human spirit.

Few men I've known understood the university better than Cal VanderWerf, and even fewer have had the skill to put these finest of educational ideals into practice, in the classroom, in the laboratory, as a department chairman, as a dean and as a president. Literally thousands of students at two major universities have studied in his classes. In each and every class there were some who got to know themselves just a little better because they'd known him. He had more confidence in them than they had in themselves, and he made them aspire to greater things. A single institution, this one, came to see itself in a completely different role through his leadership. During his presidency, Cal became an academic institution of a much more broadly reaching influence, and it was almost singularly through his devoted efforts and personal sacrifice.

Cal left behind much more than he took. Almost to the moment he passed from this earth, he was doing what he loved the most, leaving his impact on the future of American education in the sciences through the programs of two foundations, the Camille and Henry Dreyfus Foundation and Research Corporation. Cal will be sorely missed by all of us, by his family, by his friends, by his students, by his associates, by this college. But those who knew him most and knew him best know that he left just a little bit behind with us and in us. We are to carry on the work which he began and for this we must be eternally grateful, and for him we must strive to be just a little bit better because it's so.

Cal was kind, honorable, loving, and he lived his life by the greatest of God's two Commandments: Love the Lord your God with all your heart, with all your soul and especially with all your mind, and love your neighbor even more than you love yourself. Rachel, Gretchen, Glocina, Julie, Lisa, Peter, Marta, your loss is our loss. Your loss and sorrow is our sorrow, your pain is our pain. But no man is an island; no man stands alone, each man to another, each man to his own. To each of us in his own special way, Cal provided that capstone of the soprano's high C; he lived so that we might understand ourselves more. The life he lived is your life as it is ours, and he lives in those of use who remain.

"And I heard the voice of harpers, harping with their harps, and they came as it were a new song before the throne and no man could hear that song but they who were redeemed; list the cherubic hosts with golden choirs, touch their immortal hearts with golden wires for those just spirits who wear victorious crowns singing everlastingly devout and holy songs. And I saw a new heaven and a new earth for the first heaven and the first earth were passed away and there was no more seed. And I heard a great voice out of heaven saying, 'Behold the tabernacle of God as with men and they shall be His people and God himself shall be with them and be their God.' And God shall wipe away all tears from their eyes and there shall be no more death, neither sorrow nor crying nor any more pain, for the former things are passed away. Holy, holy, holy, holy is the Lord of hosts."

We've lost a dear friend but all passes, art alone endures. The places Cal has touched and all his good works will live with mankind forever and for all generations. This is the day the Lord hath made, let us rejoice in it.

About the Author

Douglas C. Neckers from SPIE, reprinted courtesy of SPIE.

Doug Neckers

By Stephen G. Anderson

Nowadays most of us take computer-generated three-dimensional visuals for granted. Aided by powerful small computers and large displays, anyone needing to work with a 3D representation can do so quite easily, be it for work or pleasure. In the early 1980s, though, such tools were not so readily available.

It was at that time, after teaching organic chemistry for many years, that Douglas Neckers (then at the Center for Photochemical Sciences in Bowling Green, Ohio) realized he could use his skills as a photopolymer chemist together with an evolving new laser-based model-making technique called stereolithography. Neckers says the "image" he created was the first-ever three-dimensional medical image, a model of a real human heart.

"I had noticed that chemistry undergraduates from all walks of life were experiencing considerable difficulty visualizing in three dimensions," he explains. "It's an important skill for all organic chemists but especially for any pre-health professionals taking the course. I wanted them to be able to see and plan what they were going to have to do surgically.

So in 1988, using input from an MRI image and a stereolithography system on loan from 3D Systems, Neckers created a polymer-based model of a real human heart.

Neckers has since retired from Bowling Green University as the McMaster Distinguished Research Professor Emeritus of Photochemical Sciences and is currently CEO of Spectra

> "At the end of the day I wanted to make sure that the students were educated properly so that they could step into a position that would lead to a very good life for them."

Group Limited (SGL), an Ohio company he founded in 1990 after creating that first medical replica.

3D prototypes help in medicine

At SGL, again working with 3D Systems and in collaboration with the Cleveland Clinic among others, Neckers used stereolithography to create medical models from a patient's CT scans that surgeons could then use to plan upcoming operations. And not incidentally his company could also provide his students with jobs when they were finished with school. As things turned out, however, because the healthcare reimbursement mechanisms were too slow for a viable business model, SGL has since evolved into the photoscience materials company that it is today.

Stereolithography is just one example of photoscience in action. Today photochemistry also has important applications in electronics, photography, printing, and manufacturing.

Early in his career Neckers realized that energy curing (more commonly called UV curing at the time) of photo-sensitized compounds offered some real benefits to industry. The process is environmentally friendly (less use of solvents) and fast. Companies whose business was rapidly turning liquids into solids — from quick-drying paint and varnishes to printing and medicine — were seriously interested and that led to the founding of the Center for Photochemical Sciences.

Almost immediately, Neckers was consulting to companies such as Mead, 3D Systems, and Ciba Geigy and sometimes giving courses in basic photochemical sciences at SPIE meetings.

"The contacts made in academia and industry brought in funds and creative energy," he said.

Over the years, that creative energy has been responsible for Neckers and his research group creating innovative plastic coatings for medicines like insulin to allow people to take it orally. The group at BGSU, with 65 patents to its credit, is also responsible for such useful items as quick-drying coatings for engines and algaecide-infused paint for ship hulls.

Industry, academic partnerships

Neckers attributes the success of the BGSU center to a continuing awareness of industry and its needs. "The collaboration with industry helped build the successful PhD program," he says. "We admitted a broad spectrum of students (not just chemists) and made sure they were trained for jobs and leadership."

It is this overt commitment, not just to teaching and education but to what it leads to, that is a constant in discussions with Neckers. "It's a matter of integrity," he said. "You can't give someone a strong education without ensuring that they then have somewhere to go when they graduate."

Neckers adds, "At the end of the day I wanted to make sure that the students were educated properly so that they could step into a position that would lead to a very good life for them."

Looking forward, Neckers offers some ideas about what may lie ahead. He notes that three-dimensional printing at home is now a reality and

wonders how long it will be before news providers offer digital links to three dimensional models that could enhance their stories.

He says that dentistry is a big growth area. Right now photochemistry offers faster curing of fillings, but maybe making teeth in a dentist's office is just around the corner. And he hopes that photoscience will play a larger role in solar energy conversion over the next 5-10 years.

And as for the medical models, Neckers believes fully functional models of, for instance, knee joints, are not too far away while noting that finding materials to behave like the various tissues involved (ligaments, cartilage, etc.) is proving a significant challenge. Nonetheless 3D Systems reports that its medical products in oral, hearing, and orthopedics represent some of its fastest growing markets.

–Stephen G. Anderson is industry and market strategist for SPIE. ■

Index